Lecture Notes in Physics

Springer

Berlin
Heidelberg
New York
Hong Kong
London
Milan
Paris
Tokyo

Physics and Astronomy

ONLINE LIBRARY

springeronline.com

The Editorial Policy for Edited Volumes

The series *Lecture Notes in Physics* (LNP), founded in 1969, reports new developments in physics research and teaching - quickly, informally but with a high degree of quality. Manuscripts to be considered for publication are topical volumes consisting of a limited number of contributions, carefully edited and closely related to each other. Each contribution should contain at least partly original and previously unpublished material, be written in a clear, pedagogical style and aimed at a broader readership, especially graduate students and nonspecialist researchers wishing to familiarize themselves with the topic concerned. For this reason, traditional proceedings cannot be considered for this series though volumes to appear in this series are often based on material presented at conferences, workshops and schools.

Acceptance

A project can only be accepted tentatively for publication, by both the editorial board and the publisher, following thorough examination of the material submitted. The book proposal sent to the publisher should consist at least of a preliminary table of contents outlining the structure of the book together with abstracts of all contributions to be included. Final acceptance is issued by the series editor in charge, in consultation with the publisher, only after receiving the complete manuscript. Final acceptance, possibly requiring minor corrections, usually follows the tentative acceptance unless the final manuscript differs significantly from expectations (project outline). In particular, the series editors are entitled to reject individual contributions if they do not meet the high quality standards of this series. The final manuscript must be ready to print, and should include both an informative introduction and a sufficiently detailed subject index.

Contractual Aspects

Publication in LNP is free of charge. There is no formal contract, no royalties are paid, and no bulk orders are required, although special discounts are offered in this case. The volume editors receive jointly 30 free copies for their personal use and are entitled, as are the contributing authors, to purchase Springer books at a reduced rate. The publisher secures the copyright for each volume. As a rule, no reprints of individual contributions can be supplied.

Manuscript Submission

The manuscript in its final and approved version must be submitted in ready to print form. The corresponding electronic source files are also required for the production process, in particular the online version. Technical assistance in compiling the final manuscript can be provided by the publisher's production editor(s), especially with regard to the publisher's own LaTeX macro package which has been specially designed for this series.

LNP Homepage (springerlink.com)

On the LNP homepage you will find:
−The LNP online archive. It contains the full texts (PDF) of all volumes published since 2000. Abstracts, table of contents and prefaces are accessible free of charge to everyone. Information about the availability of printed volumes can be obtained.
−The subscription information. The online archive is free of charge to all subscribers of the printed volumes.
−The editorial contacts, with respect to both scientific and technical matters.
−The author's / editor's instructions.

Y. Kosmann-Schwarzbach B. Grammaticos
K.M. Tamizhmani (Eds.)

Integrability of Nonlinear Systems

Springer

Editors

Yvette Kosmann-Schwarzbach
Centre de Mathématiques
École Polytechnique
91128 Palaiseau, France

Basil Grammaticos
GMPIB, Université Paris VII
Tour 24-14, 5ᵉ étage, case 7021
2 place Jussieu
75251 Paris Cedex 05, France

K. M. Tamizhmani
Department of Mathematics
Pondicherry University
Kalapet
Pondicherry 605 014, India

Y. Kosmann-Schwarzbach, B. Grammaticos, K. M. Tamizhmani (eds.), *Integrability of Nonlinear Systems*, Lect. Notes Phys. **638** (Springer-Verlag Berlin Heidelberg 2004), DOI 10.1007/b94605

Library of Congress Cataloging-in-Publication Data.

Cataloging-in-Publication Data applied for

A catalog record for this book is available from the Library of Congress.

Bibliographic information published by Die Deutsche Bibliothek
Die Deutsche Bibliothek lists this publication in the Deutsche Nationalbibliografie;
detailed bibliographic data is available in the Internet at http://dnb.ddb.de

ISSN 0075-8450
ISBN 978-3-642-05835-6   e-ISBN 978-3-540-40962-5

Springer-Verlag is a part of Springer Science+Business Media

springeronline.com

© Springer-Verlag Berlin Heidelberg 2010
Printed in Germany

Cover design: *design & production*, Heidelberg

Printed on acid-free paper
54/3141/du - 5 4 3 2 1 0

Preface

This second edition of *Integrability of Nonlinear Systems* is both streamlined and revised. The eight courses that compose this volume present a comprehensive survey of the various aspects of integrable dynamical systems. Another expository article in the first edition dealt with chaos: for this reason, as well as for technical reasons, it is not reprinted here. Several texts have been revised and others have been corrected or have had their bibliography brought up to date. The present edition will be a valuable tool for graduate students and researchers.

The first edition of this book, which appeared in 1997 as Lecture Notes in Physics 495, was the development of the lectures delivered at the International School on Nonlinear Systems which was held in Pondicherry (India) in January 1996, organized by CIMPA-Centre International de Mathématiques Pures et Appliquées/International Center for Pure and Applied Mathematics and Pondicherry University. In February 2003, another International School was held in Pondicherry, sponsored by CIMPA, UNESCO and the Pondicherry Government, dealing with *Discrete Integrable Systems*. The lectures of that school are now being edited as a volume in the Lecture Notes in Physics series by B. Grammaticos, Y. Kosmann-Schwarzbach and Thamizharasi Tamizhmani, and will constitute a companion volume to the essays presented here.

We are very grateful to the scientific editors of Springer-Verlag, Prof. Wolf Beiglböck and Dr. Christian Caron, who invited us to prepare a new edition. We acknowledge with thanks the renewed editorial advice of Dr. Bertram E. Schwarzbach, and we thank Miss Sandra Thoms for her expert help in the production of the book.

Paris, September 2003   The Editors

Contents

Analytic and Asymptotic Methods
for Nonlinear Singularity Analysis:
A Review and Extensions of Tests for the Painlevé Property

List of Contributors

Mark J. Ablowitz
Department of Applied Mathematics,
Campus Box 526,
University of Colorado at Boulder,
Boulder Colorado 80309-0526, USA
markjab@newton.colorado.edu

Paolo Casati
Dipartimento di Matematica
e Applicazioni,
Università di Milano-Bicocca,
Via degli Arcimboldi 8,
20126 Milano, Italy
casati@matapp.unimib.it

Gregorio Falqui
SISSA,
Via Beirut 2/4,
34014 Trieste, Italy
falqui@sissa.it

Basil Grammaticos
GMPIB,
Université Paris VII,
Tour 24-14, 5e étage, case 7021,
75251 Paris, France
grammati@paris7.jussieu.fr

Rod Halburd
Department of Mathematical Sciences,
Loughborough University,
Loughborough,
Leicestershire LE11 3TU,
United Kingdom
r.g.halburd@lboro.ac.uk

Jarmo Hietarinta
Department of Physics,
University of Turku,
20014 Turku, Finland
hietarin@utu.fi

Nalini Joshi
School of Mathematics
and Statistics F07,
University of Sydney,
NSW 2006, Australia
nalini@maths.usyd.edu.au

Yvette Kosmann-Schwarzbach
Centre de Mathématiques,
École Polytechnique,
91128 Palaiseau, France
yks@math.polytechnique.fr

Martin D. Kruskal
Department of Mathematics,
Rutgers University,
New Brunswick NJ 08903, USA
kruskal@math.rutgers.edu

Franco Magri
Dipartimento di Matematica
e Applicazioni,
Università di Milano-Bicocca,
Via degli Arcimboldi 8,
20126 Milano, Italy
magri@matapp.unimib.it

Marco Pedroni
Dipartimento di Matematica,
Università di Genova,
Via Dodecaneso 35,
16146 Genova, Italy
pedroni@dima.unige.it

Alfred Ramani
CPT, École Polytechnique,
CNRS, UMR 7644,
91128 Palaiseau, France
ramani@cpht.polytechnique.fr

Junkichi Satsuma
Graduate School
of Mathematical Sciences,
University of Tokyo,
Komaba, Meguro-ku,
Tokyo 153-8914, Japan
satsuma@ms.u-tokyo.ac.jp

Michael Semenov-Tian-Shansky
Laboratoire Gevrey
de Mathématique physique,
Université de Bourgogne, BP 47870,
21078 Dijon Cedex, France
semenov@mail.u-bourgogne.fr

K.M. Tamizhmani
Department of Mathematics,
Pondicherry University,
Kalapet, Pondicherry-605 014, India
tamizh@yahoo.com

Introduction

The Editors

Nonlinear systems model all but the simplest physical phenomena. In the classical theory, the tools of Poisson geometry appear in an essential way, while for quantum systems, the representation theory of Lie groups and algebras, and of the infinite-dimensional loop and Kac-Moody algebras are basic. There is a class of nonlinear systems which are integrable, and the methods of solution for these systems draw on many fields of mathematics. They are the subject of the lectures in this book.

There is both a continuous and a discrete version of the theory of integrable systems. In the continuous case, one has to study either systems of ordinary differential equations, in which case the tools are those of finite-dimensional differential geometry, Lie algebras and the Painlevé test – the prototypical example is that of the Toda system –, or partial differential equations, in which case the tools are those of infinite-dimensional differential geometry, loop algebras and the generalized Painlevé test – the prototypical examples are the Korteweg-de Vries equation (KdV), the Kadomtsev-Petviashvilii equation (KP) and the nonlinear Schrödinger equation (NLS). In the discrete case there appear discretized operators, which are either differential-difference operators, or difference operators, and the tools for studying them are those of q-analysis.

At the center of the theory of integrable systems lies the notion of a Lax pair, describing the isospectral deformation of a linear operator, a matrix differential operator, usually depending on a parameter, so that the Lax operator takes values in a loop algebra or a loop group. A Lax pair (L, M) is such that the time evolution of the Lax operator, $\dot{L} = [L, M]$, is equivalent to the given nonlinear system. The study of the associated linear problem $L\psi = \lambda\psi$ can be carried out by various methods.

In another approach to integrable equations, a given nonlinear system is written as a Hamiltonian dynamical system with respect to some Hamiltonian structure on the underlying phase-space. (For finite-dimensional manifolds, the term "Poisson structure" is usually preferred, that of "Hamiltonian structure" being more frequently applied to the infinite-dimensional case.) For finite-dimensional Hamiltonian systems on a symplectic manifold (a Poisson manifold with a non-degenerate Poisson tensor) of dimension $2n$, integrability in the sense of Liouville (1855) and Arnold (1974) is defined by the requirement that there exist n conserved quantities that are functionnally independent on a dense open set and in involution, i. e., whose pairwise Poisson brackets vanish. Geometric methods are then applied in various ways.

The Editors, Introduction, Lect. Notes Phys. **638**, 1–4 (2004)
http://www.springerlink.com/

1 Analytic Methods

The inverse scattering method (ISM), using the inverse scattering transform (IST), is closely related to the Riemann-Hilbert factorization problem and to the $\bar{\partial}$ method. This is the subject of M.J. Ablowitz's survey, "Nonlinear waves, solitons and IST", which treats IST for equations both in one space variable, $(1 + 1)$-dimensional problems, and in 2 space variables, $(2 + 1)$-dimensional problems, and whose last section contains a review of recent work on the self-dual Yang-Mills equations (SDYM) and their reductions to integrable systems.

2 Painlevé Analysis

In the Painlevé test for an ordinary differential equation, the time variable is complexified. If all movable critical points of the solutions are poles, the equation passes the test. It contributes to the determination of the integrability or non-integrability of nonlinear equations, defined in terms of their solvability by means of an associated linear problem. In the Ablowitz-Ramani-Segur method for the detection of integrability, the various ordinary differential equations that arise as reductions of a given nonlinear partial differential equation are tested for the Painlevé property.

In their survey, "Analytic and asymptotic methods for nonlinear singularity analysis", M.D. Kruskal, N. Joshi and R. Halburd review the Painlevé property and its generalizations, the various methods of singularity analysis, and recent developments concerning irregular singularities and the preservation of the Painlevé property under asymptotic limits.

The review by B. Grammaticos and A. Ramani, "Integrability", describes the various definitions of integrability, their comparison and implementation for both finite- and infinite-dimensional systems, and for both continuous and discrete systems, including some recent results obtained in collaboration with K.M. Tamizhmani. The method of singularity confinement, a discrete equivalent of the Painlevé method, is explained and applied to the discrete analogues of the Painlevé equations.

3 τ-functions, Bilinear and Trilinear Forms

Hirota's method is the most efficient known for the determination of soliton and multi-soliton solutions of integrable equations. Once the equation is written in bilinear form in terms of a new dependent variable, the τ-function, and of Hirota's bilinear differential operators, multi-soliton solutions of the original nonlinear equation are obtained by combining soliton solutions. J. Hietarinta's "Introduction to the Hirota bilinear method" is an outline of the method with examples, while J. Satsuma's "Bilinear formalism in soliton theory" develops the theory further, treats the bilinear identities satisfied by the τ-functions, and shows how the method can be generalized to a trilinear formalism valid for multi-dimensional extensions of the soliton equations, to the q-discrete and ultra-discrete cases and how it can be applied to the study of cellular automata.

4   Lie-Algebraic and Group-Theoretical Methods

When the Poisson brackets of the matrix elements of the Lax matrix, viewed as linear functions on a Lie algebra of matrices, can be expressed in terms of a so-called "r-matrix", the traces of powers of the Lax matrix are in involution, and in many cases the integrability of the original nonlinear system follows. It turns out that a Lie algebra equipped with an "r-matrix" defining a Poisson bracket, e.g., satisfying the classical or modified Yang-Baxter equation (CYBE, MYBE) is a special case of a Lie bialgebra, the infinitesimal object associated with a Lie group equipped with a Poisson structure compatible with the group multiplication, called a Poisson Lie group. Poisson Lie groups play a role in the solution of equations on a 1-dimensional lattice, and they are the ingredients of the geometric theory of the dressing transformations for wave functions satisfying a zero-curvature equation under elements of the "hidden symmetry group". The quantum version of these objects, quantum R-matrices satisfying the quantum Yang-Baxter equation (QYBE), and quantum groups are the ingredients of the quantum inverse scattering method (QISM), while the Bethe Ansatz, constructing eigenvectors for a quantum Hamiltonian by applying creation operators to the vacuum, can be interpreted in terms of the representation theory of quantum groups associated with Kac-Moody algebras.

The lectures by Y. Kosmann-Schwarzbach, "Lie bialgebras, Poisson Lie groups and dressing transformations", are an exposition, including the proofs of all the main results, of the theory of Lie bialgebras, classical r-matrices, Poisson Lie groups and Poisson actions.

The survey by M.A. Semenov-Tian-Shansky, "Quantum and classical integrable systems", treats the relation between the Hamiltonians of a quantum system solvable by the quantum inverse scattering method and the Casimir elements of the underlying hidden symmetry algebra, itself the universal enveloping algebra of a Kac-Moody algebra or a q-deformation of such an algebra, leading to deep results on the spectrum and the eigenfunctions of the quantum system. This study is preceded by that of the analogous classical situation which serves as a guide to the quantum case and utilizes the full machinery of classical r-matrices and Poisson Lie groups, and the comparison between the classical and the quantum cases is explicitly carried out.

5   Bihamiltonian Structures

When a dynamical system can be written in Hamiltonian form with respect to two Hamiltonian structures, which are compatible, in the sense that the sum of the corresponding Poisson brackets is also a Poisson bracket, this dynamical system possesses conserved quantities in involution with respect to both Poisson brackets. This fundamental idea, due to F. Magri, is the basis of "Eight lectures on integrable systems", by F. Magri, P. Casati, G. Falqui, and M. Pedroni, where they develop the geometry of bihamiltonian manifolds and various reduction theorems in Poisson geometry, before applying the results to the theory of both infinite- and finite-dimensional soliton equations. They show which reductions yield the Gelfand-Dickey and the Kadomtsev-Petviashvilii equations, and they derive the bihamiltonian structure of the Calogero system.

The surveys included in this volume treat many aspects of the theory of nonlinear systems, they are different in spirit but not unrelated. For example, there is a parallel, which deserves further explanation, between the role of q-analysis in the theory of discrete integrable sytems and that of q-deformations of algebras of functions on Lie groups and of universal enveloping algebras of Lie algebras in the theory of quantum integrable systems, while the r-matrix method for classical integrable systems on loop algebras, which seems to be purely algebraic, is in fact an infinitesimal version of the Riemann-Hilbert factorization problem.

The theory of nonlinear systems, and in particular of integrable systems, is related to several very active fields of theoretical physics. For instance, the role played in the theory of integrable systems by infinite Grassmannians (on which the τ-function "lives"), the boson-fermion correspondence, the representation theory of W-algebras, the Virasoro algebra in particular, all show links with conformal field theory.

We hope that this book will permit the reader to study some of the many facets of the theory of nonlinear systems and their integrability, and to follow their future developments, both in mathematics and in theoretical physics.

Nonlinear Waves, Solitons, and IST

M.J. Ablowitz

Department of Applied Mathematics, Campus Box 526, University of Colorado at Boulder, Boulder Colorado 80309-0526, USA
markjab@newton.colorado.edu

Abstract. These lectures are written for a wide audience with diverse backgrounds. The subject is approached from a general perspective and overly detailed discussions are avoided. Many of the topics require only a standard background in applied mathematics.

The lectures deal with the following topics: fundamentals of linear and nonlinear wave motion; isospectral flows with associated compatible linear systems including PDE's in 1+1 and 2+1 dimensions, with remarks on differential-difference and partial difference equations; the Inverse Scattering Transform (IST) for decaying initial data on the infinite line for problems in 1+1 dimensions; IST for 2+1 dimensional problems; remarks on self-dual Yang-Mills equations and their reductions. The first topic is extremely broad, but a brief review provides motivation for the other subjects covered in these lectures.

1 Fundamentals of Waves

Water waves are an interesting physical model and a natural way for us to begin our discussion. Consequently let us consider the equations of water waves for an irrotational, incompressible, inviscid fluid:

$$\bigtriangledown^2 \phi = 0 \quad \text{in} \quad -h < z < \eta \tag{1.1}$$

$$\frac{\partial \phi}{\partial z} = 0 \quad \text{on} \quad z = -h \tag{1.2}$$

$$\frac{\partial \eta}{\partial t} + \bigtriangledown \phi \cdot \bigtriangledown \eta = \frac{\partial \phi}{\partial z} \quad \text{on} \quad z = \eta \tag{1.3}$$

$$\frac{\partial \phi}{\partial t} + g\eta + \frac{1}{2} |\bigtriangledown \phi|^2 = 0 \quad \text{on} \quad z = \eta , \tag{1.4}$$

where η denotes the free surface, and, since the fluid is ideal, the velocity is derivable from a potential, $\bar{u} = \bigtriangledown \phi$. For simplicity, we shall assume waves in one dimension, $\eta = \eta(x,t)$, $\bar{u} = (u,w) = \left(\dfrac{\partial \phi}{\partial x}, \dfrac{\partial \phi}{\partial z} \right)$, $\phi = \phi(x,z,t)$. It will be convenient for us to consider the linearized equations whereby we expand the free surface conditions (1.3), the kinematic equation of a free surface, and (1.4), the Bernoulli equation, around $z = 0$:

$$\frac{\partial \eta}{\partial t} = \frac{\partial \phi}{\partial z} \quad \text{on} \quad z = 0 , \tag{1.3a}$$

M.J. Ablowitz, Nonlinear Waves, Solitons, and IST, Lect. Notes Phys. **638**, 5–29 (2004)
http://www.springerlink.com/   © Springer-Verlag Berlin Heidelberg 2004

$$\frac{\partial \phi}{\partial t} + g\eta = 0 \quad \text{on} \quad z = 0 . \tag{1.4a}$$

The linear dispersion relation is obtained by looking for a special solution of the form $\phi(x, z, t) = \text{Re}\,(\Phi(z)e^{i(kx-\omega t)})$, $\eta = \text{Re}\,(Ne^{i(kx-\omega t)})$. Equations (1.1) and (1.2) imply

$$\Phi = A\cosh k(z+h),$$

and (1.3a-1.4a) yields the linear dispersion relationship,

$$\omega^2 = gk\tanh kh. \tag{1.5}$$

We also note that one could easily add surface tension and consider two-dimensional waves in the above discussion. In this case the right-hand side of (1.4) would have the surface tension term,

$$\frac{T}{\rho} \frac{(\eta_{xx}(1+\eta_y^2) + \eta_{yy}(1+\eta_x^2) - 2\eta_{xy}\eta_x\eta_y)}{(1+\eta_x^2+\eta_y^2)^{3/2}} ,$$

where now $\eta = \eta(x, y, t)$ and T is the coefficient of surface tension. The dispersion relation is obtained by looking for wave-like solutions such as $\eta = \text{Re}\,(Ne^{i(kx+ly-\omega t)})$ and one finds

$$\begin{aligned} \omega^2 &= (g\kappa + T\kappa^3)\tanh \kappa h, \\ \kappa^2 &= k^2 + l^2, \end{aligned} \tag{1.6}$$

which the reader can verify.

Rather than proceeding with two-dimensional waves, we will first discuss the one-dimensional situation. In the case of long waves (shallow water), $|kh| << 1$, (1.5) yields,

$$\omega^2 = ghk^2(1 - \frac{1}{3}(kh)^2 + \dots). \tag{1.7}$$

The first approximation is $\omega^2 = c_0^2 k^2$, $c_0^2 = gh$ (c_0 is the long wave speed) which is the dispersion relation of the linear wave equation,

$$\eta_{tt} - c_0^2 \eta_{xx} = 0 . \tag{1.8}$$

Note that the identifications $\omega \to i\frac{\partial}{\partial t}$, $k \to -i\frac{\partial}{\partial x}$ in $\omega^2 = c_0^2 k^2$ yields (1.8). On the other hand, if one considers "unidirectional waves," by taking the square root of (1.7),

$$\omega = c_0 k(1 - \frac{1}{6}(kh)^2 \dots), \tag{1.9}$$

the above identifications for ω, k imply the equation,

$$\frac{\partial \eta}{\partial t} + c_0\frac{\partial \eta}{\partial x} + \alpha\frac{\partial^3 \eta}{\partial x^3} = 0, \tag{1.10}$$

where $\alpha = \dfrac{c_0 h^2}{6}$. In fact, a more careful asymptotic analysis allows one to derive the Korteweg-de Vries (KdV) and Kadomtsev-Petviashvili (KP) equations even with surface tension included. The asymptotic description is valid when the following conditions hold, considering two-dimensional waves, i.e., this is relevant to the derivation of the KP equation:

i)   wave amplitudes are small: $\varepsilon = |\eta|_{\max}/h \ll 1$,
ii) shallow water–long waves: $(\kappa h)^2 \ll 1$,
iii) slow transverse variations: $(m/k)^2 \ll 1$,
iv) maximal balance: $0((m/k)^2) = 0((\kappa h)^2) = 0(\varepsilon)$.

With these conditions we consider unidirectional waves; namely, since the solution of the wave equation (1.8) above has both right- and left-going waves:

$$\eta = f(x - c_0 t, y) + g(x + c_0 t, y), \qquad (1.11)$$

we consider initial values which select, say right-going waves, i.e., $g = 0$. In this case the following KP equation is found,

$$\frac{\partial}{\partial x}\left(\eta_t + c_0 \eta_x + \frac{3c_0}{2h}\eta\eta_x + \gamma\eta_{xxx}\right) + \frac{1}{2}\eta_{yy} = 0, \qquad (1.12)$$

where $\gamma = h^2(1 - \hat{T})/6$; $\hat{T} = T/3pgh^2$. The KdV equation,

$$\eta_t + c_0 \eta_x + \frac{3c_0}{2h}\eta\eta_x + \gamma\eta_{xxx} = 0, \qquad (1.13)$$

results if η is independent of y and $\eta \to 0$ as $x \to \infty$. The normalized KP equations result by rescaling η, t, x, y (we leave this to the reader to verify),

$$(u_t + 6uu_x + u_{xxx})_x + 3\sigma^2 u_{yy} = 0, \qquad (1.14)$$

$\sigma^2 = \pm 1$ ($\sigma^2 = \operatorname{sgn}(1 - \hat{T})$). We see that there are two physically interesting choices of sign depending on sgn $(1 - \hat{T})$. In the usual situation, the surface tension is taken to be negligible, hence $\sigma^2 = +1$; in the literature this is often called the KPII equation. When surface tension is large enough for sgn $(1 - \hat{T}) = -1 = \sigma^2$, (1.14) is called the KPI equation.

The normalized form of the KdV equation which follows from (1.14) is

$$u_t + 6uu_x + u_{xxx} = 0. \qquad (1.15)$$

In fact, the KdV equation is comprised of two parts, both of which are fundamentally important in the study of wave phenomena, namely first-order quasilinear hyperbolic waves,

$$u_t + 6uu_x = 0, \qquad (1.16)$$

and a linear dispersive wave equation,

$$u_t + u_{xxx} = 0. \qquad (1.17)$$

The quasilinear equation (1.16) has solutions which become multi-valued in finite time. This is due to the fact that the characteristics of the equation satisfy

$$\frac{dx}{dt} = 6u, \tag{1.18}$$

while at the same time $u = $ const along each characteristic. Thus (1.18) can be integrated

$$x = 6F(\xi)t + \xi, \tag{1.19}$$

where $x = \xi$ denotes the particular characteristic at $t = 0$, and at $t = 0$, $u(\xi, 0) = F(\xi)$. Thus the solution of (1.16) is given by

$$u(x, t) = F(\xi), \tag{1.20}$$

where $\xi = \xi(x, t)$ is given by solving the implicit equation (1.19). It follows from (1.19)–(1.20) that *any* decaying lump-like initial data will lead to crossing of characteristics, i.e., from (1.18), larger positive values of u travel faster than smaller values, and multi-valuedness of the solution.

The usual mechanism to arrest multi-valuedness, or crossing of characteristics, is to supplement (1.16) with a small term with higher-order derivatives, e.g., the Burgers equation (cf. Whitham, [1]); in this reference a general review of linear and nonlinear waves is given],

$$u_t + 6uu_x = \varepsilon u_{xx}. \tag{1.21}$$

Indeed, the Hopf-Cole transformation,

$$u = -\frac{\varepsilon}{3}\frac{\partial}{\partial x}\log\phi = -\frac{\varepsilon}{3}\frac{\phi_x}{\phi}, \tag{1.22}$$

linearizes (1.21) to

$$\phi_t = \varepsilon\phi_{xx}, \tag{1.23}$$

which can be solved by transform methods.

From the exact solution it is found that (1.16) is the main solution for (1.21) for $0 < \varepsilon << 1$ until, asymptotically speaking, the characteristics almost cross. Then an asymptotically thin shock wave is formed, which, for decaying initial data, vanishes as $t \to \infty$.

However, as shown by Zabusky and Kruskal [2], the KdV equation behaves quite differently from Burgers equation. In [2] it was shown that the KdV equation (1.15), with a small coefficient in front of the u_{xxx} term, i.e., replace u_{xxx} by $\varepsilon^2 u_{xxx}$, $0 < \varepsilon << 1$, which can be obtained from (1.15) by rescaling x, t, develops numerous special hump-like travelling waves, referred to as solitons (solitons will be discussed later in more detail), in the asymptotic region near the time of multi-valuedness. The solitons move to the right, away from the front, and, as $t \to \infty$, a dispersive tail is left behind (cf. [3]). The tail vanishes as $t \to \infty$. Years later, researchers studied the asymptotic problem of KdV with a small dispersive term in more detail (cf. [4]).

Finally, it is worth remarking upon the solution of the linear equation (1.17) with $u(x,0) = f(x)$ given, and decaying sufficiently just as $|x| \to \infty$. The solution is obtained by Fourier transforms and is found to be

$$u(x,t) = \frac{1}{2\pi} \int_{-\alpha}^{\infty} b_0(k) e^{i(kx+k^3 t)} dk, \tag{1.24}$$

where $b_0(k) = \int_{-\infty}^{\infty} f(x) e^{-ikx} dx = \hat{u}(k,0)$, and where $\hat{u}(k,t)$ denotes the Fourier transform at any time t. While the general solution establishes existence, qualitative information can be obtained by further study. Asymptotic analysis as $t \to \infty$ (stationary phase-steepest descent methods) establishes that the solution decays as

$$u(x,t) \sim \frac{1}{(3t)^{1/3}} \left(\frac{b_0(|z|) + b_0(-|z|)}{2} \right) Ai(z)$$
$$+ \frac{1}{(3t)^{2/3}} \left(\frac{b_0(|z|) - b_0(-|z|)}{2i|z|} \right) Ai'(z), \tag{1.25}$$

where $z = x/(3t)^{1/3}$ and $Ai(z)$ is the Airy function,

$$Ai(z) = \frac{1}{2\pi} \int_{-\infty}^{\infty} \exp\left(i\left(sz + \frac{s^3}{3} \right) \right) ds, \tag{1.26}$$

(cf. [3] for further details).

The method of Fourier transforms generalizes readily, e.g., it is applicable to any evolutionary PDE with constant coefficients. The scheme for solving a linear problem with a dispersion relation $\omega(k)$, e.g., (1.17) where $\omega(k) = -k^3$, by Fourier transforms is as follows:

$$u(x,0) \xrightarrow{\text{Fourier Transform}} \hat{u}(k,0)$$

$$u(x,t) \xleftarrow{\text{Inverse Fourier Transform}} \hat{u}(k,t) = \hat{u}(k,0) e^{-i\omega(k)t}$$

In fact, as we shall discuss in Sect. 3, the method for solving nonlinear wave equations, such as KdV and KP, which is referred to as the Inverse Scattering Transform (IST), is in many ways a natural generalization of Fourier transforms.

2   IST for Nonlinear Equations in 1+1 Dimensions

The KdV equation (1.15) was the first equation solved (on the infinite line with appropriately decaying data) by inverse scattering methods [5]. Subsequently Zakharov and Shabat [24] showed that the nonlinear Schrödinger equation (NLS), which arises as a centrally important equation in fluid dynamics, nonlinear optics and plasma physics,

$$iu_t + u_{xx} + 2s|u|^2u = 0, \tag{2.1}$$

$(s = +1)$ could be solved by similar methods. Shortly after this, a procedure was developed by which KdV, NLS $(s = \pm1)$, modified KdV (mKdV),

$$u_t = 6su^2u_x + u_{xxx} = 0, \tag{2.2}$$

$s = \pm1$, sine-Gordon,

$$u_{xt} = \sin u, \tag{2.3}$$

and indeed a class of nonlinear evolution equations could be solved [6]. The method was termed the Inverse Scattering Transform (IST).

The essential idea is to associate nonlinear evolution equations with the compatibility of two linear operators [3, 6],

$$v_x = Xv \tag{2.4a}$$

$$v_t = Tv, \tag{2.4b}$$

where v is an n-dimensional vector, and X, T are $n \times n$ matrices. Compatibility of (2.4a,b) implies that $v_{xt} = v_{tx}$, hence X, T satisfy

$$X_x - T_t + [X, T] = 0,$$

where $[X, T] = XT - TX$. It is easiest to consider a concrete situation where X is specified. Thus, consider the 2×2 linear equation for (2.4a),

$$v_{1,x} = -ikv_1 + qv_2 \tag{2.5a}$$

$$v_{2,x} = ikv_2 + rv_1 \tag{2.5b}$$

where q, r are functions of x, t and for (2.4b) we consider

$$v_{1,t} = Av_1 + Bv_2 \tag{2.6a}$$

$$v_{2,t} = Cv_1 + Dv_2, \tag{2.6b}$$

where A, B, C, D are scalar functions of q, r, and k; k is a parameter which is essential in the method of direct and inverse scattering. Compatibility of (2.5)–(2.6) implies $D = -A$, and

$$\begin{aligned} A_x &= qC - rB \\ B_x + 2ikB &= q_t - 2Aq \\ C_x - 2ikC &= r_t + 2Ar. \end{aligned} \tag{2.7}$$

In [3,6] it is shown that finite power series solutions for $A, B, C,$ $\quad A = \sum_{j=0}^{N} A_j k^j$
etc., lead to nonlinear equations. Important special cases are listed below. When $N = 2$, $q = \mp r^*(= u)$ results in the NLS equation (2.1) (with either choice of sign $s = \pm 1$). When $N = 3$ and if $r = -1$, $q \equiv u$, we find the KdV equation (1.15); and when $q = \mp r \equiv u$, the mKdV equation (2.2) is obtained. If $N = 1$, the sine-Gordon equation (2.3) results if $q = -r = u_x/2$ and the sinh-Gordon equation,

$$u_{xt} = \sinh u, \tag{2.8}$$

results if $q = r = u_x/2$.

In fact, in [4,6] it is further shown that the following general class of nonlinear equations is compatible with (2.5)–(2.6),

$$\begin{pmatrix} r \\ -q \end{pmatrix}_t + 2A_0(\mathcal{L}) \begin{pmatrix} r \\ q \end{pmatrix} = 0, \tag{2.9}$$

where $A_0(k) = \lim_{|x| \to \infty} A(x, t, k)$ $(A_0(k)$ may be a ratio of entire functions), and $\mathcal{L}$ is an integro-differential operator given by

$$\mathcal{L} = \frac{1}{2i} \begin{pmatrix} \partial/\partial x - 2rI_- q & 2I_- r \\ -2qI_- q & -\partial/\partial x + 2qI_- r \end{pmatrix}, \tag{2.10}$$

where $I_- \equiv \int_{-\infty}^{x}$ and $A_0(k)$ is related to the linear dispersion relation,

$$A_0(k) = \frac{i}{2}\omega_r(2k) = -\frac{-i}{2}\omega_q(-2k), \tag{2.11}$$

where $r = \exp(i(kx - \omega_r t))$ and $q = \exp(i(kx - \omega_q t))$ are linearized wave solutions.

There are numerous generalizations and extensions of these ideas. The reader is encouraged to consult the many papers and monographs related to this subject (cf. [7] for an extensive bibliography).

3   Scattering and the Inverse Scattering Transform

In this section we quote the main results in the scattering theory of the linear problem associated with the KdV equation (1.15). In fact, as discussed in Sect. 3, the KdV equation arises from (2.5)–(2.6) with $r = -1$, $q \equiv u$, and taking a finite expansion in powers of k, $A = \sum_{j=0}^{3} A_j k^j$, etc. In fact, in this case, (2.5)–(2.6) takes a simpler form when written as a scalar system,

$$v_{xx} + (u(x, t) + k^2)v = 0 \tag{3.1}$$

$$v_t = (u_x + \gamma)v + (4k^2 - 2u)v_x, \tag{3.2}$$

where γ is an arbitrary constant. The reader can verify that (3.1)-(3.2) are consistent with the KdV equation (1.15), assuming $\partial k/\partial t = 0$, i.e., it is an isospectral flow.

Scattering theory is developed on the spatial part of the compatible system – in this case (3.1). In fact, (3.1) is the well-known time-independent Schrödinger equation. Scattering and inverse scattering associated with (3.1) have a long history which we shall not review here. The results described below hold for functions $u(x,t)$ decaying sufficiently fast at infinity, e.g., satisfying

$$\int_{-\infty}^{\infty} (1 + |x|^2)|u(x)|dx < \infty. \tag{3.3}$$

The scattering analysis which we require is dependent on the analytic behavior of certain eigenfunctions defined by the following boundary conditions at infinity:

$$\begin{aligned}
\phi(x,k) &\sim e^{-ikx}, \quad x \to -\infty \\
\psi(x,k) &\sim e^{ikx}, \quad \bar{\psi}(x,k) \sim e^{-ikx}, \quad x \to +\infty.
\end{aligned} \tag{3.4}$$

Note that, for convenience, we have suppressed the variable t in ϕ, ψ. Since (3.1) is invariant under $k \to -k$, the boundary conditions (3.4) imply that

$$\psi(x,k) = \bar{\psi}(x,-k). \tag{3.5}$$

The fact that (3.1) is a second order equation means that $\phi, \psi, \bar{\psi}$ are linearly related,

$$\phi(x,k) = a(k)\bar{\psi}(x,k) + b(k)\psi(x,k). \tag{3.6}$$

Actually $a(k)$ and $b(k)$ are readily related to the reflection coefficient $r(k)$ and the transmission coefficient $t(k)$ in quantum mechanics, $k \in \mathbb{R}$,

$$r(k) = b(k)/a(k), \quad t(k) = 1/a(k), \tag{3.7}$$

and the properties of the differential equation (Wronskian relation) guarantee that

$$|r(k)|^2 + |t(k)|^2 = 1. \tag{3.8}$$

It is convenient to work with modified eigenfunctions $\mu(x,k) = v(x,k)e^{ikx}$, defined as follows:

$$\begin{aligned}
M(x,k) &= \phi(x,k)e^{ikx}, \\
N(x,k) &= \psi(x,k)e^{ikx}, \\
\bar{N}(x,k) &= \bar{\psi}(x,k)e^{ikx}.
\end{aligned} \tag{3.9}$$

Symmetry condition (3.5) translates to

$$N(x,k) = \bar{N}(x,-k)e^{2ikx} \tag{3.10}$$

and (3.6) takes the form,

$$\frac{M(x,k)}{a(k)} = \bar{N}(x,k) + r(k)e^{2ikx}\bar{N}(x,-k). \tag{3.11}$$

Indeed, (3.11) is a fundamental equation in this approach. It is a generalized Riemann-Hilbert boundary value problem (RHBVP), which is a consequence of the following facts:

i) $M(x,k)$ and $a(k)$ can be analytically extended to the upper half k-plane (UHP) and tend to unity as $|k| \to \infty$ (Im $k > 0$).

ii) $\bar{N}(x,k)$ can be analytically extended to the lower half k-plane (LHP) and tends to unity as $|k| \to \infty$ (Im $k < 0$).

These analytic properties can be proven by a careful analysis of the integral equations which represent M and $\bar{N}$, namely,

$$M(x,k) = 1 + \int_{-\infty}^{\infty} G_+(x - \xi, k)u(\xi)M(\xi, k)d\xi,$$

$$\bar{N}(x,k) = 1 + \int_{-\infty}^{\infty} G_-(x - \xi, k)u(\xi)\bar{N}(\xi, k)d\xi,$$

(3.12)

where

$$G_\pm(x,k) = \frac{1}{2\pi} \int_{C_\pm} \frac{e^{ipx}}{p(p - 2k)}$$

(3.13)

and $C_\pm$ are contours from $-\infty$ to ∞ which are indented with small semi-circles that pass just below the singularities in the case of C_+, just above in the case of C_-. The singularities are at $p = 0$ and $p = 2k$. Explicit formulae are given by

$$G_+(x,k) = \frac{1}{2ik}(1 - e^{2ikx})\theta(x)$$

(3.14a)

and

$$G_-(x,k) = -\frac{1}{2ik}(1 - e^{2ikx})\theta(-x)$$

(3.14b)

where $\theta(x)$ is the Heaviside function,

$$\theta(x) = \begin{cases} 1 \text{ if } x > 0, \\ 0 \text{ if } x < 0. \end{cases}$$

(3.15)

Expanding (3.12), which are Volterra integral equations, in a Neumann series and suitably bounding the terms of the series establishes that the series consist of uniformly convergent series of analytic functions in their regions of convergence. Moreover, relation

$$a(k) = 1 + \frac{1}{2ik} \int_{-\infty}^{\infty} u(x)M(x;k)dx$$

(3.16)

establishes the analytic behavior of $a(k)$ (recall that $M(x;k)$ is analytic for Im $k > 0$). In fact, the zeroes of $a(k)$ are relevant in what follows. It can be established that for $u(x)$ real there are a finite number of simple zeroes, all of which lie on the imaginary k axis: $a(k_j) = 0$, $k_j = ik_j$, $k_j > 0$, $j = 1, \ldots N$. Sometimes the RHBVP (3.11) is written as

$$(\mu_+(x, k) - \mu_-(x, k)) = r(k)e^{2ikx}\mu_-(x, -k), \tag{3.17}$$

where $\mu_+(x, k) = M(x, k)/a(k)$, $\mu_-(x, k) = \bar{N}(x, k)$. We also note the relationship

$$b(k) = -\frac{1}{2ik}\int_{-\infty}^{\infty} u(x)M(x, k)e^{-2ikx}dx. \tag{3.18}$$

From (3.16), (3.18) we can compute $r(k), t(k)$ (see (3.7)) in terms of initial data. The above relationships follow from the direct problem. Namely, given $u(x)$ satisfying (3.3), then $a(k)$ and $b(k)$ are given by (3.16), (3.18), hence $r(k)$ from (3.7) and the analytic properties from the integral equations (3.12)–(3.13) and the RHBVP from (3.17).

The inverse problem requires giving appropriate scattering data to uniquely determine a solution to (3.17). Equation (3.17) is a RHBVP with a shift. In fact, there is no closed form solution to (3.17). The best one can do is to transform (3.17) to an integral equation or system of integral equations determined by the scattering data.

The simplest case occurs where $a(k) \neq 0$ for Im $k > 0$. Taking a minus projection of (3.17) (after subtracting unity from both $\mu_\pm(x, k)$) where the projection operators are defined by

$$\mathcal{P}^\pm f(k) = \frac{1}{2\pi i}\int_{-\infty}^{\infty} \frac{f(\zeta)}{\zeta - (k \pm i0)}d\zeta \tag{3.19}$$

and noting that

$$\begin{align} \mathcal{P}^\pm f_\mp(k) &= 0 \\ \mathcal{P}^\pm f_\pm(k) &= \pm f_\pm(k), \end{align} \tag{3.20}$$

where $f_\pm(k)$ are analytic functions for Im$k \gtrless 0$ and $f_\pm(k) \to 0$ as $k \to \infty$, yields

$$\bar{N}(x, k) = 1 + \frac{1}{2\pi i}\int_{-\infty}^{\infty} \frac{r(\zeta)N(x, \zeta)}{\zeta - (k - i0)}d\zeta \tag{3.21}$$

or, using (3.10),

$$N(x, k) = e^{2ikx}\left\{1 + \frac{1}{2\pi i}\int_{-\infty}^{\infty} \frac{r(\zeta)N(x, \zeta)}{\zeta + (k + i0)}d\zeta\right\}. \tag{3.22}$$

Expanding (3.21) as $k \to \infty$ yields

$$\bar{N}(x, k) \sim 1 - \frac{1}{2\pi ik}\int_{-\infty}^{\infty} r(\zeta)N(x, \zeta)d\zeta. \tag{3.23}$$

On the other hand, from (3.12) (or directly from (3.1)), one establishes that

$$\bar{N}(x, k) \sim 1 - \frac{1}{2ik}\int_{x}^{\infty} u(\xi)d\xi \tag{3.24}$$

and upon comparing (3.23)–(3.24) we see that

$$u(x) = -\frac{\partial}{\partial x} \frac{1}{\pi} \int_{-\infty}^{\infty} r(\zeta) N(x, \zeta) d\zeta. \tag{3.25}$$

Thus, given the scattering data, $r(\zeta)$, we can construct a solution of the integral equation (3.22) and the potential $u(x)$.

Equations (3.22), (3.25) can be simplified by looking for solutions with a certain structure,

$$N(x, k) = e^{2ikx} \left\{ 1 + \int_{x}^{\infty} K(x, s) e^{ik(s-x)} ds \right\}. \tag{3.26}$$

Substituting (3.26) into (3.22) and operating on the result with

$$\frac{1}{2\pi} \int_{-\infty}^{\infty} dk \ e^{ik(x-y)}$$

for $y > x$, we find the so-called Gel'fand-Levitan-Marchenko equation (GLM),

$$K(x, y) + F(x + y) + \int_{x}^{\infty} K(x, s) F(s + y) \ ds = 0, \quad (y > x), \tag{3.27}$$

where $F(x) = F_c(x) = \dfrac{1}{2\pi} \displaystyle\int_{-\infty}^{\infty} r(k) e^{ikx} dk$. Similarly, substitution of (3.26) into (3.25) yields

$$u(x) = 2\frac{\partial}{\partial x} K(x, x). \tag{3.28}$$

So far we have not allowed for the possibility that $a(k)$ can vanish for Im $k > 0$. If $a(k)$ vanishes at $k_j = i\kappa_j$, $\kappa_j > 0$, $j = 1, \ldots N$, the final result is that the GLM equation is only modified by changing the function $F(x)$:

$$F(x) = F_c(x) + F_d(x) \tag{3.29}$$

where $F_c(x)$ is given below (3.27) and $F_d(x)$ is defined by

$$F_d(x) = \sum_{j=1}^{N} C_j \exp(-\kappa_j x), \tag{3.30}$$

where C_j are certain normalizing coefficients related to $\phi(x, k_j) = \hat{C}_j \psi(x, k_j)$, where $C_j = -i\hat{C}_j / a'(k_j)$.

Thus the complete solution of the inverse problem is as follows. Given the scattering data $S(k) = \{r(k), \{\kappa_j, C_j\}_{j=1}^{N}\}$, we form $F(x)$ from (3.29), solve the GLM equation (which results from the RHBVP (3.11) or (3.17)) for $K(x, y)$ and obtain the potential $u(x)$ from (3.28). In fact, the procedure for inverse scattering of the 2×2 problem (2.5) and many other scattering problems related to integrable equations in $1 + 1$ dimensions is similar. We refer the reader to [3,6,7] for more details.

The inverse scattering transform is completed when one determines the time-dependence of the scattering data. For this purpose, we consider the time-evolution equation (3.2), or, for a fixed eigenfunction, the equation satisfied by $M = \phi e^{ikx}$, where ϕ is defined by (3.4),

$$M_t = (\gamma - 4ik^3 + u_x + 2iku)M + (4k^2 - 2u)M_x \qquad (3.31)$$

with the asymptotic behaviors,

$$\begin{aligned} M &\to 1 & \text{as} \quad x \to -\infty, \\ M &= a(k,t) + b(k,t)e^{2ikx} & \text{as} \quad x \to +\infty. \end{aligned} \qquad (3.32)$$

Note that we now denote all functions with explicit time-dependence. The latter equation of (3.32) follows from (3.4), (3.6), (3.9). These asymptotic relations imply

$$\gamma = 4ik^3, \qquad a_t = 0, \qquad b_t = 8ik^3 b. \qquad (3.33)$$

Thus

$$a(k,t) = a(k,0) \qquad (3.34a)$$

$$b(k,t) = b(k,0)e^{8ik^3 t} \qquad (3.34b)$$

$$v(k,t) = r(k,0)e^{8ik^3 t} = r_0(k)e^{8ik^3 t}, \qquad (3.34c)$$

and from (3.34a), the discrete eigenvalues, $k_j = i\kappa_j$, $j = 1,\ldots N$, are clearly constants of the motion. It can also be readily shown that the normalization constants satisfy simple evolution equations,

$$C_j(t) = C_j(0)e^{8ik_j^3 t} = C_j(0)e^{8\kappa_j^3 t}. \qquad (3.34d)$$

Thus, $F(x,t)$ in the GLM equation (3.27) is given by

$$F(x,t) = \frac{1}{2\pi} \int_{-\infty}^{\infty} r_0(k)e^{ikx+8ik^3 t} dk + \sum_{j=1}^{N} C_j(0)e^{-\kappa_j x + 8\kappa_j^3 t} . \qquad (3.35)$$

This kernel fixes the integral equation (3.27), and hence the solution $u(x,t)$ to the KdV equation follows from (3.28) in terms of initial data. The scheme of solution is similar to that of Fourier transforms:

$$u(x,0) \xrightarrow{\quad \text{Direct Scattering} \quad} S(k,0) = \{r_0(k), \{k_j, C_j(0)\}_{j=1}^{N}\}$$

$$\big\downarrow \quad \text{Time evolution of scattering data}$$

$$u(x,t) \xleftarrow{\quad \text{Inverse Scattering} \quad} S(k,t) = \{r(k,t), \{k_j, C_j(t)\}_{j=1}^{N}\}.$$

There are a number of results which follow from the above developments. a) For scattering data that correspond to potentials satisfying (3.3), the solution to the GLM equation exists. With suitable conditions on u and its derivatives, global solutions to the KdV equation can be established. b) Long-time asymptotic analysis of the KdV equation can be ascertained. The solution is comprised of a discrete part consisting of N soliton (see below) waves moving to the right, and a dispersive tail which decays algebraically as $t \to \infty$. c) The discrete part of the spectrum can be solved in terms of a linear algebraic system. In the GLM equation, the discrete spectrum corresponds to a degenerate kernel. From the RHBVP, the following linear system results,

$$N_l(x,t) + \sum_{p=1}^{N} \frac{C_p(0)}{i(\kappa_p + \kappa_l)} \exp(-2\kappa_p x + 8\kappa_p^3 t) N_p(x,t) = \exp(-2\kappa_l x) \qquad (3.36)$$

where $N_l(x,t) \equiv N(x, k = i\kappa_l, t)$, and from the solution of (3.36) we reconstruct the solution of KdV, $u(x,t)$, via

$$u(x,t) = 2i \frac{\partial}{\partial x} \sum_{p=1}^{N} C_p(t) N_p(x,t). \qquad (3.37)$$

A one-soliton solution $(N = 1)$ is given by

$$u(x,t) = 2\kappa_1^2 \operatorname{sech}^2 \kappa_1(x - 4\kappa_1^2 t - x_1) \qquad (3.38)$$

where $C_1(0) = 2\kappa_1 \exp(2\kappa_1 x_1)$, and a two-soliton solution $(N = 2)$ is given by

$$u(x,t) = \frac{4(\kappa_2^2 - \kappa_1^2)[(\kappa_2^2 - \kappa_1^2) + \kappa_1^2 \cosh(2\kappa_2 \xi_2) + \kappa_2^2 \cosh(2\kappa_1 \xi_1)]}{[(\kappa_2 - \kappa_1) \cosh(\kappa_1 \xi_1 + \kappa_2 \xi_2) + (\kappa_2 + \kappa_1) \cosh(\kappa_2 \xi_2 - \kappa_1 \xi_1)]^2}, \qquad (3.39)$$

where $\xi_i = x - 4\kappa_i^2 t - x_i$, $C_i(0) = 2\kappa_i \exp(2\kappa_i x_i)$, $i = 1,2$. The two-soliton solution shows that the sum of two solitary waves of the form given by (3.38) is the asymptotic state of (3.39), but there is a phase shift due to the interaction.

We also note that knowledge that the function $a(k,t)$ is a constant of the motion can be related to the infinite number of conservation laws of KdV (cf. [3, 7]).

It should also be noted that discretizations of (2.5), (3.1) lead to interesting discrete nonlinear evolution equations which can be solved by IST. The best known of these equations is the Toda lattice,

$$\frac{\partial^2 u_n}{\partial t^2} = \exp(-(u_n - u_{n-1})) - \exp(-(u_{n+1} - u_n)), \qquad (3.40)$$

which is related to the linear discrete Schrödinger scattering problem,

$$\alpha_n v_{n+1} + \alpha_{n-1} v_{n-1} + \beta_n v_n = k v_n, \qquad (3.41)$$

where

$$\alpha_n = \frac{1}{2} \exp\left(-\frac{1}{2}(u_n - u_{n-1})\right), \qquad \beta_n = -\frac{1}{2}\frac{\partial u_{n-1}}{\partial t},$$

and the integrable discrete NLS equation

$$i\frac{\partial u_n}{\partial t} = (u_{n+1} + u_{n-1} - 2u_n) + s|u_n|^2(u_{n+1} + u_{n-1}), \tag{3.42}$$

which is related to the 2×2 discrete scattering problem,

$$\begin{aligned}
v_{1_{n+1}} &= zv_{1_n} + r_n v_{2_n} \\
v_{2_{n+1}} &= \frac{1}{z}v_{2_n} + q_n v_{1_n},
\end{aligned} \tag{3.43}$$

where

$$q_n = u_n, \quad r_n = -su_n^*, \quad s = \pm 1, \tag{3.44}$$

and u^* is the complex conjugate of u. There are also double discretizations of the NLS equation which can be solved by IST. These discretizations are obtained by discretizing the temporal equation (2.6). One example of a doubly discrete NLS equation is given by

$$\frac{i\Delta^m u_n^m}{\Delta t} = \delta^2\left(\frac{(u_n^m + u_n^{m+1})}{2\Delta x^2}\right) + u_{n-1}^m(L_{n-1} - 1)$$

$$+ u_{n+1}^{m+1}(L_n - 1) + \frac{1}{4}\left[u_n^m(u_n^{m*}u_{n+1}^m + u_n^{m+1*}u_{n+1}^{m+1}) + \right.$$

$$u_n^{m+1}(u_{n-1}^m u_n^{m*} + u_{n-1}^{m+1}u_n^{m+1*}) + 2|u_n|^2 u_n^{m+1} \left. L_n\right]$$

$$+ 2|u_n^{m+1}|^2 u_{n-1}^m L_{n-1} - u_n^m T_n - u_n^{m+1}T_{n-1}^*, \tag{3.45}$$

where

$$\delta^2(u_n^m) = (u_{n+1}^m + u_{n-1}^m - 2u_n^m),$$

$$L_{k-1} = \frac{1 \pm |u_k^{m+1}|^2}{1 \pm |u_k^m|^2}L_k,$$

$$L_k \to 1, \quad k \to -\infty \tag{3.46}$$

$$T_k - T_{k-1} = (u_k^m u_{k-1}^{m*} + u_{k+1}^m u_k^{m*}), \qquad T_k \to 0, \quad k \to -\infty. \tag{3.47}$$

Equation (3.45)–(3.47) is an integrable nonlinear analogue of the Crank-Nicholson scheme. A review of the scattering and inverse scattering theory associated with discrete equations can be found in [3,7].

It should also be mentioned that there is another class of nonlinear evolution equations in 1+1 dimensions that are solvable by IST. This is the class of singular integro-differential equations. The paradigm equation is

$$u_t + \frac{1}{\delta}u_x + 2uu_x + Tu_{xx} = 0, \tag{3.48}$$

where δ is a constant and Tu is the singular integral operator,

$$(Tu)(x) = \frac{1}{2\delta} \fint_{-\infty}^{\infty} \cosh\left(\frac{\pi}{2\delta}(y - x)\right) u(y)dy, \qquad (3.49)$$

and $\fint_{-\infty}^{\infty}$ denotes the principal value integral. (3.48) is referred to as the Intermediate Long Wave (ILW) equation. Indeed, it has two well known limits:
a) $\delta \to 0$ ILW reduces to the KdV equation,

$$u_t + 2uu_x + \frac{\delta}{3}u_{xxx} = 0, \qquad (3.50)$$

b) $\delta \to \infty$ ILW reduces to the Benjamin-Ono (BO) equation.

$$u_t + 2uu_x + Hu_{xx} = 0, \qquad (3.51)$$

where Hu is the Hilbert transform,

$$(Hu)(x) = \frac{1}{\pi} \fint_{-\infty}^{\infty} \frac{u(y)}{y - x} dy. \qquad (3.52)$$

The BO equation was derived in the context of long internal gravity waves in a stratified fluid [14–16], whereas the ILW equation was derived in a similar context in [17,18]. In [7] the IST analysis associated with the ILW equation and BO equation is reviewed. The unusual aspect of the IST scheme is the fact that the scattering operator is a differential RHBVP. Related generalizations are also discussed in [7].

4   IST for 2+1 Equations

In Sect. 2 we discussed the relevance of the KP equation in two-dimensional water waves. The normalized KP equation is given by (1.14). In this section, a broad outline of the main results of IST for the KP equation will be outlined. Just as the KdV equation was the first 1+1 equation linearized by IST methods, the KP equation was the first nontrivial equation linearized by 2+1 IST methods. After the methods were established for the KP equation, they were quickly generalized to other equations such as the Davey-Stewartson equation and the 2+1 N wave equation (cf. [7]).

The compatible linear system for KP is given by

$$\sigma v_y + v_{xx} + uv = 0 \qquad (4.1a)$$

$$v_t + 4v_{xxx} + 6uv_x + 3u_x v - 3\sigma(\partial_x^{-1}u_y)v + \gamma v = 0, \qquad (4.1b)$$

where $\partial_x^{-1} = \int_{-\infty}^{x} dx'$, and γ is an arbitrary constant. In the case when $\sigma^2 = -1$, i.e., KPI, then (4.1a) is the nonstationary Schrödinger equation. When $\sigma^2 = +1$, i.e., KPII, then (4.1a) is a "reverse" heat equation which is well-known to be ill-posed as an initial value problem. However, as a scattering problem, the case when $\sigma^2 = +1$ can be analyzed effectively. However the situation when $\sigma^2 = +1$ is very different from $\sigma^2 = -1$.

A) KPI: $\sigma^2 = -1$

In this case it is convenient to make the transformation,

$$v(x, y, k) = m(x, y, k) \exp(i(kx - k^2 y)), \tag{4.2}$$

whereupon (4.1a) is transformed to ($\sigma = i$),

$$im_y + m_{xx} + 2ikm_x = -um. \tag{4.3}$$

We want to find an eigenfunction which is bounded for all x, y, k. Such an eigenfunction satisfies the following integral equation,

$$m = 1 + \tilde{G}(um), \tag{4.4}$$

where

$$\tilde{G}f = \int_{-\infty}^{\infty} \int_{-\infty}^{\infty} G(x - x', y - y', k) f(x', y') dx' dy' \tag{4.5}$$

and Green's function, $G(x, y, k)$, satisfies

$$iG_y + G_{xx} + 2ikG_x = -\delta(x)\delta(y). \tag{4.6}$$

Taking a Fourier transform in x, y yields

$$G(x, y, k) = \frac{1}{(2\pi)^2} \int_{-\infty}^{\infty} \int_{-\infty}^{\infty} \frac{e^{ipx+iqy}}{(q + p(p + 2k))} dp\, dq. \tag{4.7}$$

It is clear that if $k = k_R + ik_I$, then $G(x, y, k)$ is not well defined for $k_I = 0$. But there are natural analytic functions for Im $k \gtrless 0$. In fact, by contour integration,

$$G_{\pm}(x, y, k) = \frac{i}{2\pi} \int_{-\infty}^{\infty} e^{ipx - ip(p+2k)y} \{\theta(y)\theta(\mp p) - \theta(-y)\theta(\pm p)\} dp, \tag{4.8}$$

where $G_{\pm}$ stands for the limit $k \to k_R \pm i0$. A study of the properties of the integral equation (cf. [8,9]) shows that there are solutions $m_{\pm}(x, y, k)$ to (4.4). It is natural to ask how these functions are related. In [10; see also 7], the following nonlocal generalized RHBVP is derived,

$$(m_+ - m_-)(x, y, k) = \int_{-\infty}^{\infty} f(k, l) e^{\beta(x, y, k, l)} m_-(x, y, l) dl, \tag{4.9a}$$

where

$$\beta(x, y, k, l) = i(l - k)x - i(l^2 - k^2)y, \tag{4.9b}$$

$$f(k, l) = \frac{i \, \text{sgn} \, (k - l)}{2\pi} \int_{-\infty}^{\infty} \int_{-\infty}^{\infty} u(x', y') N(x', y', k, l) dx' dy' \tag{4.9c}$$

$$N(x, y, k, l) = \exp(\beta(x, y, k, l) + \tilde{G}_-(uN)). \tag{4.9d}$$

An essential difference between (4.9) and the result for KdV is that $m_\pm(x, y, k)$ may have poles since, in the derivation of (4.9), we employed the fact that $m_\pm$ satisfies a Fredholm integral equation (4.4). In KdV, the underlying integral equations were of Volterra type, and consequently their solutions have no poles. However, in the form (3.17), we see that $\mu_+(x, k) = M(x, k)/a(k)$, and poles are introduced through the zeroes of the scattering data $a(k)$. It turns out that the poles of the eigenfunction, $m_\pm(x, y, k)$, suitable normalizing coefficients and the "reflection" coefficients $f(k, l)$ complete the IST picture. We shall assume that the eigenfunctions $m_\pm(x, y, k)$ have only simple poles – indeed, in recent work it is demonstrated that this is not necessary; the poles can be of any order [11]. The functions $m_\pm$ are assumed to have the representations

$$m_\pm(x, y, k) = 1 + \sum_{j=1}^{N} \frac{i\phi_{j\pm}(x, y)}{(k - k_{j\pm})} + \mu_\pm(x, y, k), \tag{4.10}$$

where $\mu_\pm$ are analytic functions for Im $k \gtrless 0$ and $\mu_\pm \to 0$ as $|k| \to \infty$. The following important relation holds,

$$\lim_{k \to k_{j\pm}} \left(m_\pm(x, y, k) - \frac{i\phi_{j\pm}(x, y)}{k - k_{j\pm}} \right) = (x - 2k_{j\pm}y + \gamma_{j\pm})\phi_{j\pm}(x, y). \tag{4.11}$$

In fact, if $\mu_\pm = 0$, i.e., the case of pure poles, then (4.11) is a linear system of equations and the potential is obtained from

$$u(x, y) = 2\frac{\partial}{\partial x} \sum_{j=1}^{N} (\phi_{j+} + \phi_{j-}). \tag{4.12}$$

In deriving (4.11) via the integral equation (4.4) with Im $k_{j\pm} \gtrless 0$, a constraint appears; namely

$$Q(k_{j+}, \phi_{j+}) = -Q(k_{j-}, \phi_{j-}) = 1, \tag{4.13}$$

where

$$Q(k, \phi) = \frac{1}{2\pi} \int_{-\infty}^{\infty} \int_{-\infty}^{\infty} u(x, y)\phi(x, y)dx\, dy. \tag{4.14}$$

In recent work [11], it has been shown that values of the integral $Q(k, \phi)$ can be an integer and that Q is actually an underlying index of the problem. The time-dependence of the data is obtained from (4.1b) and it is found that

$$\frac{\partial f(k, l, t)}{\partial t} = 4i(l^3 - k^3)f(k, l, t) \tag{4.15a}$$

$$\frac{\partial}{\partial t}k_{j\pm} = 0 \tag{4.15b}$$

$$\gamma_{j\pm}(t) = 12(k_{j\pm})^2 t + \gamma_{j\pm}(0). \tag{4.15c}$$

The time dependence (4.15) with (4.9), (4.10) and (4.11) complete the IST framework, assuming only simple poles for m as expressed in (4.10). The class of lump solutions which are real, nonsingular and decay as $0(1/r^2)$, $r^2 = x^2 + y^2$, as $r \to \infty$, is obtained from (4.11)–(4.12) where $k_{j-} = \bar{k}_{j+}$, $\gamma_{j-} = \bar{\gamma}_{j+}$ (bar stands for the complex conjugate). A one-lump solution ($N = 1$) is found to be

$$u = 2\frac{\partial^2}{\partial x^2} \log F, \tag{4.16a}$$

where

$$F(x, y, t) = (x' - 2k_R y')^2 + 4k_I^2 y'^2 + \frac{1}{4k_I^2}$$
$$x' = x - 12(k_R^2 + k_I^2)t - x_0, \quad y' = y - 12k_R t - y_0 \tag{4.16b}$$
$$x_0 = \gamma_R(0) + \gamma_I(0), \quad y_0 = -\frac{\gamma_I}{2k_I}$$
$$k_1 = k_R + ik_I.$$

B) KPII: $\sigma^2 = 1$
  In this case we transform (4.1a) ($\sigma = -1$) via

$$v(x, y, k) = m(x, y, k) \exp(ikx - k^2 y)$$

to

$$-m_y + m_{xx} + 2ikm_x = -um. \tag{4.17}$$

The direct problem involves a study of the properties of a particular solution of (4.17). This solution is expressed in terms of the following integral equation,

$$m = 1 + \tilde{G}(um), \tag{4.18}$$

where, using the notations (4.5–4.7),

$$-G_y + G_{xx} + 2ikG_x = -\delta(x)\delta(y) \tag{4.19}$$

$$G(x, y, k) = \frac{1}{(2\pi)^2} \int \int \frac{e^{ipx+iqy}}{p^2 + 2pk + iq} dp\, dq. \tag{4.20}$$

In (4.20) and hereafter all double integrals are taken from $-\infty$ to ∞ in both variables of integration. Thus, unlike (4.7), Green's function expressed by (4.20) has *no* regions of analyticity. In fact, $G = G(x, y, k_R, k_I)$ where $k = k_R + ik_I$; i.e., G depends on the real and imaginary parts of k. Note that by use of contour integration,

$$G(x, y, k_R, k_I) = \frac{\text{sgn}(-y)}{2\pi} \int_{-\infty}^{\infty} \theta(y(p^2 + 2k_R p)) \exp(ipx - p(p + 2k)y)dp. \tag{4.21}$$

This implies that the function m via (4.18) is also a function of k_R, k_I and therefore is analytic nowhere in the k-plane. By taking the anti-holomorphic derivative of (4.18), i.e. operating on (4.18) by $\frac{\partial}{\partial \bar{k}} = \frac{1}{2}\left(\frac{\partial}{\partial k_R} + i\frac{\partial}{\partial k_I}\right)$, and using, as we sometimes do, the notation $G(x, y, k)$ to denote $G(x, y, k_R, k_I)$ etc., we find that

$$\frac{\partial G}{\partial \bar{k}}(x, y, k) = \frac{\text{sgn }(-k_R)}{2\pi}e^{i(p_0 x + i q_0 y)} \tag{4.22a}$$

$$G(x, y; -\bar{k}) = G(x, y, k)\exp(-i(p_0 x + q_0 y)) \tag{4.22b}$$

$$p_0 = -2k_R, \quad q_0 = 4k_R k_I . \tag{4.22c}$$

We find that m satisfies the following $\bar{\partial}$ equation,

$$\frac{\partial m}{\partial \bar{k}}(x, y, k) = R(k_R, k_I)e^{i p_0 x + i q_0 y}m(x, y, -\bar{k}), \tag{4.23}$$

where $R(k_R, k_I)$ plays the role of the scattering data, and is related to the potential via

$$R(k_R, k_I) = \frac{\text{sgn }(-k_R)}{2\pi}\iint u(x, y)m(x, y, k)e^{-i p_0 x - i q_0 y}dx\, dy. \tag{4.24}$$

Thus, given an appropriate potential $u(x, y)$ vanishing sufficiently fast at infinity (cf. [12]), the direct problem establishes relations (4.23)–(4.24). The inverse problem is fixed by giving $R(k_R, k_I)$ (there are no discrete state solutions known which lead to real, nonsingular, decaying states for KPII) in order to determine $m(x, y, k)$ and then $u(x, y)$. The inverse problem is developed by using the generalized Cauchy integral formula (cf. [13]),

$$m(x, y, k) = \frac{1}{2\pi i}\int\int_{R_\infty}\frac{\partial m}{\partial \bar{z}}(x, y, z)\frac{dz \wedge d\bar{z}}{z - k} + \frac{1}{2\pi i}\int_{C_\infty}\frac{m(x, y, z)}{z - k}dz, \tag{4.25}$$

where R_∞ is the entire complex plane, C_∞ is a circular contour at infinity, $z = z_R + iz_I$, and $dz \wedge d\bar{z} = 2idz_Rdz_I$. As $k \to \infty$, we can establish that $m \sim 1$, hence the second term on the right-hand side of (4.26) is unity. Using (4.23) we find that

$$m(x, y, k) = 1 + \frac{1}{2\pi i}\iint R(z_R, z_I)e^{i(p_0 x + q_0 y)}\frac{m(x, y, -\bar{z})}{z - k}dz \wedge d\bar{z}. \tag{4.26}$$

Once $m(x, y, k)$ is found, the potential is reconstructed from

$$u(x, y) = -\frac{\partial}{\partial x}\left(\frac{2i}{\pi}\iint R(z_R, z_I)e^{i(p_0 x + q_0 y)}m(x, y, -\bar{z})dz_Rdz_I\right). \tag{4.27}$$

The latter formula is obtained by comparing the limit as $k \to \infty$ in (4.26) and (4.18).

Finally, the time-dependence of the scattering data is shown from (4.1b) to be

$$\frac{\partial R}{\partial t} = -4i(k^3 + \bar{k}^3)R. \tag{4.28}$$

Thus, the IST framework for KPII is complete; namely at $t = 0$, $u(x, y, 0)$ determines $R(k_R, k_I, 0)$; (4.28) gives $R(k_R, k_I, t)$ and from (4.26)–(4.27) the eigenfunction $m(x, y, t, k)$ and solution $u(x, y, t)$ are obtained.

5   Remarks on Related Problems

In the previous sections we have discussed the integrability of a class of nonlinear equations in 1+1and in 2+1 dimensions. It is natural to ask whether there are any 4-dimensional integrable systems. Indeed, there is at least one such important system, the self-dual Yang-Mills (SDYM) equations. They are the result of the compatibility of the following linear pair,

$$\left(\frac{\partial}{\partial\alpha} + \zeta\frac{\partial}{\partial\bar{\beta}}\right)\Psi = (\gamma_\alpha + \zeta\gamma_{\bar{\beta}})\Psi \tag{5.1a}$$

$$\left(\frac{\partial}{\partial\beta} - \zeta\frac{\partial}{\partial\bar{\alpha}}\right)\Psi = (\gamma_\beta - \zeta\gamma_{\bar{\alpha}})\Psi, \tag{5.1b}$$

where $\gamma_\alpha, \gamma_{\bar{\alpha}}, \gamma_\beta, \gamma_{\bar{\beta}}$ are four dependent variables, often called gauge potentials, and $\alpha, \bar{\alpha}, \beta, \bar{\beta}$ are four independent variables which can be written in terms of the usual Cartesian coordinates as

$$\begin{array}{ll} \alpha = t + iz & , \quad \beta = x + iy \\ \bar{\alpha} = t - iz & , \quad \bar{\beta} = x - iy. \end{array} \tag{5.2}$$

Compatibility of (5.1a,b) is effected by operating with $\left(\frac{\partial}{\partial\beta} - \zeta\frac{\partial}{\partial\bar{\alpha}}\right)$ on (5.1a) and setting this equal to the equation formed by operating with $\left(\frac{\partial}{\partial\alpha} + \zeta\frac{\partial}{\partial\bar{\beta}}\right)$ on (5.1b). The result is

$$\left(\frac{\partial}{\partial\beta} - \zeta\frac{\partial}{\partial\bar{\alpha}}\right)(\gamma_\alpha + \zeta\gamma_{\bar{\beta}}) - \left(\frac{\partial}{\partial\alpha} + \zeta\frac{\partial}{\partial\bar{\beta}}\right)(\gamma_\beta - \zeta\gamma_{\bar{\alpha}}) + [\gamma_\alpha + \zeta\gamma_{\bar{\beta}}, \ \gamma_\beta - \zeta\gamma_{\bar{\alpha}}] = 0, \tag{5.3}$$

where $[A, B] = AB - BA$. The SDYM equations result from (5.3) by equating all powers of ζ to zero, i.e., powers ζ^2, ζ, ζ^0. One finds the following equations. Define

$$F_{\alpha\beta} \equiv \frac{\partial\gamma_\beta}{\partial\alpha} - \frac{\partial\gamma_\alpha}{\partial\beta} - [\gamma_\alpha, \gamma_\beta] = 0. \tag{5.4}$$

Then

$$\begin{array}{l} F_{\alpha\beta} = 0 \quad , \\ F_{\bar{\alpha}\bar{\beta}} = 0 \quad , \\ F_{\alpha\bar{\alpha}} + F_{\beta\bar{\beta}} = 0 \ . \end{array} \tag{5.5}$$

Indeed, there are three equations for the four gauge potentials. There is a gauge freedom, namely the transformation

$$\gamma_a = \left(f\hat{\gamma}_a - \frac{\partial f}{\partial a}\right)f^{-1} \tag{5.6}$$

which leaves SDYM invariant (here "a" can be $\alpha, \bar{\alpha}, \beta$, or $\bar{\beta}$).

There has been significant interest in the SDYM equations as a "master" integrable system. Ward [19] has conjectured that perhaps all "integrable" equations, e.g. soliton equations, may be obtained as a reduction of the SDYM equations. The reduction process has three aspects (see [7]).

i)   Employ the gauge freedom (5.6) of the equations. Frequently the choice of gauge can simplify the analysis and make the search for integrable reductions considerably easier.

ii) Reduction of independent variables, i.e., $\gamma_a(\alpha, \bar{\alpha}, \beta, \bar{\beta})$ can be functions of α, or α, β, etc.

iii) Choice of the underlying gauge group (algebra) in which one carries out the analysis. Sometimes it is a matrix algebra, e.g., $su(n), gl(n)$; but in many interesting cases the gauge algebra is infinite dimensional, e.g., sdiff(S^3).

It is often easiest to make identifications via the linear pair of SDYM. For example, suppose $\gamma_a \in gl(N)$, $\gamma_a = \gamma_a(\alpha, \beta)$, $\gamma_{\bar{\beta}} = iJ = $ diagonal matrix, $\gamma_{\bar{\alpha}} = iA_0 = $ diagonal matrix. Then, calling $\gamma_\alpha = Q$, $\gamma_\beta = A_1$ (5.1a–5.1b) reduce to

$$\frac{\partial \Psi}{\partial \alpha} = (Q + i\zeta J)\Psi \tag{5.7a}$$

$$\frac{\partial \Psi}{\partial \beta} = (A_1 - i\zeta A_0)\Psi. \tag{5.7b}$$

In fact, (5.7) is the linear pair associated with the N wave system (when $N = 3$ it is the 3 wave system). We need go no further in writing the equations (cf. [7]) except to point out that once the spatial part of the linear system is known, then actually the entire hierarchy can be ascertained. Other special cases include KdV, NLS, sine Gordon etc.

It is also worth remarking that the well-known 2+1 dimensional soliton system can be obtained from SDYM if we assume that the gauge potentials are elements of the infinite dimensional gauge algebra of differential polynomials. For example, suppose $\gamma_{\bar{\alpha}} = \gamma_{\bar{\beta}} = 0$, $Q = Q(\alpha, y, \beta)$, $A_1 = A_1(\alpha, y, \beta)$, J, A_0 are diagonal matrices and

$$\gamma_\alpha = Q + J\frac{\partial}{\partial y}$$

$$\gamma_\beta = A_1 + A_0\frac{\partial}{\partial y}. \tag{5.8}$$

Then (5.1) reduces to

$$\frac{\partial \Psi}{\partial \alpha} = \left(Q + J\frac{\partial}{\partial y}\right)\Psi$$

$$\frac{\partial \Psi}{\partial \beta} = \left(A_1 + A_0\frac{\partial}{\partial y}\right)\Psi. \tag{5.9}$$

Compatibility of (5.9) yields the N wave equations in 2+1 dimensions (here the independent variables are α, y, β; i.e., β plays the role of time). Again the hierarchy is generated from the spatial part of the linear system. Other special cases include the KP and Davey-Stewartson systems. A detailed discussion of 2+1 reductions can be found in [20].

It should also be mentioned that SDYM reduces to the classical 0+1 dimensional Painlevé equations. In [21] it is shown that all of the six Painlevé equations can be obtained from SDYM with finite-dimensional Lie Groups (matrix gauge algebra).

In [22] it was shown that using an infinite-dimensional gauge algebra, sdiff (S^3), SDYM could be reduced to the system

$$\begin{aligned}\dot{w}_1 &= w_2w_3 - w_1(w_2 + w_3)\\\dot{w}_2 &= w_3w_1 - w_2(w_3 + w_1),\\\dot{w}_3 &= w_1w_2 - w_3(w_1 + w_2)\end{aligned} \tag{5.10}$$

which was proposed by Darboux in 1878 in his study of triply orthogonal surfaces, and for which solutions were obtained by Halphen in 1881. Indeed, if we let $y = -2(w_1 + w_2 + w_3)$, then it can be shown that y satisfies

$$\dddot{y} = 2y\ddot{y} - 3\dot{y}^2, \tag{5.11}$$

which was studied by Chazy in 1909–1911. The properties of Chazy's equation imply that two solutions are related as follows. Call

$$y = \frac{1}{2}\frac{d}{dt}\log \Delta(t). \tag{5.12}$$

Then two functions $\Delta(t)$ are related by

$$\Delta_{II}(t) = \frac{\Delta_I(\gamma t)}{(ct + d)^{12}}, \quad \text{where } \gamma t = \frac{at + b}{ct + d}, \quad ad - bc = 1. \tag{5.13}$$

Indeed (5.13) yields a well known functional equation when $\Delta_{II} = \Delta_I = \Delta(t)$,

$$\Delta(t) = \frac{\Delta(\gamma t)}{(ct + d)^{12}} \quad \Delta \to 0, \ \text{Im } t \to \infty, \tag{5.14}$$

a, b, c, d integers. Such a function $\Delta(t)$ is called the discriminant modular form; explicit formulae representing the function are

$$\Delta(t) = q\prod_{n=1}^{\infty}(1 - q^n)^{24} = \sum_{n=1}^{\infty}\tau_n q^n, \quad q = e^{2\pi it}. \tag{5.15}$$

The coefficients τ_n of the Fourier series of $\Delta(t)$ are the well-known Ramanujan coefficients. From (5.12), an alternative form for $y(t)$ is

$$y(t) = \pi i E_2(t) = \pi i(1 - 24\sum_{n=1}^{\infty}\sigma_1(n)q^n), \tag{5.16}$$

where $E_2(t)$ is the Eisenstein series, and $\sigma_1(n)$ are particular number-theoretic coefficients. From (5.15) or (5.16) the general solution is obtained from (5.13) where a, b, c, d are taken to be arbitrary coefficients with $ad - bc = 1$. If we use (5.12), then a convenient "Bäcklund" type transformation is

$$y_{II}(t) = \frac{y_I(\gamma t)}{(ct + d)^2} - \frac{6c}{ct + d} .$$ (5.17)

Equations (5.12)-(5.17) demonstrate that the solutions to Chazy's equation (5.11) and the solution to the Darboux-Halphen system (5.10) (by finding w_1, w_2, w_3 in terms of $y, \dot{y}, \ddot{y}$) are expressible in terms of automorphic functions.

An interesting question to ask is whether the solution of (5.10–5.11) can be obtained via the inverse method. In fact, in a recent paper [23] it has been shown that the linear compatible system of SDYM can be reduced to a monodromy problem. The novelty is that in this case the monodromy problem has evolving monodromy data+– unlike those associated with the Painlevé equation where the monodromy is fixed (isomonodromy). Then the linear problem can be used to find the solutions of (5.10) which are automorphic functions, and via (5.10), to solve Chazy's equation (5.11). Generalizations of the Darboux-Halphen system are also examined and solved in [23]. It is outside the scope of this article to go into those details.

Acknowledgments

This work was partially supported by the Air Force Office of Scientific Research, Air Force Materials Command, USAF under grant F49620-97-1-0017, and by the NSF under grant DMS-9404265. The US Government is authorized to reproduce and distribute reprints for governmental purposes notwithstanding any copyright notation thereon. The views and conclusions contained herein are those of the author and should not be interpreted as necessarily representing the official policies or endorsements, either expressed or implied, of the Air Force Office of Scientific Research or the US Government.

References

1. G.B. Whitham: *Linear and Nonlinear Waves* (Wiley, New York 1974)
2. N.J. Zabusky, M.D. Kruskal: Interactions of solitons in a collisionless plasma and the recurrence of initial states, *Phys. Rev. Lett.* **15**, 240–243 (1965)
3. M.J. Ablowitz, H. Segur: *Solitons and the Inverse Scattering Transform,* SIAM Studies in Applied Mathematics, 425 pp. (SIAM, Philadelphia, PA 1981)
4. P. Lax, D. Levermore: The small dispersion limit of the Korteweg-de Vries equation, I, II and III, *Commun. Pure Appl. Math.* **37**, 253–290 (1983); 571–593; 809–830; S. Venakides: The zero dispersion limit of the KdV equation with nontrivial reflection coefficient, *Commun. Pure Appl. Math.* **38**, 125–155 (1985); The generation of modulated wavetrains in the solution of the KdV equation, *Commun. Pure Appl. Math.* **38**, 883–909 (1985)

5. C. Gardner, J. Greene, M. Kruskal, R. Miura: Method for solving the Korteweg-de Vries equation, *Phys. Rev. Lett.* **19**, 1095–1097 (1967); Korteweg-de Vries and generalizations. VI. Methods for exact solution, *Commun. Pure Appl. Math.* **27**, 97–133 (1974)

6. M.J. Ablowitz, D. Kaup, A.C. Newell, H. Segur: The inverse scattering transform – Fourier analysis for nonlinear problems, *Stud. Appl. Math.* **53**, 249–315 (1974)

7. M.J. Ablowitz, P. Clarkson: *Solitons, Nonlinear Evolution Equations and Inverse Scattering*, London Mathematical Society Lecture Notes Series #149, 516 pp. (Cambridge University Press, Cambridge, UK 1991)

8. H. Segur: Comments on IS for the Kadomtsev-Petviashvili eqauation, in *Mathematical Methods in Hydrodynamics and Integrability in Dynamical Systems,* Proceedings, La Jolla, eds. M. Tabor and Y.M. Treve, AIP Conf. Proc. **88**, 211–228 (1981)

9. X. Zhou: Inverse scattering transform for the time dependent Schrödinger equation with applications to the KPI equation, *Commun. Math. Phys.* **128**, (1990) 551–564

10. A.S. Fokas, M.J. Ablowitz: On the inverse scattering of the time-dependent Schrödinger equation and the associated Kadomtsev-Petviashvili equation (I), *Stud. Appl. Math.* **69**, 211–228 (1983)

11. M.J. Ablowitz, J. Villarroel: Solutions to the time dependent Schrödinger and the Kadomtsev-Petviashvili Equations, *Phys. Rev. Lett.* **78**, 570–573 (1997)

12. M.V. Wickerhauser: Inverse scattering for the heat operator and evolutions in 2+1 variables, *Commun. Math. Phys.* **108**, 67–89 (1987)

13. M.J. Ablowitz, A.S. Fokas: *Complex Variables, Introduction and Applications* (Cambridge University Press, Cambridge, UK 1996)

14. T.B. Benjamin: Internal waves of permanent form in fluids of great depth, *J. Fluid Mech.* **29**, 559–592 (1967)

15. R.E. Davies, A. Acrivos: Solitary internal waves in deep water, *J. Fluid Mech.* **29**, 593–607 (1967)

16. H. Ono: Algebraic solitary waves in stratified fluids, *J. Phys. Soc. Japan* **39**, 1082–1091 (1975)

17. R.I. Joseph: Solitary waves in a finite depth fluid, *J. Phys. A: Math. Gen.* **10**, L225–L227 (1977)

18. T. Kubota, D.R.S. Ko, L.D. Dobbs: Weakly nonlinear long internal gravity waves in stratified fluids of finite depth, *AIAA J. Hydronautics* **12**, 157–165 (1978)

19. R.S. Ward: Integrable and solvable systems, and relations among them, *Phil. Trans. R. Soc. Lond. A* **315**, 451–457 (1985)

20. M.J. Ablowitz, S. Chakravarty, L.A. Takhtajan: *Commun. Math. Phys.* **158**, 289–314 (1993)

21. R. Maszczyk, L.J. Mason, N.M.J. Woodhouse: *Classical Quantum Gravity* **11**, 65 (1994)

22. S. Chakravarty, M.J. Ablowitz, P.A. Clarkson: *Phys. Rev. Lett.* **65**, 1085–1087 (1990)

23. S. Chakravarty, M.J. Ablowitz: Integrability, Monodromy Evolving Deformations and the Self-Dual Bianchi IX Systems, *Phys. Rev. Lett.* **76**, 857–860 (1996)

24. V.E. Zakharov, A.B. Shabat: Exact theory of two-dimensional self-focusing and one-dimensional waves in nonlinear media, *Sov. Phys. JETP* **34**, 62–69 (1972)

References Added in the Second Edition

The book: *Discrete and Continuous Nonlinear Schrödinger Systems,* M.J. Ablowitz, B. Prinari, and D. Trubatch, 258 pages, will be published by Cambridge University Press, Cambridge, UK in 2003. This book will be useful for students and researchers who wish to study the inverse scattering transform and applications. This book extends the techniques used in the monograph: *Solitons, Nonlinear Evolution Equations and Inverse Scattering,* M.J. Ablowitz and P.A. Clarkson, London Mathematical Society Lecture Notes Series #149, 516 pages, Cambridge University Press, Cambridge, UK, 1991.

In the new book readers will be able to find a careful discussion of the inverse scattering transform via Riemann-Hilbert methods associated with discrete and continuous scalar and vector nonlinear Schrödinger equations.

Integrability – and How to Detect It

B. Grammaticos[1] and A. Ramani[2]

[1] GMPIB, Université Paris VII, Tour 24-14, 5ᵉ étage, case 7021, 75251 Paris, France
`grammati@paris7.jussieu.fr`
[2] CPT, École Polytechnique, CNRS, UMR 7644, 91128 Palaiseau, France
`ramani@cpht.polytechnique.fr`

Abstract. We present a physicist's approach to integrability and its detection. Starting from specific examples we present a working definition of what is meant by "integrability". The integrability detector on which this whole course in based is the "Painlevé method" which links the integrable character of a (differential) system to the singularity structure of its solutions. Recent results on integrable discrete systems are also discussed here. They are, for the major part, obtained through the application of the "singularity confinement" approach that is the discrete equivalent of the Painlevé method. Foremost among these results are the discrete Painlevé equations that generalize in the discrete domain the transcendental functions introduced by Painlevé and which have so many interesting applications in the domain of nonlinear physics.

1 General Introduction: Who Cares about Integrability?

It would seem fit for a course entitled "Integrability" to start with the definition of this notion. Alas, this is not possible. There exists a profusion of integrability definitions and where you have two scientists you have (at least) three different definitions of integrability. It is not our aim to present here a precise, rigorous definition of this notion. We shall rather present our intuitive arguments leading to a 'working' definition of integrability that lies at the heart of our work [1].

The word integrability, coming from "integral", immediately evokes differential equations. Why do differential equations play an important role in physics? This is a philosophical question that cannot be answered within the strict framework of science. The fact is that centuries of investigations have established beyond any doubt the validity of the deterministic description of physical phenomena through differential equations. Classical Mechanics is *par excellence* a domain of application of differential equations. The nonlinearity inherent in most classical equations of motion makes the question of stability and the prediction of long-term behaviour all the more interesting. While the modern theories of Quantum Fields and of Gravitation have complicated matters a little, differential equations have retained their significance.

Poincaré gave a definition of integrability that captures the essence of the term [2]. According to him, to integrate a differential equation is to find for the general solution a finite expression, possibly multivalued, in a finite number of functions. The word "finite" indicates that integrability is related to a *global* rather than *local* knowledge of the solution. However, this definition is not very

B. Grammaticos and A. Ramani, Integrability – and How to Detect It, Lect. Notes Phys. **638**, 31–94 (2004)
`http://www.springerlink.com/`   © Springer-Verlag Berlin Heidelberg 2004

useful unless one defines more precisely what is meant by "function". We shall come back to this point in the next section, but let us point out here that the most important feature that characterizes a function is its singlevaluedness.

Integrability is a rare phenomenon. The typical dynamical system is nonintegrable. How does one study such generic systems? Until very recently this was practically impossible. Except for some very particular cases, the only way to study a generic dynamical system was through the use of computers: "without computers you cannot visualize randomness in real systems" [3]. On the other hand, integrable systems can be studied in much greater detail than generic, nonintegrable ones. Algebraic and analytic methods are operative here. However, an arbitrarily small change in an integrable equation can destroy its integrability. Still, some structure of the integrable system persists under (not too large) perturbations. Near-integrable systems can be studied through special analytical techniques [4] which allow one to dicover the qualitative behaviour of the system.

Given that integrability is structurally unstable, one may worry as to the pertinence of integrable systems in the description of physical phenomena. Segur points out [5] that "if a given problem can be approximated by an integrable model then it is likely that it can also be approximated to the same accuracy by a model that is not integrable". Thus the worry is that results that depend fundamentally on integrability cannot be very important. Still, this is not the feeling shared by the integrability community. Calogero offers a basis for this optimistic attitude [6]. He has pointed out that some integrable partial differential equations (PDE) are both "universal" and "widely applicable". His argument is that a limiting (usually asymptotic) procedure applied to a large class of nonlinear PDE's leads, to the same limit, to a universal equation which is integrable. If this limiting procedure is physically reasonable this guarantees the wide applicability of the integrable equation. More recently, Fokas has shown that the use of nonlinear transforms allows one to extend the class of universal integrable equations [7]. Thus one expects integrable equations to play a non negligible role in the description of realistic physical systems, even though they are expected to describe some limiting, asymptotic situation.

Novikov [3] takes this argument one step further: "Physicists and mathematical philosophers of science for the most part do not believe that the laws of nature are to be expressed by arbitrarily chosen, general equations. Most of them somehow believe *de facto* in a higher reason". Indeed one is amazed at the simple mathematical form of physical laws. What is still more amazing is that the values of the fundamental physical constants are so finelly tuned as to make the appearance of sentient life in the universe possible. This means not only that the laws are simple but that the initial conditions of the universe are appropriate.

At this point we should be able to answer the question of the title of this section. However, in order to make things even clearer, let us present some definitions, proposed by Segur [5], that will help our argumentation. According to Segur:

Mathematics is the study of abstract structure and relationships. "Abstract" means that the study is done without concern whether the structure studied helps build a better widget, even though that may have motivated the study originally.

Physics is the study of the structure of the universe we inhabit. According to these definitions, a subject can move from mathematics to physics, or vice versa, depending on current evidence about whether the structure in question can be observed experimentally. Movement in each direction has been known to happen.

Sciences is the search for structure which, when found, is encoded in "laws".

He points out that these definitions may look unusual, but they certainly contain some truth.

Thus, the answer to the question "who cares about integrability?" is: "physicists should and mathematicians may, if they wish". Physicists just cannot ignore the rich structure that is present in integrable systems. Problems with unexpected structure often turn out to be related to integrable systems. Mathematicians, operating in a more abstract world, may well ignore integrability. However, the richness of this field is attracting the interest of a steadily increasing number of mathematicians, who volunteer for the exploration of the wide, uncharted regions of the integrability domain.

2   Historical Presentation: From Newton to Kruskal

Newton was the first to solve a fundamental problem in the domain of differential dynamical systems. It was the problem of the motion of two massive points under the action of their mutual gravitational attraction. This was certainly one of the most important discoveries in physics. At the same time Newton invented the mathematical tools for the formulation of his model, guessed the law of gravitation and solved the equations of motion so as to derive Kepler's laws. Of course, the 2-body problem, solved by Newton, is only an approximation of the physical reality. Several effects must be neglected in order to simplify the problem and bring it down to the tractable 2-body model. In fact, even today, it is not clear whether the system of gravitating masses that compose our Solar System is stable or not, but this only adds more value to Newton's reductionist approach. The 2-body Newtonian problem is also a nice example of a *super*integrable system [8]. Indeed for a 2-body problem with an interaction potential $V(|\boldsymbol{x}_1 - \boldsymbol{x}_2|)$ depending only on the distance between the particles is always integrable. The invariant of the total linear momentum,

$$\boldsymbol{P} = \boldsymbol{p}_1 + \boldsymbol{p}_2, \qquad (2.1)$$

where $\boldsymbol{p}_i = m_i \dot{\boldsymbol{x}}_i$, $i = 1, 2$ are the momenta of each particle always exists. This allows the equations of motion to be reduced to a 3-dimensional system for the relative coordinate $r = x_1 - x_2$. The reduced Hamiltonian now reads,

$$E = \frac{1}{2m}p^2 + V(r), \qquad (2.2)$$

where $p = m(p_1/m_1 - p_2/m_2)$ and m is the reduced mass $1/m = 1/m_1 + 1/m_2$. Hamiltonian (2.2) always has two conserved quantities, the total energy and the angular momentum:

$$M = r \times p \qquad (2.3)$$

Because the Poisson bracket of any two of the components of M does not vanish, one can define only two quantities 'in involution': the total angular momentum, $M^2 = M_x^2 + M_y^2 + M_z^2$, and one of the projections, say M_z, as independent invariants.

However the Newtonian potential $V = g|x_1 - x_2|^{-1}$ possesses an extra, dynamical, symmetry leading to an additional invariant, known as the Laplace-Runge-Lenz vector,

$$L = g\frac{r}{r} + \frac{1}{m}p \times (r \times p). \qquad (2.4)$$

(We shall not go into any details here concerning the involution properties of L with itself and with M.) It would not be exaggerated to state that if the Solar System displays its secular stablity this is due to the special structure of the $\frac{1}{r}$ potential and is thus intimately related to its superintegrability.

The domain of differential equations became from the outset the centre of intense activity but soon most of the methods of formal integration known today were established and, somehow, the interest in explicit integrability diminished. In the years that followed there was a shift of interest towards a number of linear differential equations that play an important role in mathematical physics. The consideration of these equations led naturally to the complex domain and to modern analytic theory. Now, one can ask, if we consider an ordinary differential equation (ODE) with real coefficients and seek only the solution for real values of the independent variable, why should we consider the extension of the solution to the complex plane of the independent variable? Although questions that begin by "why" are proverbially difficult to answer (and often lead to circular reasonings or philosophical considerations), we shall try here to furnish some elements of an answer [9].

- Algebraic equations with real coefficients have solutions defined only in the complex plane.
- The exponential function e^z has the nice property of being periodic (with period $2\pi i$) only if it is defined on the complex plane.
- The Weierstrass elliptic function $\wp(z, a, b)$ is doubly periodic for complex z. If one restricts the parameters a, b to real values and z on the real axis, $\wp$ becomes simply periodic.

One last example [10] is given by the power spectrum of a (purely real) signal $x(t)$. The high-frequency behaviour of the Fourier transform depends on the location and nature of the singularities of $x(t)$ in the complex-time plane.

The extension to the complex plane gives the possibility, instead of asking for a global solution for an ODE, of looking for solutions locally and obtaining a more global result by analytic continuation. Before making a more precise statement, let us introduce some useful notions. A singular point is is one in the neighbourhood of which the solution of the ODE is not analytic. We call critical point ('branch point' is an equivalent name) any point in the neighbourhood of which at least two different determinations of the solution of the ODE exist. The point may be isolated or not. An essential singularity is not necessarily a critical point, and a noncritical essential singularity is not an obstacle to single-valuedness.If we wish to define a function, we must find a way to treat the critical points and obtain a singlevalued application. There are two ways to do this, called uniformisations. The first method is by removing lines, called cuts, so as to avoid local loops around critical points. The second is by introducing a Riemann surface, made of several copies of the initial Riemann sphere, cut and pasted together. A most important result is that the procedure defined here can always be applied to the solutions of linear ODE's. This is possible because the location of the critical singularities of the solutions of a linear ODE are determined entirely by the coefficients of the ODE. In modern (singularity-analysis) terminology, we say that the critical singularities of a linear ODE are *fixed*. The consequence of the above result is immediate: the solution af any linear ODE defines a function. Thus, following Poincaré's definition, every linear ODE is integrable.

This discovery led to the obvious question: can one play the same game with *nonlinear* equations and define new functions? Nonlinear equations present considerable difficulties because of the structure of their singularities. While in the linear case the singularities are fixed, in the case of nonlinear differential equations there exist singularities the location of which (in the complex plane) depends on the initial conditions (or, equivalently, on the integration constants). These singularities are called movable. The movable character of critical singularities makes uniformisation impossible. The situation is even worse since nonlinear ODE's possess solutions beyond the general one (with the right number of integration constants) and the special solutions (recovered from the general solution for particular values of the constants). These solutions are called singular and can have a structure of singularities totally different from the one of the general solution. The problem of defining new functions through nonlinear equations was addressed in its simplest form, i.e. first-order equations, by Fuchs and Painlevé [11,12]. They found that the only first order equation without movable critical singularities is the Riccati equation,

$$w' = aw^2 + bw + c, \tag{2.5}$$

which contains, as a special case the linear equation. However, the Riccati equation is linearizable through the substitution $w = F/G$ (Cole-Hopf transformation) so no new functions are introduced. More general equations of the form $w'^n = f(w, z)$, where f is polynomial in w and analytical in z, were also considered, and the integrable cases were integrated in terms of elliptic functions.

At that time one of the small miracles that are often associated with integrability, occured [13]. Kovalevskaya, who was Fuchs's student, set out to study the integrability of a physical problem using singularity-analysis techniques. If she had just confirmed the integrability of the already known integrable cases her work would have been, at best, interesting and soon forgotten. What happened was that Kovalevskaya discovered a new, highly nontrivial, case. She set out to study the motion of a heavy top, spinning around a fixed point. The equations of motion with respect to a moving Cartesian coordinate system based on the principal axes of inertia with origin at its fixed point, known as Euler's equations, are:

$$A\frac{dp}{dt} = (B - C)qr + Mg(\gamma y_0 - \beta z_0)$$

$$B\frac{dq}{dt} = (C - A)pr + Mg(\alpha z_0 - \gamma x_0)$$

$$C\frac{dr}{dt} = (A - B)pq + Mg(\beta x_0 - \alpha y_0)$$

$$(2.6)$$

$$\frac{d\alpha}{dt} = \beta r - \gamma q$$

$$\frac{d\beta}{dt} = \gamma p - \alpha r$$

$$\frac{d\gamma}{dt} = \alpha q - \beta p,$$

where (p, q, r) are the components of angular velocity, (α, β, γ) the directions cosines of the direction of gravity, (A, B, C) the moments of inertia, (x_0, y_0, z_0) the position of the centre of mass of the system, M the mass of the top, and g the acceleration due to gravity. The complete integrability of the system requires the knowledge of four integrals of motion. Three such integrals are straightforward, the geometric constraint,

$$\alpha^2 + \beta^2 + \gamma^2 = 1, \tag{2.7}$$

the total energy,

$$Ap^2 + Bq^2 + Cr^2 - 2Mg(\alpha x_0 + \beta y_0 + \gamma z_0) = K_1, \tag{2.8}$$

and the projection of the angular momentum on the direction of gravity,

$$A\alpha p + B\beta q + C\gamma r = K_2. \tag{2.9}$$

A fourth integral was known only in three cases:
Spherical: $A = B = C$ with integral $px_0 + qy_0 + rz_0 = K$,
Euler: $x_0 = y_0 = z_0 = 0$ with integral $A^2p^2 + B^2q^2 + C^2r^2 = K$, and
Lagrange: $A = B$ and $x_0 = y_0 = 0$ with integral $Cr = K$.

In each of these cases the solutions of the equations of motion were given in terms of elliptic functions and were thus meromorphic in time t. Kovalevskaya set out to investigate the existence of other cases with solutions meromorphic in t. It turned out that such a case exists, provided that

$$A = B = 2C \quad \text{and} \quad z_0 = 0 \tag{2.10}$$

This case has been dubbed the *Kowalevski top*, in her honour. The fourth integral in this case may be written,

$$[C(p+iq)^2 + Mg(x_0+iy_0)(\alpha+i\beta)][C(p-iq)^2 + Mg(x_0-iy_0)(\alpha-i\beta)] = K. \tag{2.11}$$

Using (2.11) Kovalevskaya was able to show that the solution can be expressed as the inverse of a combination of hyperelliptic integrals. Hyperelliptic functions are not meromorphic in general, but it turned out that the symmetric combinations of hyperelliptic integrals involved in the solution of the Kowalevski top do have meromorphic inverses.

Another major discovery in the domain of the integrability took place at the end of the 19th century although at the time it was not perceived as such. Following the discovery of the solitary wave by Scott Russel, intense activity led to the formulation of the Korteweg-de Vries equation [14]. This equation is a nonlinear evolution equation describing the propagation of long, one-dimensional, small amplitude, surface gravity waves in a shallow water channel,

$$\frac{\partial \eta}{\partial \tau} = \frac{3}{2}\sqrt{\frac{g}{h}}\frac{\partial}{\partial \xi}\left(\frac{\eta^2}{2} + \frac{2\alpha\eta}{3} + \frac{\sigma}{3}\frac{\partial^2\eta}{\partial \xi^2}\right), \tag{2.12}$$

with $\sigma = h^3/3 - Th/(\rho g)$ and where h is the surface elevation of the wave above the equilibrium level h, α a small arbitrary constant related to the uniform motion of the liquid, g the acceleration due to gravity, T the surface tension, and ρ the density. The terms "long" and "small" are meant in comparison to the depth of the channel. Equation (2.12), known today as the KdV equation, can be brought in a nondimensional form through $t = \sqrt{g/\sigma}h\tau/2$, $x = \xi\sqrt{h/\sigma}$ and $u = (\eta/2 + \alpha/3)/h$,

$$u_t + 6uu_x + u_{xxx} = 0, \tag{2.13}$$

where subscripts denote partial differentiation. The KdV equation possesses a solitary wave solution of the form,

$$u(x,t) = 2\kappa^2 \operatorname{sech}^2(\kappa(x - 4\kappa^2 t - x_0)), \tag{2.14}$$

with κ, x_0 constants, thus providing the theoretical background for the interpretation of Scott Russell's observations.

The domain of nonlinear ODE's with fixed critical singularities continued to attract the interest of mathematicians so Painlevé [15] set out to classify all the second-order equations that belong to this class. In particular he examined equations of the form

$$w'' = f(w', w, z), \tag{2.15}$$

with f polynomial in w', rational in w and analytic in z. This classification was completed by Gambier [16] who presented the complete list of equations that satisfy the requirement of an absence of movable critical singularities. The most interesting result of this analysis was the discovery of six equations that define new functions. They are known today under the name of Painlevé equations:

$$w'' = 6w^2 + z$$

$$w'' = 2w^3 + zw + a$$

$$w'' = \frac{w'^2}{w} - \frac{w'}{z} + \frac{1}{z}(aw^2 + b) + cw^3 + \frac{d}{w}$$

$$w'' = \frac{w'^2}{2w} + \frac{3w^3}{2} + 4zw^2 + 2(z^2 - a)w - \frac{b^2}{2w} \qquad (2.16)$$

$$w'' = w'^2(\frac{1}{2w} + \frac{1}{w-1}) - \frac{w'}{z} + \frac{(w-1)^2}{z^2}(aw + \frac{b}{w}) + c\frac{w}{z} + \frac{dw(w+1)}{w-1}$$

$$w'' = \frac{w'^2}{2}(\frac{1}{w} + \frac{1}{w-1} + \frac{1}{w-z}) - w'(\frac{1}{z} + \frac{1}{z-1} + \frac{1}{w-z})$$
$$+ \frac{w(w-1)(w-z)}{2z^2(z-1)^2}(a - \frac{bz}{w^2} + c\frac{z-1}{(w-1)^2} + \frac{(d-1)z(z-1)}{(w-z)^2}),$$

where a, b, c, d are arbitrary constants. The functions defined by these equations are known as the *Painlevé transcendents*. Although the Painlevé equations are integrable in principle, their integration could not be performed with the methods available at that time. Still, the argument based on the singularity structure is compelling (in particular since Painlevé took particular care to prove the absence of movable essential singularities) and Painlevé's intuition was fully justified almost a century later.

Several attempts have been made to extend Painlevé's results to higher-order equations, in particular by Chazy and Garnier [17,18]. However, these attempts did not meet with the success of Painlevé despite the fact that they yielded some interesting results. Subsequently the activity in the domain of integrability dwindled and by the time of the scientific boom that followed the Second World War the subject had become of (at best) marginal interest. In the years that followed the War, scientists became engrossed with the new tool at their disposal, the high-speed electronic computer. The domain of numerical simulations of physical processes came into being and has been growing ever since. It was a discovery in this area that would lead to the renaissance of integrability.

Fermi, Pasta and Ulam [19] tried to obtain an indication of energy equipartition by studying a lattice of coupled anharmonic oscillators with fixed ends,

$$m\frac{d^2x_n}{dt^2} = k(x_{n+1} + x_{n-1} - 2x_n)[1 + \alpha(x_{n+1} - x_{n-1})]. \qquad (2.17)$$

The surprising result was that, instead of spreading over the modes towards equilibrium, the energy eventually recurred. This unexpected result motivated a study by Kruskal and Zabusky [20] that would turn out to be the cornerstone of the modern integrability theory. They considered the continuous limit of the FPU model and this turned out to be the KdV equation! Through numerical simulation they discovered that, under periodic boundary conditions, the initial profile was recovered. But, what is more important, they observed that in the intermediate time, before recurence, the initial profile separated into several solitary waves that interacted elastically. Zabusky and Kruskal named these special waves "solitons". The solitons preserve (asymptotically) their shape and velocity upon nonlinear interactions with other solitons. This was a remarkable discovery but it was made numerically. Thus, a detailed study was needed, a study promptly undertaken by Kruskal and his team [21]. Soon several special properties of the KdV equation were obtained:

- the existence of an infinite number of conservation laws
- its transformation (discovered by Miura [22] and bearing his name since) to another remarkable equation the modified KdV,

$$v_t + 6v^2 v_x + v_{xxx} = 0, \qquad (2.18)$$

- the existence of an arbitrary number of solitons, as was proven by Hirota [23] who developed his powerfull bilinear formalism for this purpose,

and, finally,

- its linearization.

For the latter the important step was to relate KdV to a linear time-independent Schrödinger problem,

$$\Phi_{xx} + u\Phi = \lambda\Phi, \qquad (2.19)$$

with u, the solution of KdV, playing the role of the potential. Thus the idea was to use the methods of Quantum Mechanical Inverse Scattering (IST) that allow the reconstruction of the potential from scattering data. The time evolution of the wave function Φ is given by a second equation,

$$\Phi_t = (\gamma + u_x)\Phi + (4\lambda + 2u)\Phi_x. \qquad (2.20)$$

The compatibility of (2.19) and (2.20), under the assumption $\lambda_t = 0$, yields the KdV equation. Since for the reconstruction of the potential only linear, integrodifferential, equations are involved, the IST procedure indeed linearizes the KdV equation. Lax put the IST method for solving the KdV into a more general framework [24]. Rewriting (2.19) and (2.20) as $L\Phi = \lambda\Phi$, $\Phi_t = M\Phi$, the compatibility condition for the linear operators L, M is obtained as $L_t + [L, M] = 0$. The *Lax pair* technique was soon to become the key for the treatment of integrable nonlinear PDE's.

Although exhibiting recurrence, the FPU model is not integrable until one has taken its continuous limit. Thus an important question was whether one can find a nonlinear lattice that was completely integrable. Toda showed that this was indeed possible [25]. The lattice with exponential interactions between nearest neighbours that bears his name,

$$\frac{d^2 x_n}{dt^2} = e^{x_{n+1} - x_n} + e^{x_n - x_{n-1}}, \tag{2.21}$$

is indeed integrable, has an infinite number of conservation laws, possesses a Lax pair and satisfies every condition for integrability.

The KdV equation and its integrability might have been an exception, but this soon turned out not to be the case. Zakharov and Shabat [26] discovered another integrable nonlinear PDE, the nonlinear Schrödinger equation: $iu_t + u_{xx} + \kappa|u|^2 u = 0$. Ablowitz, Kaup, Newell and Segur [27], motivated by important observations of Kruskal, solved the Sine-Gordon equation: $u_{xt} = \sin u$. Soon the domain of integrable PDE's was blossoming. It was not astonishing that the Painlevé equations started making their appearance in connection with integrable evolution equations. Thus, Ablowitz and Segur [28] showed that the IST techniques could be used to *linearize* the Painlevé equations. The interesting point is that this linearization was in terms of integrodifferential equations. This may explain why this solution was not obtained earlier although the 'Lax pairs' for the Painlevé equations were known since the works of Garnier and Schlesinger. The appearance of the Painlevé equations as well as of other equations belonging to the Painlevé-Gambier classification, as reductions of integrable PDE's, was not fortuitous. Ablowitz and Segur realized that integrability was the key word and soon the ARS conjecture was proposed [29] (in collaboration with one of us, A.R.): "Every ODE which arises as a reduction of a completely integrable PDE is of Painlevé type (perhaps after a transformation of variables)". This conjecture provided a most useful integrability detector. In the years that followed the ARS approach, which is very close in spirit to that of Kovalevskaya, turned out to be a most powerful tool for the investigation of integrability. Several new integrable systems were discovered through the singularity analysis approach.

Improvements to this approach were proposed. Weiss and collaborators managed to treat PDE's directly without the constraint of considering reductions [30]. This was significant progress because, according to the ARS conjecture, one had to treat every reduction before being able to assert anything about the given PDE (and, of course, it is very difficult to make sure that *every* reduction has been considered). Kruskal extended the singularity-analysis approach in a nonlocal way through his poly-Painlevé method [31]. While, in the traditional approach, one is concerned whether the solutions are multivalued, in the poly-Painlevé approach the distinction is made between nondense and dense multivaluedness. The former is considered to be compatible with integrability while the latter is not. Apart from these innovative approaches, considerable progress has been made in the 'mainstream' singularity-analysis domain but, a major open question still remains: "what are the acceptable transformations?" This, and the absence of a certain rigor, reduce the singularity analysis approach to a good heuristic tool for the study of integrability. It is undoubtedly powerful but one must bear in mind that it is not infallible and it does not have the rigor of a theorem.

At this point our historical review of integrability would have ended, were it not for a series of discoveries that opened a new domain, that of integrable discrete systems. In the past few years there has been a growing interest in these systems and the situation mirrors the euphoria of the early era of integrable evolution equations. This is understandable. The reason has been summarized by Kruskal [32], in his usual, 'right on target', way, "For years we have been thinking that the integrable evolution equations were the fundamental ones. It is becoming clear now that the fundamental objects are the integrable *discrete* equations." In the sections that follow the reasons for this statement will become evident.

3 Towards a Working Definition of Integrability

As we have already pointed out, integrability is a term widely used in the domain of dynamical systems, and despite this fact (or because of it) the various practitioners do not seem to agree on its definition. In the first part of this course we will deal almost entirely with dynamical systems described by differential equations and the question of their integrability [33]. The simplest aspect of integrability is solvability, meaning the existence of solution expressed in terms of elementary functions. Still, this is not a clear-cut definition since there is tremendous arbitrariness in what is considered to be an 'elementary' function. As an example consider of solvability a simplified version of the Rikitake system,

$$\dot{x} = yz$$

$$\dot{y} = -xz \qquad (3.1)$$

$$\dot{z} = -xy.$$

Its solutions are given as elliptic functions,

$$x = A \ \text{sn}(p(t - t_0))$$

$$y = A \ \text{cn}(p(t - t_0)) \qquad (3.2)$$

$$z = p \ \text{dn}(p(t - t_0)),$$

with parameter $m = A^2/p^2$.

While explicit knowledge of the solutions of the system may be very useful, it is clear that most times we must contend ourselves with less. In fact, one of the main uses of integrability lies in the fact that it allows us to obtain global information on the long-time behaviour of the system, usually through the existence of conserved quantities, i.e. quantities whose value is constant throughout the time-evolution of the system. Thus, integrability is characterized by the existence of 'constants of the motion', 'integrals', or 'invariants'. For instance, the Hamiltonian system,

$$H = \frac{1}{2}(p_x^2 + p_y^2) + y^4 + 3x^2y^2/4 + x^4/8, \qquad (3.3)$$

has two constants of motion, the Hamiltonian itself, and a second invariant,

$$I = p_x^4 + \frac{1}{2}(6x^2y^2 + x^4)p_x^2 - 2x^3yp_xp_y + \frac{1}{2}x^4p_y^2 + \frac{1}{4}(x^4y^4 + x^6y^2) + \frac{x^8}{16}. \qquad (3.4)$$

Integrals are used to reduce equations of motion. A particularly simple case is that of one-dimensional Hamiltonian systems,

$$H = \frac{1}{2}p_x^2 + V(x) \qquad (3.5)$$

The energy, H_0, is the first constant of motion, and this allows us to obtain the second constant,

$$t - t_0 = \int \frac{dx}{\sqrt{2(H_0 - V(x))}}. \qquad (3.6)$$

Still t_0 is not so useful as H_0. In order to obtain it we must integrate the equations of motion. Moreover, t_0 almost always depends on the integration path and not only on the value of x, while the energy is a local function in phase-space. In [34], we have attempted a classification of the various aspects of integrability. We have distinguished three different situations.

a) The system can be solved by quadratures. (In [35], we have used the term "explicit integrability" for this case). For instance, the two-dimensional Hamiltonian system,

$$H = \frac{1}{2}(p_x^2 + p_y^2) + F(\rho) + \frac{1}{\rho^2}G(\phi), \qquad (3.7)$$

where $\rho = \sqrt{x^2 + y^2}$ and $\phi = \operatorname{atan}(y/x)$, has the second integral,

$$I = (xp_y - yp_x)^2 + 2G(\phi). \qquad (3.8)$$

This allows the equations of motion to be reduced first, to a quadrature for ρ,

$$\dot{\rho}^2 = 2H_0 - 2F(\rho) - \frac{1}{\rho^2}I_0 \qquad (3.9)$$

where H_0 and I_0 are the conserved values of H and I respectively, then $\rho(t)$ is obtained from (3.9), the equation for ϕ can also be reduced to a quadrature

$$\int \frac{d\phi}{\sqrt{I_0 - 2G(\phi)}} = \pm \int \frac{dt}{\rho^2(t)}. \qquad (3.10)$$

b) The equations of motion can be reduced to a system of linear equations which are considered to be integrable, as we explained in the previous section. The simplest example is that of the well known Riccati equation,

$$\dot{x} = a(t)x^2 + b(t)x + c(t), \qquad (3.11)$$

which linearizes to

$$\ddot{w} + (\frac{\dot{a}}{a} - b)\dot{w} + acw = 0 \qquad (3.12)$$

through the transformation

$$x = -\frac{\dot{w}}{aw}. \qquad (3.13)$$

Some PDE's are also integrable through linearization, Burger's equation being the archetype,

$$u_t + u_{xx} + 2uu_x = 0. \qquad (3.14)$$

The Cole-Hopf transformation,

$$u = \frac{v_x}{v}, \qquad (3.15)$$

reduces its solution to that of the heat equation,

$$v_t + v_{xx} = 0. \qquad (3.16)$$

c) The system can be linearized in terms of integrodifferential equations. This is, for example, the case of the Painlevé transcendental equations. As we have seen, six are known at order two, but more surely exist at higher orders. The very idea of linearization through integro-differential equations comes from the Inverse Scattering Transform (IST) technique which was developed in the context of PDE's.

3.1   Complete Integrability

The notion of integrability, meaning the existence of integrals of motion, is so vague as to be almost useless. Thus, we must refine it further in order to obtain a rigorous definition, particularly in view of the relation which we will study between integrability and the singularity structure of the solutions of the equations of motion. As far as the singularity analysis practitioners are concerned, the term integrability implies the existence of complex analytic (functionally independent) integrals of motion. Thus the kind of integrability in which we are interested could have been dubbed "complex analytic integrability".

Complete integrability means that these integrals exist in sufficient number. For a system of N first-order autonomous ordinary differential equations, sufficient means $N - 1$ time-independent invariants (whereupon the system can be reduced to a single quadrature) or N time-dependent ones (in which case the solutions can be obtained by solving an algebraic problem). In general terms, it seems reasonable to ask for integrals respecting the invariance of the initial system. For example, for autonomous systems we should look for integrals which are either time-independent or form-invariant under time translation (see 3.29 below).

Hamiltonian systems are special because, for the complete integrability of a Hamiltonian system with M degrees of freedom, the existence of $(M - 1)$ single-valued first integrals I_i ('actions'), in involution, i.e. with vanishing Poisson bracket, in addition to the Hamiltonian itself, allows the construction of $(M-1)$ additional integrals Ω_i ('angles'), following the Hamilton-Jacobi procedure. A system of N first-order ODE's may sometimes be similar to a Hamiltonian system [36]. Indeed, complete integrability can be interpreted as the existence of k $(1 \leq k \leq N-1)$ first integrals I_i, provided that $(N-k-1)$ more, say Ω_j, can be computed by integration of closed differential forms obtained from the I_i. This includes of course the Hamiltonian case for which $k = M$. Moreover, if a differential system of order N admits an invariant density measure, then it also falls into that category for $k = N-2$, since a last invariant can be obtained through the Jacobi Last Multiplier theorem [8]. For instance, the three-dimensional Lotka-Volterra system,

$$\dot{x} = x(Cy + z + \lambda)$$

$$\dot{y} = y(x + Az + \mu) \tag{3.17}$$

$$\dot{z} = z(Bx + y + \nu),$$

always admits an invariant density measure $1/xyz$ [37]. Therefore, in the special case $A = B = 1$, $\lambda = \mu = \nu = 0$, where there exists the invariant,

$$I = (x - Cy)(1 - \frac{z}{y}), \tag{3.18}$$

we obtain the second invariant,

$$\Omega = \frac{xz}{y}(1 - \frac{y}{z})^{C+1}. \tag{3.19}$$

Of course, analytic integrability is not the only possibility. In [36], in order to relate integrability to the rationality of the Kovalevskaya exponents (KE), Yoshida has introduced the notion of "algebraic integrability". What he meant by that is that the constants of motion should be rational functions. However he imposed rationality on both the 'action'- and the 'angle'-type invariants. Requiring rationality of the latter is a bit awkward, although it is satisfied for some well-known systems. For instance, the Newtonian Hamiltonian in two dimensions,

$$H = \frac{1}{2}(p_x^2 + p_y^2) + \frac{1}{\rho}, \tag{3.20}$$

where $\rho = \sqrt{x^2 + y^2}$, has three additional single-valued integrals

$$I_1 = xp_y - yp_x$$

$$I_2 = p_x(xp_y - yp_x) - \frac{y}{\rho} \tag{3.21}$$

$$I_3 = p_y(xp_y - yp_x) + \frac{x}{\rho},$$

but not all of them are in involution. Still, it is clear that in general the 'angle' invariants are multivalued (see (3.19) for irrational C). If Yoshida's hypothesis regarding the 'angles' is relaxed, one can construct counterexamples with irrational (or complex) Kovalevskaya exponents [10].

Algebraic integrability has also been the object of a series of studies by Adler and van Moerbecke [38]. Their notion of "algebraic complete integrability" is related not only to the existence of a sufficient number of invariants in involution but to a further demand that the solutions be expressible in terms of Abelian integrals. The advantage of this restriction was that they were able to prove some fundamental theorems relating the complex structure of the solutions with integrability and it explained the relation between Painlevé analysis and the dynamics on complex algebraic tori. The two types of algebraic integrability presented above (Yoshida and Adler-van Moerbecke) clearly imposed more restrictions on the system than analytic integrability but, in return, they insure that the solutions do not behave too badly.

Going in the other direction, we can introduce less restrictive types of integrability. The simplest generalization one can think of is to ask for integrals that are analytic only within a given domain rather than globally analytic integrals. Such integrals are not uncommon. For instance, the Lotka-Volterra system (3.17) has many subcases where such integrals exist. For example, when $ABC + 1 = 0$ and $\lambda = \mu = \nu$, the conserved quantity,

$$I = x^{AB}y^{-B}z(x - Cy + ACz)^{-AB+B-1}, \qquad (3.22)$$

is defined in any domain not containing the plane $x - Cy + ACz = 0$ and the coordinate planes $x = 0$, $y = 0$. In fact, for a strictly real-time evolution of the system, integrals which are analytic in some such domains may well suffice.

'Integrable' systems of another kind are those which describe a particle in a potential decreasing sufficiently fast for the particle to be asymptotically free [39]. Thus, the asymptotic momenta are constants of the motion (here "asymptotic" means as time goes to infinity). However, though this is true for real trajectories, it ceases to be so when one considers complex time. Then the asymptotic momenta are not single-valued functions of the initial point and asymptotically free systems are not "analytically" integrable. The generalized Toda potential we will encounter later in this course is an example of asymptotically free potential, and it is generically nonintegrable, as proven by the application of Ziglin's theorem [40].

Going one step further we could argue that every differential equation is integrable in a trivial sense. In fact, let us consider the system of ODE's,

$$\dot{x}_i = F_i(t, x_1, \ldots, x_n) \qquad i = 1, \ldots, n, \qquad (3.23)$$

with initial conditions $x_i(t_0) = c_i$. These initial conditions can be taken to be the constants of motion of the system. The general solution of the system is,

$$x_i = F_i(t, c_1, \ldots, c_n), \qquad (3.24)$$

and, by inverting (3.24), which is equivalent to integrating back in time from t to t_0, we can write the n constants as,

$$c_i = I_i(t, x_1, \ldots, x_n). \tag{3.25}$$

However the inversion is not at all guaranteed to be single-valued. Indeed this will be the case if the integration path wanders in the complex plane around 'bad' singularities.

Whether real-time information on the system is sufficient, is essentially a philosophical question. While it is clear that the physical time is real, it is equally clear that the structure of the solution in complex time heavily influences the real-time behaviour [10]. Singularity analysis techniques are based exclusively on complex time and it is at this price only that one can obtain results on analytic (or algebraic) integrability. The enormous difference between real and complex time, as far as integrability is concerned, can be easily grasped in the following example due to Kruskal. Consider the simple ODE in the complex domain,

$$\dot{x} = \frac{\alpha}{t-a} + \frac{\beta}{t-b} + \frac{\gamma}{t-c} \tag{3.26}$$

Its integration (by quadratures) is straightforward:

$$I = x - \alpha \log(t-a) - \beta \log(t-b) - \gamma \log(t-c) \tag{3.27}$$

However, since the logarithm of a complex number is defined only up to an integer multiple of $2i\pi$, the right-hand side of (3.27), and thus the value of I as well, is determined only up to an additive term, $2i\pi(k\alpha + m\beta + n\gamma)$, with k, m, n, arbitrary integers. Now if one (resp., two) of the α, β, γ are zero, one can construct a two- (resp., one-) dimensional lattice, and define I in a unique way, within the interior of an elementary lattice cell. But, if $\alpha\beta\gamma \neq 0$ and α, β, γ are linearly independent over the integers the indeterminacy in (3.27) is very high: for given x and t, the possible values of I fill the plane densely. Conversely, the knowledge of t and I does not suffice to determine x in any useful way. Thus, the integral (3.27) is useful only in the presence of 'discrete' multivaluedness while 'dense' multivaluedness will be viewed as incompatible with integrability. These are the arguments that lie at the origin of Kruskal's poly-Painlevé approach.

Another interesting concept is that of "pseudo-integrable" systems. They are integrable(?) systems in which the motion takes place on surfaces more complicated than simple tori [41]. Examples have been given of such systems, where the orbits lie on surfaces of genus higher than two, i.e. 'multiply holed tori'. Perhaps the simplest example one can present is two-dimensional polygonal billiards with angles equal to rational multiples of π when reflections are not uniquely defined everywhere. This is the case of reflections at an angle larger than π [42]. Whether some chaotic behaviour is possible in the case of pseudo-integrable systems is not known, but it is clear that these systems are not fully integrable.

3.2   Partial and Constrained Integrability

In the previous subsection, we have examined the various forms of complete integrability. While in some cases the conditions on the integrals were weak to the point of allowing the question of chaotic behaviour of the system to be raised, it was always assumed that the system possessed a complete set of integrals. Relaxing this assumption introduces naturally the notion of partial integrability. Thus one possible form of partial integrability is to have an insufficient number of integrals of motion. In the case of Hamiltonian systems with M degrees of freedom, insufficient means less than $M - 1$ 'action'-like integrals in involution, in addition to the Hamiltonian. Consider, for instance, the Hamiltonian for the asymmetric spinning top with gravity we encountered in the previous section. For general values of the moments of inertia only one additional invariant exists, namely the component of the angular momentum in the direction of gravity. Thus the Hamiltonian of the top is, in general, only partially integrable.

Partial integrability can also be associated with the existence of integrals of inadequate form. For a system of N first-order differential equations the existence of $N-1$ time-dependent first integrals is not sufficient for complete integrability: the system can be reduced to a first-order, nonautonomous differential equation, $\dot{x} = f(x,t)$, which is in general nonintegrable. Still, time-dependent invariants offer some long-term global information about the system. In the case of the Lorenz system,

$$\dot{x} = \sigma(y - x)$$

$$\dot{y} = -y + \rho x - xz \tag{3.28}$$

$$\dot{z} = -bz + xy,$$

for $b = 2\sigma$, one time-dependent integral exists,

$$x^2 - 2\sigma z = Ce^{-2\sigma t}. \tag{3.29}$$

We readily see that, for $\sigma > 0$, as time goes to (positive) infinity the motion is attracted, to the paraboloidal surface $x^2 - 2\sigma z = 0$. The integral (3.29) is precisely of the "form-invariant under time translation" type to which we alluded previously. In fact, a shift in time, $t \to t + t'$, conserves the form of the integral, the new value of the constant being just $C' = Ce^{-2\sigma t'}$.

Hamiltonian systems are, as always, special. Time-dependent Hamiltonians do not, of course conserve energy, so even one-dimensional ones are generally not integrable. However, the existence of just one conserved quantity, even an explicitly time-dependent one, suffices for integrability. Indeed, for any Hamiltonian, $H(x, p_x, t)$, one can introduce a new variable z, and define the time-independent, two-degrees-of-freedom Hamiltonian, $H' = H(x, p_x, p_z) - z$, whereupon the time derivative of p_z is just unity. If H has a conserved quantity, $I(x, p_x, t)$, then H' has a second invariant, $I(x, p_x, p_z)$, and so is integrable.

Another type of incomplete integrability is "constrained" integrability. One well-known example is the fixed-energy integrability of Hamiltonian systems. For instance, when $H = 0$, the Hamiltonian,

$$H = \frac{1}{2}(p_x^2 + p_y^2) + 4\big(a(x^6 + y^6) + (4b - a)(x^2 - y^2)x^2y^2\big),\qquad (3.30)$$

possesses the second invariant

$$I = \frac{x^2 - y^2}{x^2 + y^2}(p_x^2 - p_y^2) - \frac{4xy}{x^2 + y^2}p_xp_y + 8a(x^2 - y^2)^2 - 32bx^2y^2\qquad (3.31)$$

In [43], Hietarinta has investigated the question in detail and presented several examples of two-dimensional Hamiltonian systems which are integrable at zero energy only.

4   Integrability and How to Detect It

The most conclusive proof of integrability is the explicit construction of a sufficient number of first integrals. This is, of course, not the final step in the integration of the problem (unless one finds a sufficient number of invariants that allow the reduction of the solution to a nondifferential, algebraic problem). For Hamiltonian systems one must procede to the construction of the angles associated with the invariants, taken as actions. In some cases the problem turns out to be separable, although the separation may be highly nontrivial. For instance, in [44] we have examined the integrable Hamiltonian,

$$H = \frac{1}{2}\left(p_x^2 + p_y^2\right) + x^4 + 6x^2y^2 + 8y^4,\qquad (4.1)$$

with second invariant

$$C^2 = p_x^4 + 4x^2(x^2 + 6y^2)p_x^2 - 16x^3p_xp_y + 4x^4p_y^2 + 4x^4(x^4 + 4x^2y^2 + 4y^4),\quad (4.2)$$

from the point of view of separability. We have found that the introduction of the variables:

$$u = \frac{p_x^2 + c}{x^2} + 2x^2 + 4y^2$$

and

$$v = \frac{p_x^2 - c}{x^2} + 2x^2 + 4y^2,\qquad (4.3)$$

where c is the square root of the value of the constant of motion (4.2) allows one to separate the equations of motion into

$$\dot{u}^2 = 2u^3 - 8u(2h + c)$$

and

$$\dot{v}^2 = 2v^3 - 8v(2h - c),\qquad (4.4)$$

where h is the value of the energy. The integration of (4.4) is in terms of elliptic functions and, once u, v are known, one can construct x and y in a straightforward way. The interesting (and highly nontrivial) feature of the separation variables (4.3) is that they contain the momenta explicitly and cannot thus be obtained through a simple transformation of coordinates.

In the case of PDE's one considers that integrability is ensured once the Lax pair for the equation in question is obtained. It remains, of course, to reduce the PDE to a Riemann-Hilbert problem and then to solve it, but this is considered to be a technicality (sometimes, admittedly, presenting considerable difficulties). Another, perhaps less conclusive criterion of integrability, is the construction of multi-soliton solutions. The Hirota bilinear formalism and its multilinear extensions are particularly useful, since they transform the construction of multi-soliton solutions into a purely algebraic problem. Still, the proof of the existence of solutions with an arbitrary number of solitons presents considerable difficulties. Moreover, the application of the multi-soliton criterion is delicate, since several types of solitons may exist and one must consider the interaction of all possible combinations.

Another approach to the detection of integrability is through the numerical study of the behaviour of the solutions in real time [45]. Since chaotic behaviour is incompatible with integrability, numerically detected chaos is a clear indication that the system cannot be integrable. However, regular behaviour is not synonymous with integrability: nonintegrable systems do not necessarily exhibit large-scale chaos. What makes the situation even worse is the observation that chaos may appear even in the simulation of an integrable system if one is careless with the numerical implementation. What one needs is a reliable integrability detector so that, given a differential system, one can determine *a priori* whether or not the system is integrable. ("A priori" means without first finding the solutions or a sufficient number of integrals).

This course focuses on singularity analysis also known as the "Painlevé method" for the detection of integrability [46]. The version we shall present in detail is the one known as the ARS approach. This is not the only way to implement singularity analysis. In the next section we shall give a brief account of Painlevé's original method as well as the most recent extensions of the ARS approach. Our preference for this method is based on obvious reasons but, objectively, the ARS criterion is particularly easy to implement and has led to some nice discoveries in the domain of integrability.

4.1   Fixed and Movable Singularities

As we have already hinted, singularities play an important role in determining the integrability of a given nonlinear ODE. Linear equations have only fixed singularities. Let us consider the second-order linear ODE,

$$\frac{d^2w}{dz^2} + p(z)\frac{dw}{dz} + q(z)w = 0. \tag{4.5}$$

A point z_0 in the neighbourhood of which p and q are analytic is called a regular point of the ODE and the solution $w(z)$ can be expressed as a Taylor series in the neighbourhood of z_0. The singular points of the solutions of the equation are located at the singular points of the coefficients p and q. A singular point z_0 is called regular if $(z - z_0)p(z)$ and $(z - z_0)^2 q(z)$ are analytic in the neighbourhood of z_0 [47]. Otherwise it is called irregular. An equation is called Fuchsian if every singular point is regular. The generalisation of these notions to a n-th order equation is straightforward.

Nonlinear equations have not only fixed singularities but movable singularities as well, i.e. singularities whose location depends on the integration constants. Various kinds of movable singularities can exist. Let us illustrate this through specific examples. (In what follows we are going to concentrate on equations that are mostly analytic, so our examples will be chosen from this class). We have:

$w' + w^2 = 0$ with solution $w = (z - z_0)^{-1}$,

$2w' + w^3 = 0$ with solution $w = (z - z_0)^{-1/2}$,

$ww'' - w' + 1 = 0$ with solution $w = (z - z_0)\ln(z - z_0) + \alpha(z - z_0)$,

$\mu ww'' - (1 - \mu)w'^2 = 0$ with solution $w = \alpha(z - z_0)^\mu$,

$(ww'' - w'^2)^2 + 4zw'^3 = 0$ with solution $w = \alpha e^{(z-z_0)^{-1}}$,

$(1 + w^2)w'' + (1 - 2w)w'^2 = 0$ with solution $w = \tan[\alpha + \ln(z - z_0)]$.

where z_0 and α are the integration constants. Thus we have here as movable singularities a pole, an algebraic branch point, a logarithmic branch point, a transcendental singular point (for irrational μ), an isolated essential singularity and a nonisolated essential singularity. Everything but a pole is called a critical singularity.

4.2   The Ablowitz-Ramani-Segur Algorithm

The ARS algorithm [48] was originally developed in order to determine whether a nonlinear ODE (or system of ODE's) admits movable branch points, either algebraic or logarithmic. It is important to keep in mind that this algorithm provides a necessary condition for the absence of such movable branch points. Thus, the (somewhat atypical) occurence of movable essential singularities cannot be detected by this procedure. Let us consider a system of ODE's of the form

$$w_i' = F_i(w_1, w_2, \ldots, w_n; z) \qquad i = 1, \ldots, n. \tag{4.6}$$

The main assumption on which the ARS algorithm rests is that the dominant behaviour of the solutions in the neighbourhood of a movable singularity is of the form

$$w_i \sim a_i(z - z_0)^{p_i}, \qquad z \to z_0. \tag{4.7}$$

Dominant logarithmic branches can also exist, of course, and, while apparently excluded by the ansatz (4.7), they are also taken into consideration in our study.

Most often, the system at hand is not written in the form (4.6) of first order ODE's but includes higher derivatives as well. In any case, not all the functions w_i need to tend to infinity as $z \to z_0$, for $p_i > 0$ in (4.7). In some cases all w_i's go to finite values and only some of the higher derivatives become singular. This seemingly innocent technical point can be of capital importance when it comes to finding all possible singular behaviour of the system. One more remark before we start concerns the case where the complex conjugate w_i^*'s of the dependent variables w_i's explicitly appear in the equations. Our approach is straightforward: we treat the w_i^*'s as new variables, v_i's say, and apply the algorithm without assuming any relation between the w_i's and the v_i's (the original system (4.6) will, of course, have to be augmented by the equations for the v_i's). The reason for doing this is easy to understand: if one starts from the equations for the v_i's , which are formally complex conjugates of those for the w_i's and assumes initial conditions, at a point on the real z-axis, where the v_i's are indeed complex conjugate of the w_i's, then one gets $v_i(z) = [w_i(z^*)]^*$ for all z. Thus the behaviour of v_i at a singular point z_0 is related to the behaviour of w_i at z_0^*, not at z_0 (unless z_0 is real). Thus, in general, the singular behaviour of v_i and w_i at a given complex z_0 may be different.

The ARS algorithm proceeds in three steps, dealing with the dominant behaviour, the resonances and the constants of integration, respectively.

Step 1: Dominant Behaviours
Let us look for a solution of (4.6) of the form,

$$w_i = a_i(z - z_0)^{p_i}, \tag{4.8}$$

where some $Re(p_i) < 0$ and z_0 is arbitrary. Substituting (4.8) in (4.6) one finds all possible p_i's for which two or more terms in each equation balance, while the rest can be ignored, as arising at higher orders in powers of $(z - z_0)$. For each such choice of p_i's, the balance of these so called leading terms also determines the corresponding values of the a_i's .

This first step is also the most delicate of the ARS algorithm. In order to arrive at the correct conclusions, one must find and examine separately all possible dominant behaviour. Extra care is needed here, as omissions can easily lead to erroneous results. First of all, note that several choices of p_i's are possible. If one of the p_i's turns out not to be an integer, then z_0 is an algebraic branch point at the leading order, and that would appear to discourage any further application of the algorithm. It may turn out, however, that a simple change of variables suffices to turn the system into one with no movable branch points. Even if this is not the case, if the p_i in question is a rational though of a special type, the algorithm is still applicable and may be related to the so-called 'weak Painlevé' concept [49]. If all possible p_i's are integers, then, for each of them, the leading behaviour can be viewed as the first term of a Laurent series around a movable pole, i.e. our ansatz now becomes:

$$w_i = (z - z_0)^{p_i} \sum_0^\infty a_i^{(m)} (z - z_0)^m, \tag{4.9}$$

where $a_i^{(0)} = a_i$, the location z_0 of the singularity being the first free (integration) constant of the system (4.6). For an nth-order system there are still $n - 1$ such arbitrary constants to be sought among the $a_i^{(m)}$'s in (4.9). If they are all found to be present there, expansion (4.9) will be referred to as *generic*. The powers m at which these constants arise are termed "resonances", or, sometimes, "Fuchs indices ", of the series (4.9), and it is to their determination that we now turn.

Step 2: Resonances
At this step, we start by retaining only the leading terms in the original equations and we substitute in every w_i the simplified expression:

$$w_i = a_i(z - z_0)^{p_i}(1 + \gamma_i(z - z_0)^r), \quad r > 0, \quad i = 1, \ldots, n \tag{4.10}$$

We then retain in (4.6) only the terms linear in γ_i , which we write as,

$$Q(r)\gamma = 0, \qquad \gamma = (\gamma_1 \ldots, \gamma_n) \tag{4.11}$$

where $Q(r)$ is an $n \times n$ matrix, with r entering only in its diagonal elements, and at most linearly. Clearly then, some of the γ_i's will be arbitrary, and hence free constants will enter in (4.9), at the n roots of the algebraic equation,

$$\det Q(r) = (r + 1)(r^{n-1} + A_2 r^{n-2} + \cdots + A_n) = 0, \tag{4.12}$$

where $r = -1$ is related to the one free constant we have *ab initio*, namely the location z_0 of the singularity. Some general remarks are in order here:

(a) A resonance $r = 0$ corresponds to the case where the coefficient of one of the leading terms is arbitrary
(b) As we have already seen, $r = -1$ is always a root of (4.12), as can be seen by perturbing $z - z_0$ to $z - z_0 + \epsilon$ at a leading order in (4.8), expanding in powers of ϵ, and observing that the first contribution enters at order $(z - z_0)^{p_i - 1}$.
(c) Any resonance with $Re(r) < 0$ (except $r = -1$), must be ignored, since they violate the hypothesis that the p_i's are the powers of the leading terms in series (4.9). Such resonances imply that the corresponding singular expansions are not generic, in the sense defined above. (However, see the remarks in Sect. 5.)
(d) Any resonance with $Re(r) > 0$, but where r is not an integer, indicates that $z = z_0$ is a movable branch point. The algorithm terminates at this stage. If r is real and rational, it must still be checked, whether this algebraic branch point falls into the weak-Painlevé class already mentioned in Step 1.
(e) In the case where p_i is itself rational, the appearance of a rational r, with the same denominator as p_i indicates a finite branching with multiplicity determined

by the leading singularity and is directly related to the so-called 'weak-Painlevé' concept. This special case was found sometimes to yield integrable systems.

Thus a singular expansion (4.9) will also be called *generic* if it is associated with $(n-1)$ non negative integer resonances. If, for every leading behaviour, one finds less than $n-1$ such resonances, then all the solutions found are non-generic. This usually indicates that the ansatz (4.8) misses an essential part of the solution, most probably, a leading logarithmic singularity.

If for every leading singular behaviour of step 1, all the resonances with non-negative real part are integer numbers and provided that *at least one* leading behaviour is generic [50], i.e. involves $(n-1)$ non-negative resonances, we may then proceed to

Step 3: The Constants of Integration
In this step, we shall check for the occurrence of non-dominant logarithmic branch points. To do this, we substitute into the full equation (4.6), for every different leading behaviour (4.7) the truncated expansion

$$w_i = \alpha_i \tau_i^{p_i} + \sum_1^{r_s} a_i^{(m)} \tau^{p_i+m}, \tag{4.13}$$

where we take $\tau = z - z_0$, where r_s is the largest positive root of (4.12). We then identify the terms order by order in powers of τ. We obtain equations reminiscent of (4.11) but with, in general, terms of lower-order appearing on the right-hand-side, to get,

$$Q(m)a^{(m)} = R^{(m)}(z_0; a^{(j)}), \qquad j = 1,\ldots,m-1, \tag{4.14}$$

with $m = 1,\ldots,r_s$, $R = (R_1,\ldots,R_n)$. Then:
(i) for $m < r_1$, where r_1 is the smallest positive resonance, (4.14) determines $a^{(m)}$.
(ii) At $m = r_1$, for (4.14) to have a solution, i.e for $a^{(r_1)}$ to have one arbitrary component, assuming r_1 is a simple root of (4.12), the following compatibility condition must be satisfied,

$$\det Q^{(1)}(r_1) = 0, \tag{4.15}$$

where $Q^{(1)}(r_1)$ is the matrix $Q(r_1)$ with its first column replaced by $R^{(r_1)}$.
(iii) If (4.15) is satisfied, then for $r_1 < m < r_2$, the next smallest positive resonance, (4.14) again determines $a^{(m)}$.
(iv) The same procedure must be repeated successively at each higher resonance up to the largest one. The case of multiple roots does not present any particular difficulty: one must ensure that the number of arbitrary components of $a^{(r)}$ be equal in number to the multiplicity of the resonance r.

However it may turn out that for some resonance r condition (4.15) is not satisfied. Then one or more of the expansions (4.9) will have to be altered in the following way,

$$w_i = \sum_{0}^{r-1} a_i^{(m)} \tau^{p_i+m} + (a_i^{(r)} + b_i^{(r)} \ln\tau)\tau^{p_i+r} + \dots, \tag{4.16}$$

with $\ln\tau$, $(\ln\tau)^2$ etc. possibly entering at higher orders. The logarithms introduce new terms in the expansion: we determine the coefficients $b_i^{(r)}$ by requiring that the coefficient of the appropriate power of τ vanishe while the $a_i^{(r)}$ are free.

In summary, we shall say that a system of ODE's (4.6) satisfies the necessary conditions for the Painlevé property, i.e. for having no movable critical points other than poles, if its solutions can be expanded in pure Laurent series (4.9) near every one of their movable singularities at $z = z_0$. In other words, following the ARS algorithm as outlined above we must come across no algebraic branch points and no logarithmic singularities. This turns out to be a rare occurence. In those special cases where all the necessary conditions for the Painlevé property are fulfilled, there remains the examination of the sufficiency of the conclusion of the conjecture: are there indeed algebraic integrals that can be used to reduce the dimensionality of the system and sometimes even to solve it explicitly by quadratures? To date, the most effective way of accomplishing this has been the direct integration of the corresponding equations of motion.

Still, the ARS approach is not infallible. Since it does not test for movable essential critical singularities, it may turn out that a system passes the ARS test and is still nonintegrable because of such a bad singularity. However this situation is very rare. Only recently has a real physical system rather than an *ad hoc* construction been found where the Painlevé property is violated through an essential singularity. The details will be presented in the next section.

5   Implementing Singularity Analysis: From Painlevé to ARS and Beyond

The first application of singularity analysis was on first-order equations. Painlevé proved [12] that for equations of the form,

$$F(x', x, t) = 0, \tag{5.1}$$

with F polynomial in x' and x, and analytic in t, the movable singularities of the solutions are poles and/or algebraic branch points. Fuchs showed [11] that the only equation of the form,

$$x' = f(x, t), \tag{5.2}$$

where f is rational in x and analytic in t, with critical points that are all fixed, is the Riccati equation,

$$x' = a(t)x^2 + b(t)x + c(t). \tag{5.3}$$

Its integration is straightforward. If $a = 0$ (5.3) is linear. Otherwise, the transformation

$$x = -\frac{u'}{au} \tag{5.4}$$

reduces the equation to a linear one of the second-order,

$$au'' - (a' + ab)u' + a^2cu = 0. \tag{5.5}$$

Binomial equations of the form $x'^n = f(x, t)$ have also been analysed by Briot and Bouquet [51]. They found that for $n > 1$ the following equations that have the Painlevé property:

$x'^2 = 4x^3 + \lambda x + 1$, integrable in terms of elliptic functions
$x'^2 = x(q(t)x + r(t))^2$, reducible to a Riccati
$x'^3 = x^2(x - 1)^2$, (elliptic function)
$x'^4 = x^3(x - 1)^3$, (elliptic function)   (5.6)
$x'^6 = x^4(x - 1)^3$, (elliptic function)
$x'^n = q(t)x^{n-1}$, integrable by quadratures.

The problem treated by Kovalevskaya [13] has already been mentioned in Sect. 2. The solution of the top equations (2.6) have a singular behaviour:

$$p, q, r \sim \frac{1}{\tau}$$

$$\alpha, \beta, \gamma \sim \frac{1}{\tau^2}. \tag{5.7}$$

The computation of the resonances is quite complicated but, in the integrable classes, including Kovalevskaya's, $r = -1, 0, 1, 2, 3, 4$. Substituting into the full equation we find that the resonance conditions are satisfied and thus the integrability of the Kovalevskaya case $A = B = 2C$, $z_0 = 0$ is confirmed by singularity analysis.

Although Kovalevskaya's result was very interesting and proved the usefulness of singularity analysis, it was not as important as the one due to Painlevé and his school [15,16]. Painlevé not only provided a systematic classification of equations of the form,

$$x'' = f(x', x, t), \tag{5.8}$$

with f a polynomial in x', rational in x and analytic in t, but obtained new transcendents that appear regularly in physical applications. The starting point for Painlevé's approach was the observation that critical singularities of second-order equations can be branch points, both algebraic and logarithmic, as well as essential singularities. Painlevé developed his method (known as α-method) that made it possible to test an equation for the existence of all of these singularities in the solution. Moreover, since Painlevé was concerned by the integrability of his equations, he proposed his approach as a "double method". The first part (based on the α-method) was the local study giving the necessary conditions for the absence of critical singularities. The second part was the proof of the sufficiency of the conditions and either the integration or the proof of the irreducibility of the equations. In order to illustrate the Painlevé's α-method we will examine the derivation of the first transcendental equation that bears his name and consider an equation of the form,

$$x'' = x^2 + f(t), \tag{5.9}$$

where $f(t)$ is analytic. This is the simplest nontrivial form of (5.8). In the spirit of ARS we can say that (5.9) does not have algebraic branch points and one need only investigate the existence of logarithmic singularities. Painlevé introduces a small parameter α by a scaling, $x = X/\alpha^2$, $t = t_0 + \alpha T$. We thus find:

$$\frac{d^2 X}{dT^2} = 6X^2 + \alpha^4 f(t_0) + \alpha^5 f'(t_0) + \frac{1}{2}\alpha^6 f''(t_0) + \mathcal{O}(\alpha^7) \tag{5.10}$$

and seek a solution in the form of a power series in α,

$$X(T) = X_0(T) + \alpha^4 X_4(T) + \alpha^5 X_5(T) + \alpha^6 X_6(T) + \mathcal{O}(\alpha^7). \tag{5.11}$$

(There is no need to introduce terms proportional to $\alpha, \alpha^2, \alpha^3$ [9]). We find,

$$\frac{d^2 X_0}{dT^2} = 6X_0^2 \tag{5.12}$$

and

$$\frac{d^2 X_{r+4}}{dT^2} - 12X_0 X_{r+4} = \frac{T^r}{r!}\frac{d^r f}{dt^r}(t_0), \tag{5.13}$$

for $r = 0, 1, 2$. The general solution of (5.12) is the Weierstrass elliptic function, $X_0 = \wp(T - T_0; 0, h)$, with h and T_0 as constants of integration. Thus the homogeneous part of (5.13) is a Lamé equation,

$$\frac{d^2 Y}{dT^2} - 12\wp(T - T_0; 0, h)Y = 0, \tag{5.14}$$

and its general solution is

$$Y(T) = a(T\frac{d\wp}{dT} + 2\wp) + b\frac{d\wp}{dT}, \tag{5.15}$$

with a, b integration constants. The solution of the full (5.13) is obtained by the method of variation of parameters,

$$X_{r+4} = U_{r+4}(T\frac{d\wp}{dT} + 2\wp) + V_{r+4}\frac{d\wp}{dT}, \tag{5.16}$$

and the coefficients U, V are given by

$$\frac{dU_{r+4}}{dT} = \frac{T^r}{24r!}\frac{d^r f}{dt^r}(t_0)\frac{dX_0}{dT} \tag{5.17}$$

$$\frac{dV_{r+4}}{dT} = \frac{T^r}{24r!}\frac{d^r f}{dt^r}(t_0)(T\frac{dX_0}{dT} + 2X_0).\tag{5.18}$$

Integrating (5.17) and (5.18) we find that U and V are given in terms of elliptic functions for $r = 0, 1$. For $r = 2$, expanding the solution X_0 around the movable singularity at T_0, where $X_0 \sim (T - T_0)^{-2}$, we find that a logarithm appears. For the solution to be free of movable critical points it is necessary for the coefficient of the logarithm to vanish and the explicit calculation leads to

$$\frac{d^2 f}{dt^2}(t_0) = 0.\tag{5.19}$$

Since t_0 is arbitrary, this means that, for integrability, f must be linear in t. Apart from cases that are integrable in terms of elementary functions, one finds the P_I equation,

$$x'' = 6x^2 + t.\tag{5.20}$$

Painlevé also showed that (5.20) is free of movable essential singularities, thus completing the proof that P_I has no movable critical points. In practice, the Painlevé α-method requires the exact solution of a nonlinear ODE as well as that of inhomogeneous linear ODE's with the same homogeneous part and different inhomogeneous parts at each order. Thus, a particular solution is needed at each order for the integration. As a result the whole approach is somewhat cumbersome. (This is probably the reason why Painlevé was not able to produce the total classification of second order ODE's and had to leave this task to Gambier who tackled this problem with the method of Kovalevskaya.) Gambier obtained all the equations of the Painlevé type and in particular produced a list of 24 fundamental ones [16]: if one knows the solution of these 24 equations, then one can construct the solution of any other equation of the Painlevé type at order two. Here is the Gambier (t) list, where a, b, c, d, e are constants, q, r are free functions of t, and f_n, ϕ_n, ψ_n are definite functions of q and r:

(G1) $x'' = 0$

(G2) $x'' = 6x^2$

(G3) $x'' = 6x^2 - \dfrac{1}{24}$

(G4) $x'' = 6x^2 + t$

(G5) $x'' = -3xx' - x^3 + q(x' + x^2)$

(G6) $x'' = -2xx' + qx' + q'x$

(G7) $x'' = 2x^3$

(G8) $x'' = 2x^3 + ax + b$

(G9) $x'' = 2x^3 + tx + a$

(G10) $x'' = \dfrac{x'^2}{x}$

(G11) $x'' = \dfrac{x'^2}{x} + ax^3 + bx^2 + c + \dfrac{d}{x}$

(G12) $x'' = \dfrac{x'^2}{x} - \dfrac{x'}{t} + \dfrac{1}{t}(ax^2 + b) + cx^3 + \dfrac{d}{x}$

(G13) $x'' = \dfrac{x'^2}{x} + q\dfrac{x'}{x} - q' + rxx' + r'x^2$

(G14) $x'' = (1 - \dfrac{1}{n})\dfrac{x'^2}{x} + qxx' - \dfrac{nq^2}{(n+2)^2}x^3 + \dfrac{nq'}{n+2}x^2$

(G15) $x'' = (1 - \dfrac{1}{n})\dfrac{x'^2}{x} + (f_n x + \phi_n - \dfrac{n-2}{nx})x'$

$\qquad - \dfrac{nf_n^2}{(n+2)^2}x^3 + \dfrac{n(f_n' - f_n\phi_n)}{n+2}x^2 + \psi_n x - \phi_n - \dfrac{1}{nx}$

(G16) $x'' = \dfrac{x'^2}{2x} + \dfrac{3x^3}{2}$

(G17) $x'' = \dfrac{x'^2}{2x} + \dfrac{3x^3}{2} + 4ax^2 + 2bx - \dfrac{c^2}{2x}$

(G18) $x'' = \dfrac{x'^2}{2x} + \dfrac{3x^3}{2} + 4tx^2 + 2(t^2 - a)x - \dfrac{b^2}{2x}$

(G19) $x'' = \dfrac{x'^2 - 1}{2x}$

(G20) $x'' = x'^2(\dfrac{1}{2x} + \dfrac{1}{x-1})$

(G21) $x'' = x'^2(\dfrac{1}{2x} + \dfrac{1}{x-1}) + (x-1)^2(ax + \dfrac{b}{x}) + cx + \dfrac{dx}{x-1}$

(G22) $x'' = x'^2(\dfrac{1}{2x} + \dfrac{1}{x-1}) - \dfrac{x'}{t} + \dfrac{(x-1)^2}{t^2}(ax + \dfrac{b}{x}) + c\dfrac{x}{t} + \dfrac{dx(x+1)}{x-1}$

(G23) $x'' = \dfrac{x'^2}{2}(\dfrac{1}{x} + \dfrac{1}{x-1} + \dfrac{1}{x-a})$

$\qquad + x(x-1)(x-a)\big(b + \dfrac{c}{x^2} + \dfrac{d}{(x-1)^2} + \dfrac{e}{(x-a)^2}\big)$

(G24) $x'' = \dfrac{x'^2}{2}(\dfrac{1}{x} + \dfrac{1}{x-1} + \dfrac{1}{x-t}) - x'(\dfrac{1}{t} + \dfrac{1}{t-1} + \dfrac{1}{x-t})$

$\qquad + \dfrac{x(x-1)(x-t)}{2t^2(t-1)^2}\big(a - \dfrac{bt}{x^2} + c\dfrac{t-1}{(x-1)^2} + \dfrac{(d-1)t(t-1)}{(x-t)^2}\big).$

Gambier's original article is really worth reading. Its approach looks very modern, even now, almost a century later and contains some very interesting remarks. Here is one that we find really fundamental:

> «*Je rencontrais des systèmes de conditions différentielles dont l'intégration était, quoiqu'au fond bien simple, assez difficile à apercevoir. Par un mécanisme qui est général, mais qui était difficile à prévoir, la résolution*

de ce premier problème, intégration des conditions, est intimement liée à l'intégration de l'équation différentielle elle-même».

In other words, the integration of the (integrability) conditions is intimately related to the integration of the nonlinear equation itself.

Let us illustrate the derivation of P_I by Gambier's method. Starting with (5.9) we look for the dominant behaviour in the neighbourhood of a singularity t_0. We assume that

$$x \sim a\tau^p, \tag{5.21}$$

where $\tau = t - t_0$. Substituting into (5.9) we find $p = -2$ and $a = 1$, corresponding to x'' and x^2 being dominant. Since p is an integer, we can proceed further and look for the second integration constant (t_0 being the first). We look in particular for the power of τ, called the "index" according to Fuchs, or the resonance in the ARS terminology, at which this second constant appears. We introduce

$$x = \tau^{-2} + \gamma\tau^{r-2} \tag{5.22}$$

into the dominant part of (5.9). Linearizing for γ we find that

$$(r - 2)(r - 3) - 12 = 0, \tag{5.23}$$

with roots $r = -1$, corresponding to the arbitrariness of t_0, and $r = 6$. Since this second resonance is integer we can proceed to a check of compatibility that will guarantee the absence of logarithmic branch points. We expand

$$x = \tau^{-2} \sum_{r=0}^{6} a_r \tau^r \tag{5.24}$$

with $a_0 = 1$. The calculations are straightforward and we find as a condition $d^2 f/dt^2 = 0$, i.e. f must be linear.

Third and fourth order equations were treated by Chazy and Garnier [17], [18] who attempted to obtain a Painlevé-Gambier classification at orders three and four. However, the difficulties are considerably higher and only partial classifications were obtained. Bureau was the only one who, before the appearance of integrable PDE's, pursued the singularity analysis approach. His method neither resembled Kovalevkaya's nor Painlevé's. We shall not go into these details here. The two major problems that Bureau set out to solve were the analysis of the system:

$$x' = P(x, y, t)$$
$$y' = Q(x, y, t), \tag{5.25}$$

where P and Q are polynomial in x, y [52]. The rather disappointing result was that no new transcendents were found. The more general, and also more interesting, problem with P and Q rational was, unfortunately, not treated. Bureau's second problem was that of binomial equations [53],

$$x''^2 = f(x', x, t). \tag{5.26}$$

Although he obtained very interesting results the complete classification had to wait. In a recent work, Cosgrove [54,55] has given a classification of all the integrable binomial equations of the form

$$x''^n = f(x', x, t) \tag{5.27}$$

We list below his results, where we have tried to follow the same conventions as in the Gambier case:

(SD-I) $x''^2 = q(t)R_3(tx' - x, x')$

(SD-II) $x''^2 = (q(t)x' + r(t)x + s(t))^2 R_1(tx' - x, x')$

(SD-III) $x''^2 = (q(t)x + r(t))^2 R_2(tx' - x, x')$

(SD-IV) $x''^2 = (q(t)x^2 + r(t)x + s(t))^2 R_1(tx' - x, x')$

(SD-V) $x''^2 = (q(t)x + r(t))^2 R_1(tx' - x, x')$

(SD-VI) $x''^2 = q(t)R_2(tx' - x, x')$

(BP-VII) $x''^3 = q(t)(R_2(tx' - x, x'))^2$

(BP-VIII) $x''^3 = (q(t)x + r(t))^3 (R_2(tx' - x, x'))^2$

(BP-IX) $x''^4 = q(t)(R_2(tx' - x, x'))^3$

(BP-X) $x''^4 = q(t)(R_1(tx' - x, x'))^2 (\tilde{R}_1(tx' - x, x'))^3$

(BP-XI) $x''^6 = q(t)(R_1(tx' - x, x'))^4 (\tilde{R}_1(tx' - x, x'))^5$

(BP-XII) $x''^6 = q(t)(R_1(tx' - x, x'))^3 (\tilde{R}_1(tx' - x, x'))^5$

(BP-XIII) $x''^6 = q(t)(R_1(tx' - x, x'))^3 (\tilde{R}_1(tx' - x, x'))^4$

(BP-XIV) $x''^n = q(t)(R_1(tx' - x, x'))^{n+1}$

(BP-XV) $x''^n = q(t)(R_1(tx' - x, x'))^{n-1}$

The R_i's correspond to the following expressions, where $a, b, c, \ldots$ are constants:

$$R_1 = a(tx' - x) + bx' + c$$

$$\tilde{R}_1 = d(tx' - x) + ex' + f \tag{5.28}$$

$$R_2 = a(tx' - x)^2 + bx'(tx' - x) + cx'^2 + d(tx' - x) + ex' + f$$

$$R_3 = a(tx' - x)^3 + bx'(tx' - x)^2 + cx'^2(tx' - x) + dx'^3$$
$$+ e(tx' - x)^2 + fx'(tx' - x) + gx'^2 + h(tx' - x) + kx' + l.$$

The integration of these equations led to the following results.

SD-I: This is the 'master' equation. Its solution can be given in terms of all Painlevé transcedents from PVI to PI depending on the parameter values.
SD-II: can be reduced to a second-order linear equation,
SD-III: integrated in terms of PV or PIII,

SD-IV: integrated in terms of PIV,

SD-V: integrated in terms of PI,

SD-VI: can be reduced to second-order linear equation,

BP-VII: integrated in terms of PIV or PI,

BP-VIII: integrated in terms of PII or Airy functions,

BP-IX: integrated in terms of elliptic functions,   .

BP-X: integrated in terms of PII or Airy functions,

BP-XI: integrated in terms of elliptic functions,

BP-XII: integrated in terms of elliptic functions,

BP-XIII: integrated in terms of PI,

BP-XIV: solved by quadratures,

BP-XV: solved by quadratures.

The important remark here is that no new transcendent was found in this generalisation of Painlevé's work.

With the exception of Cosgrove, our presentation till now has respected the chronological order, from Painlevé to ARS. In fact, the main volume of work on (mostly physical) systems in the 80's was done using the ARS method. However, it soon became clear that the shortcomings of ARS could be bypassed in some cases and, thus, extensions were proposed.

The application of the ARS criterion to PDE's has the drawback of being applied to reductions of the PDE to ODE's. Weiss and collaborators (WTC) [30] did away with this limitation by proposing a Painlevé test that could be applied directly to the PDE. They introduced the notion of a singularity manifold and claimed that a PDE possess the Painlevé property if its solutions are singlevalued in the neighbourhood of a *noncharacteristic* [56] movable singularity manifold. For simplicity let us consider a 2-dimensional PDE. The WTC approach consists of seeking a Laurent series expansion of the solution

$$u(x,t) = \phi^{-k}(x,t) \sum_{r=0}^{\infty} u_r(x,t)\phi^r(x,t) \tag{5.29}$$

where k is integer, the u_r are analytic functions ($u_0 \neq 0$) in the neighbourhood of the singularity manifold defined by $\phi(x,t) = 0$. Requiring that the singularity manifold be movable means that ϕ is only determined by the initial (and/or boundary) data, without any other constraint (besides being noncharacteristic). The practical implementation of the test follows closely the ARS algorithm. One substitutes the expansion in the PDE, determines k and the u_r's by the recursion,

$$Q(r)u_r = R_r(u_0, u_1, \ldots, u_{n-1}, \phi, x, t) \tag{5.30}$$

where $Q(r)$ is a polynomial of order equal to the order of the PDE. The equation $Q(r) = 0$ defines the resonances and the compatibility condition $R_r = 0$ must be satisfied at every resonance. In order to turn this approach into an easily applied algorithm, Kruskal [57] proposed the following simplification. Instead of defining the singularity manifold, one can solve for x and write $\phi = x + \psi(t)$ where ψ is an arbitrary analytic function and the u_r's are now functions of t only.

Another technical improvement, that simplifies the expansion (5.29) without undue simplifications of the singularity manifold is the method proposed by Conte and Musette [58]. It is based on the observation that the expansion variable χ need not be the singularity manifold ϕ. The only requirement is that χ must vanish where ϕ does and be a singlevalued function of ϕ and its derivatives. It turns out that one possible optimal expansion variable is:

$$\chi = \left(\frac{\phi_x}{\phi} - \frac{\phi_{xx}}{2\phi_x}\right)^{-1}.$$

(5.31)

Introducing the Schwartzian S,

$$S \equiv \{\phi; x\} = \frac{\phi_{xxx}}{\phi_x} - \frac{3}{2}\frac{\phi_{xx}^2}{\phi_x},$$

(5.32)

and the auxiliary quantity,

$$C = -\frac{\phi_t}{\phi_x},$$

(5.33)

one obtains

$$\chi_t = -C + C_x\chi - \frac{1}{2}(CS + C_{xx})\chi^2$$

(5.34)

$$\chi_x = 1 + \frac{S}{2}\chi^2$$

(5.35)

$$\frac{\phi_t}{\phi} = -C\chi^{-1} - \frac{1}{2}C_x = -C\frac{\phi_x}{\phi} + \frac{1}{2}C_x.$$

(5.36)

As in Kruskal's approach one must privilege some variable x. (Kruskal's choice corresponds to $\chi_x = 1$). The advantage of this approach is that it is invariant under a general homographic transformation of ϕ.

Kruskal's poly-Painlevé method [31] is also an important extension of the Painlevé test. Instead of performing the analysis locally, Kruskal analyzes the singularities and their interaction. In particular a multivaluedness of the 'dense' type is considered to be incompatible with integrability. The method is asymptotic and a parameter ϵ must be introduced, which can be taken to be arbitrarily small. The main idea is the following: if one can show that the same trajectory can be characterized by two values of an integration constant c which differ by a quantity proportinal to ϵ^n (for some n) this will mean that c, for this trajectory, is densely multivalued. The practical application of the method presents considerable difficulties and very few results have so far been yielded by this approach.

Still another extension of the Painlevé approach was first proposed by Kruskal in his work entitled "Flexibility in applying the Painlevé test" [59]. This whole work was motivated by what Kruskal calls a "worrisome example", a time-independent reduction of some continuous Heisenberg spin chains. Although the singularity analysis of this example is, for an experienced practitioner, straightforward, Kruskal's deep reflections have inspired at least one line of research and much more is contained in this article. Kruskal's approach has to do with the treatment of negative resonances. Negative resonances have played a rather obscure role, at least in the modern tradition of Painlevé analysis through the ARS algorithm. In fact the ARS recommendation concerning negative resonances is explicit [29]: "ignore any roots [of the indicial equation $Q(r) = 0$] with $Re(r) < 0$". (In practice, though, one asks for negative resonances to be integral too, although no compatibility at these resonances was ever considered). We must point out here that Kruskal has been drawing attention to this subtlety in private exchanges with most of us Painlevé practitioners ever since the beginning of the modern singularity analysis era. He in fact insists that a negative resonance may induce multivaluedness: directly if it is noninteger and through logarithmic terms if a compatibility condition is not satisfied at some negative integral resonance value. Kruskal's main point is that the Painlevé Laurent series expansion can be considered to be the lowest-order term of a perturbation series in the coefficient ϵ of the negative-resonance term,

$$x(\tau) = \sum_{n=0}^{\infty} \epsilon^n x_n(\tau), \tag{5.37}$$

where $x_0(\tau)$ is the Painlevé series and the higher $x_n(\tau)$ terms are generalized power series determined successively for $n = 1, 2, \ldots$ to satisfy the differential equation. This ϵ-series should be valid for small ϵ and $\tau - \tau_0$ small but not too small, i.e. in an annulus around τ_0 with inner radius depending on ϵ. Kruskal's ideas were developed up by Conte, Fordy and Pickering [60] and implemented algorithmically. They have shown that there exist cases (Chazy's equations being prominent among them) where the negative resonances play an important role in (non)integrability.

We will illustrate the importance of Kruskal's approach by an application to the Mixmaster Universe model (MUM) [61]. The equations of motion in the appropriate (noncanonical) variables are:

$$\dot{X} = X(p_x - p_y - p_z)$$

$$\dot{Y} = Y(p_y - p_z - p_x)$$

$$\dot{Z} = Z(p_z - p_x - p_y)$$

$$\dot{p}_x = X(Y + Z - X) \tag{5.38}$$

$$\dot{p}_y = Y(Z + X - Y)$$

$$\dot{p}_z = Z(X + Y - Z),$$

where the 'dot' denotes differentiation with respect to time. The ARS singularity analysis for (5.38) has yielded two different singular behaviours [62].
i) X, p_x alone diverge while Y, Z, p_y, p_z are finite, or any other circular permutation,

$$X = \pm \frac{i}{t - t_0}, \quad p_x = -\frac{1}{t - t_0}$$

$$Y = y_{01}(t - t_0) \,, Z = z_{01}(t - t_0), \quad p_y = p_2, \quad p_z = p_3.$$

The resonances in this case are: r: $-1, 0, 0, 0, 0, 2$.
The resonance -1 is related, as usual, to the freedom of the location t_0 of the singularity, while the quadruple 0-resonance is related to the free y_{01}, z_{01}, p_2, p_3 parameters. We have also verified that the $r=2$ resonance indeed satisfies the compatibility condition. Thus this expansion is generic, i.e. it has 6 free parameters, and it is of the Painlevé type.
ii) All X, Y, Z, p_x, p_y, p_z diverge as simple poles,

$$X, Y, Z = \pm \frac{i}{t - t_0}, \quad p_x, p_y, p_z = \frac{1}{t - t_0}.$$

The resonances in this case are: r= -1,-1,-1,2,2,2.
This nongeneric case is intriguing since it possesses a triple (-1) resonance, a feature that might indicate a dominant logarithmic singularity. However this was not the case and thus we concluded in [62] that type ii) singularities passed the ARS test.

However it turned out that this was not true [63]. In order to follow Kruskal's approach we consider the type ii) singular expansion as part of a perturbation expansion (in the coefficient ϵ of the negative resonance terms). In order to simplify the presentation we introduce a second small parameter η related to the positive resonances and thus propose the following expansion for X,

$$X = \pm \tfrac{i}{t-t_0}\{ \, 1 + \eta x_{01}(t - t_0)^2 + \eta^2 x_{02}(t - t_0)^4 + \ldots$$
$$\frac{\epsilon}{t - t_0}(x_{10} + \eta x_{11}(t - t_0)^2 + \eta^2 x_{12}(t - t_0)^4 + \ldots)$$
$$\frac{\epsilon^2}{(t - t_0)^2}(x_{20} + \eta x_{21}(t - t_0)^2 + \ldots)\}, \tag{5.39}$$

and similarly for Y, Z, p_x, p_y, p_z. The x_{10}, y_{10}, z_{10} and x_{01}, y_{01}, z_{01} are free because -1 and 2 are triple resonances; the corresponding coefficients of the momenta p_i are determined from those of the coordinates X, Y, Z. Substituting the expansion (5.39) in the equations of motion (5.38), we can compute the coefficients order by order in ϵ. However, now we get a resonance condition at every order whenever the power of $t - t_0$ is -1 or $+2$. This means that a resonance condition will occur at the $\epsilon^n \eta^m$ terms whenever $n = 2m + 1$ or $n = 2m - 2$. We have started by checking the first few conditions. None is automatically satisfied but necessitate that some resonance compatibility condition holds. The first condition we encountered was $x_{10} = y_{10} = z_{10}$. A look at the expansion of the coordinates in this case indicates the true nature of this condition; it corresponds to taking an arbitrary shift in the singularity location t_0 but *the same shift for all X, Y, Z* and expanding around it. We have checked that this condition suffices in order to satisfy the higher-order conditions.

So, although in the classical ARS approach the Mixmaster Universe model has a valid Painlevé expansion, the perturbative singular expansion (5.39) does not satisfy the compatibility condition at every order (unless a special expansion $x_{10} = y_{10} = z_{10}$ is considered). Now, what does this mean? A possible interpretation of the incompatibility of negative resonances may be given in the light of the results of [64,65]. In [64] we have argued that the only singular behaviour of the solutions of the MUM are the (i) and (ii) given above. In particular, we have stated that no singular behaviour can exist where two of the X's are divergent. However, it turns out that such a situation can exist (although, admittedly, it is more complicated than we initially thought). Let us, thus, assume that two of the X's, say Y and Z, are more singular than X. From (5.38) it results that p_x, p_y, p_z diverge like $\mathcal{O}(1/\tau)$. Then the fourth equation shows that Y, Z must diverge like $1/\tau^2$ while X is regular and starts as a constant. However this behaviour would be incompatible with the remaining two equations *unless* a cancellation occurs in $Y - Z$. In fact, Y and Z must be equal, not only at the level of the dominant term, but also at the level of the subdominant terms of order $1/\tau$. It is then easy to compute the leading singularity, $X = A$, $Y = B/\tau^2$, $Z = B/\tau^2$ and $p_x = 2/\tau$, $p_y = 1/\tau$, $p_z = 1/\tau$ with $AB = -1$. The cancellation of the difference $Y - Z$ (and also of $p_y - p_z$) suggests a change of variables where this difference appears explicitly: $Y - Z = \delta$, $Y + Z = \sigma$, $p_y - p_z = q$, $p_y + p_z = p$. Equations (5.38) may now be written:

$$\dot{X} = X(p_x - p) \tag{5.40a}$$

$$\dot{\sigma} = -p_x \sigma + \delta q \tag{5.40b}$$

$$\dot{\delta} = -p_x \delta + \sigma q \tag{5.40c}$$

$$\dot{p}_x = X(\sigma - X) \tag{5.40d}$$

$$\dot{p} = X\sigma - \delta^2 \tag{5.40e}$$

$$\dot{q} = (X - \sigma)\delta \tag{5.40f}$$

The leading behaviour is $p_x \sim 2/\tau$, $X \sim A$, $p \sim 2/\tau$, $\sigma \sim 2B/\tau^2$. As we have seen previously, δ and q cannot diverge like $1/\tau^2$ nor like $1/\tau$. In fact, the cancellation argument can be pushed further by examining equations (14c) and (14f) more closely. Let us first assume that the dominant term in δ is of order τ^n. Then, from (14c), we find that the dominant behaviour of q is $\mathcal{O}(\tau^{n+1})$, and using (14f) we find that the dominant behaviour in δ is $\mathcal{O}(\tau^{n+2})$. Thus δ and q vanish at all orders! Going back to the X, p variables we have $Y = Z$, $p_y = p_z$ at all orders.

In order to compute the beyond-all-orders behaviour of δ and q we start by obtaining the singular expansion for the reduced, $\delta = q = 0$, system (5.40). Next, we remark that (14c) and (14f) are linear equations in terms of δ and q. Combining the two we can obtain a single second-order equation for δ. Dropping the subdominant term $X\delta$ in (5.40f) we finally obtain:

$$\ddot{\delta} - 2\dot{\delta}\frac{\dot{\sigma}}{\sigma} + (\sigma^2 + 2(\frac{\dot{\sigma}}{\sigma})^2 - \frac{\ddot{\sigma}}{\sigma})\delta = 0. \tag{5.41}$$

Retaining the dominant terms in the expansion of σ we obtain

$$\ddot{\delta} + \frac{4}{\tau}\dot{\delta} + \delta(\frac{4}{A^2\tau^4} - \frac{8C}{A\tau^3}) = 0 \tag{5.42}$$

The general solution of (5.42) is

$$\delta = c_1 \tau^{2iC-1} e^{\frac{2i}{A\tau}} + c_2 \tau^{-2iC-1} e^{-\frac{2i}{A\tau}}. \tag{5.43}$$

Thus the difference of Y, Z is indeed a quantity beyond all orders of perturbation and contains essential singularities. Moreover, since C is a *free* constant, irrational in general, solution (5.43) has a transcendental branching point. What is worse, since the general solution contains both terms in (5.43), there is no possibility of bypassing the transcendental essential singularity by some choice of appropriate Stokes sectors. The conclusion is that the singularity considered leads to critical branching and thus violates the Painlevé property; it is expected to be nonintegrable. This result is particularly interesting since it shows that the local singularity analysis can be extended so as to deal with essential singularities.

An alternate way to deal with essential singularities of exponential character was presented by Kruskal in his seminal article [59]. We do not know of any systematic application of these ideas of Kruskal: if they could be implemented algorithmically they could solve the major difficulty of the detection of essential singularities.

6   Applications to Finite and Infinite Dimensional Systems

In what follows we shall present applications of the ARS approach which have served in discovering new integrable systems.

6.1   Integrable Differential Systems

a. The Lorenz System
One of the very first applications of the Painlevé analysis on systems of ODE's was the study of the Lorenz equations,

$$\dot{x} = \sigma(y - x)$$

$$\dot{y} = \rho x - xz - y \tag{6.1}$$

$$\dot{z} = xy - bz,$$

which arises in simple models of hydrodynamic turbulence. For general values of b, ρ, σ, the solutions present a chaotic behaviour. Still, there exist values for which the behaviour of the system becomes regular. Segur [66] has studied the system from the point of view of Painlevé analysis. Starting from the leading singular behaviour of the type $x \sim 1/\tau$, $y \sim 1/\tau^2$, $z \sim 1/\tau^2$, (with $\tau = t - t_0$), he found that the resonances were $r = -1, 2$ and 4. Two compatibility conditions resulted. Apart from the solution at infinite Reynolds number ρ, where the system is integrated in terms of elliptic functions, the following parameter values also ensured the Painlevé property for the system:

i) $\sigma = 0$
ii) $\sigma = 1/2, \ b = 1, \ \rho = 0$
iii) $\sigma = 1, \ b = 2, \ \rho = 1/9$ $\qquad\qquad\qquad\qquad\qquad$ (6.2)
iv) $\sigma = 1/3, \ b = 0, \ \rho = free.$

All of them are integrable. In case i), the equations are linear. In case ii), two time-dependent integrals may be found,

$$y^2 + z^2 = C_1 e^{-2t} \qquad \text{and} \qquad x^2 - z = C_2 e^{-t}, \qquad (6.3)$$

and thus the system can be reduced to a quadrature, and the solutions can be expressed in terms of elliptic functions. In case iii), one integral exists,

$$x^2 - 2z = Ce^{-2t}, \qquad (6.4)$$

and, after a change of variables, one obtains the second Painlevé transcendent. Finally, in case iv), the equations can be combined to a single third-order equation for x which can be integrated once to give

$$x\ddot{x} - \dot{x}^2 + x^4/4 = Ce^{-4t/3}. \qquad (6.5)$$

A simple change of variables, $X = xe^{t/3}, \ T = e^{-t/3}$, suffices to transform this equation into the third Painlevé transcendent.

In their analysis of the Lorenz system, Tabor and Weiss [67] made the following remark. When $b = 2\sigma$, the first resonance condition is satisfied but not the second one. Interestingly enough, one integral exists in this case,

$$x^2 - 2\sigma z = Ce^{-2\sigma t}, \qquad (6.6)$$

but no other integral is known. Thus in this case, partial integrability is related to the partial fulfillment of the Painlevé conditions. Still, the same condition is satisfied also for the value $b = 1 - 3\sigma$, but no integral is known in this case. Thus, for the system under study, it appears clearly that complete integrability is related to the Painlevé property, while partial integrability bears only some remote relation to the latter.

b. The Rikitake System

The Rikitake two-disk dynamo model, proposed for the description of the time variation of the earth's magnetic field [68] is

$$\dot{x} = -\mu x + \beta y + yz$$

$$\dot{y} = -\mu y + \beta x + xz \qquad (6.7)$$

$$\dot{z} = -xy + \alpha.$$

The analysis of (6.6) is particularly simple since it has only one type of singularity, near which the leading order behaviour of its solutions is $x \sim i/\tau$, $y \sim -i/\tau$, $z \sim 1/\tau$. Developing the asymptotic series to higher orders one easily find that two more free constants are expected simultaneously at the second higher order. The compatibility conditions for these free constants to enter with only integral powers of τ yield

$$\alpha = 0 \text{ and either } \beta = 0 \text{ or } \mu = 0.$$

The two cases possessing the Painlevé property can be easily integrated. With $\alpha = 0$ and $\beta = 0$ there is one integral,

$$x^2 - y^2 = C^2 e^{-2\mu t}, \tag{6.8}$$

and after the variable transformation $x + y = Cwe^{-\mu t}$, $T = e^{-\mu t}$, the Rikitake system leads to a special case of the third Painlevé transcendental equation for w. On the other hand, the case $\alpha = 0$ and $\mu = 0$ is even simpler. Multiplying the first and second equations (6.6) by $\dot{x}$ and $\dot{y}$ respectively, adding and subtracting, yields two integrals,

$$x^2 + y^2 + 2z^2 = C$$

$$x^2 - y^2 + 4\beta z = D. \tag{6.9}$$

Using (6.8), the complete integration of the system can be performed in terms of elliptic functions.

It is now interesting to ask whether the partial fulfillment of conditions for the Painlevé property yields partially integrable models. The answer is not a systematic "yes". Take, for example, the case $\beta = 0$ (with $\alpha\mu \neq 0$); integral (6.8) still exists. If $\mu = 0$, one integral also exists for all α and β,

$$x^2 - y^2 + 4\beta(z - \alpha t) = C, \tag{6.10}$$

but no further integration appears possible. Finally, for $\alpha = 0$ and for any β, μ, no simple integral appears to exist.

6.2   Integrable Two-Dimensional Hamiltonian Systems

In the following paragraphs we shall review our analysis of two-degrees of freedom Hamiltonians from the point of view of the Painlevé property and integrability. As a starting point, we shall discuss the cases of cubic and quartic potentiels which are simpler to study, and admit generalizations to N degrees of freedom more easily.

a. Cubic Potentials
Let us begin, therefore, by examining Hamiltonians of the form

$$H = \frac{1}{2}\left(p_x^2 + p_y^2\right) + y^3 + ay^2x + byx^2 + cx^3. \tag{6.11}$$

By an appropriate 'rotation' of our dependent variables it is possible, without loss of generality, to remove the $y^2 x$ term from (6.11) and consider only the family of potentials,

$$V(x, y) = y^3 + byx^2 + cx^3 \qquad (6.12)$$

The Painlevé analysis begins by looking for singular solutions of the equations of motion,

$$\ddot{x} = -2bxy - 3cx^2 \qquad \ddot{y} = -3y^2 - bx^2, \qquad (6.13)$$

which turn out to have two types of singularities:

    (i) $x \propto \alpha \tau^{-2}$, $y \propto \beta \tau^{-2}$

    (ii) $x \propto \tau^s$, $y \propto -2\tau^{-2}$ with $s(s - 1) = 4b$.

In this last case s must be an integer in order to satisfy the Painlevé property. In the special case $c = 0$, s may also be a half-integer, since the Painlevé property can be recovered by changing variables to $X \equiv x^2$ and $Y \equiv y$. Turning to singularity type (i), on the other hand, we find at leading order

$$6 = -2b\beta - 3c\alpha \qquad 6\beta = -3\beta^2 - b\alpha^2. \qquad (6.14)$$

and, for the resonances, the equation

$$(N + 2\beta b + 6\alpha c)(N + 6\beta) - 4b^2 a^2 = 0, \qquad (6.15)$$

where $N = (r - 2)(r - 3)$. We find either $N = 12$ or $N = (2b - 6)\beta$. The former case gives $r = -1$ and $r = 6$, while the latter leads to two possibilities: N_1 and N_2 corresponding to the solutions β_1, β_2 of (6.14). We now observe that

$$N_1 + N_2 = -2(2b - 6) + \frac{1}{9}N_1 N_2 b \qquad (6.16)$$

and, introducing $N_3 = s(s - 1)$, we obtain the symmetric relation,

$$36(N_1 + N_2 + N_3 - 12) = N_1 N_2 N_3, \qquad (6.17)$$

where all N_i's must be the products of two consecutive integers (or half-integers), from which we can extract all the choices compatible with the Painlevé property more conveniently. After excluding the case $N_1 = N_2 = 6$, which leads to logarithmic singularities, these cases are (up to a permutation of N_1, N_2, N_3 which amounts to a rotation in the x, y plane):

a) $N_1 = 0$, $N_2 = 12$, $N_3 = 0$, which gives a separable potential $V = y^3 + \lambda x^3$,

b) $N_1 = 90$, $N_2 = 90$, $N_3 = 3/4$, with $c = 0$ and $s = -1/2$ yielding $V = y^3 + \frac{3}{16}yx^2$,

c) $N_1 = 30$, $N_2 = 30$, $N_3 = 2$ giving $V = y^3 + \frac{1}{2}yx^2$,

d) $N_1 = 20$, $N_2 = 90$, $N_3 = 2$, with $V = y^3 + \frac{1}{2}yx^2 + \frac{i}{6\sqrt{3}}x^3$.

All these potentials have been shown to be integrable [49].

b. Quartic Potentials

The case of quartic potentials can also be treated in the same manner as the cubic potentials. Working with a form even in both x and y to keep the calculations manageable,

$$V = y^4 + ax^2y^2 + bx^4, \qquad (6.18)$$

we immediately single out three integrable cases:
a) $a = 0$: V separable in x and y,
b) $a = 6$, $b = 1$: V separable in $x \pm y$,
c) $a = 2$, $b = 1$: V rotationally symmetric, hence separable in polar coordinates.
Looking for nontrivial integrable cases of (6.18) we start again with the equations of motion,

$$\ddot{x} = -2axy^2 - 4bx^3 \qquad \ddot{y} = -4y^3 - 2ax^2y \qquad (6.19)$$

and distinguish three singularity types with leading behaviour (as $t \to t_0$):

i) $x \propto \tau^s$, $\quad y \propto \gamma\tau^{-1}$, $\qquad s > -1$,
ii) $x \propto \delta\tau^{-1}$, $\quad y \propto \tau^q$, $\qquad q > -1$,
iii) $x \propto \alpha\tau^{-1}$, $\quad y \propto \beta\tau^{-1}$.

Singularity types (i) and (ii) yield $s(s-1) = a$, $q(q-1) = a/b$ while type (iii) at leading order yields $2\beta^2 + a\alpha^2 + 1 = 0$, $a\beta^2 + 2b\alpha^2 + 1 = 0$. A resonance analysis similar to that of the cubic potentials, leads here to the equation

$$(N - 6)(N + 6 + (12 + 2a)\beta^2 + (2a + 12b)\alpha^2) = 0, \qquad (6.20)$$

where $N = (r - 1)(r - 2)$. $N = 6$ leads to the resonances $r = -1, 4$. The second possibility is that $N = -(6 + (12 + 2a)\beta^2 + (2a + 12b)\alpha^2)$ is the product of two consecutive integers. Implementing this yields the following integrable cases:
d) $s = 3/2$, $q = 4$ ($r = 5, -2$), leading to an integrable potential with parameters $a = 3/4$, $b = 1/16$, encountered in Sect. 3, (3.3), separable in parabolic coordinates [69],
e) $s = 3/2$ $q = 3$, ($r = 8, -5$), leading to the integrable potential, with parameters $a = 3/4$, $b = 1/8$ that we encountered in Sect. 4, (4.1).

c. The Generalized Toda Hamiltonian

We end our discussion of two-dimensional Hamiltonians by listing the integrable cases of one more example, the Toda Hamiltonian,

$$H = \frac{1}{2}(p_x^2 + p_y^2) - e^{x+\alpha y} - e^{\beta x+\gamma y}. \qquad (6.21)$$

The coefficient of the exponentials in (6.21) case can always be set equal to 1 by appropriately translating x and y. Then, by a simple scaling, the coefficient of x in the first exponential is also set at unity. Finally, rotating the x, y coordinates (except if $\alpha = \pm i$) allows us to set $\alpha = 0$ and to obtain the equations of motion,

$$\ddot{x} = e^x + \beta e^{\beta x+\gamma y}$$

$$\ddot{y} = \gamma e^{\beta x+\gamma y}. \qquad (6.22)$$

Clearly, x and y are not the right variables to use in a Painlevé analysis. One may wish to introduce exponentials as new variables, or equivalently, accept leading logarithmic singularities, but with integral coefficients (so as to recover integral powers of t, upon exponentiation). Following the latter approach we find three possible divergent behaviours: (i) either the $e^{\beta x + \gamma y}$ is dominant in both equations, or (ii) it is dominant only in the second, or (iii) both terms are dominant in the first equation. In all these cases, x and y diverge like $\ln t$ multiplied by some integer. In (i) and (ii) the argument of the dominant exponential is $-2\ln t$, while that of the other is $k\ln t$, where $k > -2$. Thus one derives from the leading order equations

$$-2\beta = -1, 0, 1, 2, \ldots, \qquad -2\beta/(\beta^2 + \gamma^2) = -1, 0, 1, 2, \ldots \qquad (6.23)$$

In the last case, (iii), $x \sim -2\ln t$, $y \sim -((1-\beta)/\gamma)\ln t$, and the only constraint comes from the resonance condition

$$2 - 2\beta\left(1 + \frac{(1-\beta)^2}{\gamma^2}\right) = n(n-1), \qquad (6.24)$$

where n is an integer, with the other two resonances being $r = -1, 2$. The case $\beta = 0$ is trivially separable. Now combining conditions (6.23) and (6.24) one easily identifies three cases which satisfy the necessary conditions for exhibiting the Painlevé property:
  a) $\beta = -1/2$, $\gamma^2 = 3/4$,
  b) $\beta = -1$, $\gamma^2 = 1$ (and its equivalent $\beta = -1/2$, $\gamma^2 = 1/4$),
  c) $\beta = -3/2$, $\gamma^2 = 3/4$ (and its equivalent $\beta = -1/2$, $\gamma^2 = 1/12$)
  That these three cases also satisfy sufficient conditions for integrability is shown in [46]. An important observation, concerning these three integrable cases is that their corresponding Hamiltonians are related to the root systems of certain Lie algebras. This observation has been proven seminal in the extension of integrability to the N-dimensional case.

From what we have said so far, one easily realizes that the singularity analysis of higher dimensional systems is, in general, a very tedious affair. As the dimensionality increases the number of possible leading behaviours, resonances and conditions to check, grows very rapidly. Thus an exhaustive analysis of a general N-degree of freedom Hamiltonian system is a practically impossible task. However, interesting, nontrivial examples, for which this analysis has been performed, do exist (quartic oscillators, generalized Toda hamiltonians etc.). The interested reader will find these details in our review article [46].

6.3   Infinite-Dimensional Systems

We turn now to infinite-dimensional systems i.e. systems defined by PDE's. We shall not present any result based on the ARS approach: once the equation is reduced to an ODE, the application of the algorithm is straightforward. We shall rather present two examples that will illustrate the Weiss-Kruskal method.

We start with Weiss's original approach applied to Burger's equation,

$$u_t + uu_x + u_{xx} = 0 \tag{6.25}$$

assume a leading singularity of the form $u = u_0\phi^\alpha$ where ϕ is a function of both x and t, and balance the last two terms in (6.25). We thus find

$$\alpha = -1 \quad \text{and} \quad u_0 = 2\phi_x. \tag{6.26}$$

Next, we substitute an expansion $u = \phi^{-1}\sum u_n\phi_n$ into (6.25) and establish a recursion relation for the u_n's. We readily obtain

$$\phi_x^2(2 - n)(n + 1)u_n = R(u_0, ...u_{n-1}, \phi_t, \phi_x, ...). \tag{6.27}$$

Thus the resonances are -1 and 2, with -1 being related to the arbitrariness of ϕ. There exists one compatibility condition at order $n = 2$ which must be verified. Order by order, we obtain, at $n = 1$,

$$\phi_t + u_1\phi_x^2 + \phi_{xx} = 0, \tag{6.28}$$

which defines u_1, and at $n = 2$,

$$0 = \frac{\partial}{\partial x}\left(\phi_t + u_1\phi_x^2 + \phi_{xx}\right), \tag{6.29}$$

which, given (6.28), is satisfied. Thus Burger's equation possesses the Painlevé property for partial differential equations in the sense of Weiss *et al.* and in fact is integrable by linearization.

Next, we perform the Painlevé analysis for the modified Korteweg-de Vries equation,

$$u_t + 6u^2u_x + u_{xxx} = 0, \tag{6.30}$$

using the simplified Kruskal ansatz for the singularity manifold $\phi \equiv x + \psi(t)$. Balancing the most singular terms we find,

$$u \sim u_0\phi^{-1} \quad \text{with} \quad u_0^2 = 1, \tag{6.31}$$

and the resonances turn out to be -1, 3 and 4. So we expand u up to order four in ϕ,

$$u = \phi^{-1}\sum_{n=0}^{4} u_n\phi_n, \tag{6.32}$$

where $u_n = u_n(t)$. The calculations are much more simple than Weiss's original formulation since $\phi_x = 1$ and $\phi_{xx} = 0$. We thus obtain, order by order,

$$u_1 = 0 \quad \text{and} \quad u_2 = -u_0\psi_t/6. \tag{6.33}$$

The resonance condition at order three is identically satisfied while the resonance at order four yields the compatibility condition,

$$6u_2^2 + u_2u_0\psi_t = 0, \tag{6.33}$$

which, in view of (6.33) is also satisfied. Thus, the modified Korteweg-de Vries equation does possess the Painlevé property for partial differential equations in the sense of Weiss and Kruskal.

We are going to limit ourselves to just these two cases. The existing literature abounds in examples of the application of the algorithm to integrable partial differential equations. The wealth of the integrable examples discovered through singularity analysis shows the power of this method. Still one must bear in mind that the ARS approach can only be considered as a useful indicator of integrability. It does not furnish rigorous proofs and it is not infallible. One must complement it with intuition, insight and inspiration. Its blind-eyed, mechanical, application can sometimes lead to disaster and (at least to the authors' taste) does not constitute an exciting prospect.

7 Integrable Discrete Systems Do Exist!

The fundamental assumption since the beginning of the mathematical description of the physical reality (going back to Galilei and Newton) is that space-time is continuous. Thus it is believed that the physical equations of motion can be formulated in terms of differential equations. How can we be sure that this assumption is correct? The answer is that we cannot! From a physical point of view the only thing that we can say is that if space-time has a discrete, lattice-like, structure, then the lattice constant must be fine enough so as to lie beyond the detection capacity of present-day experiments [70]. The fact that we, with our senses, perceive the world as continuous does not mean anything. Matter does indeed look and feel continuous in spite of its discrete, atomic or molecular, structure. We shall not go here into any details concerning the physical implications of a discrete space-time. One thing is certain: if space-time were discrete then we would be compelled to use discrete equations for the formulation of the physical equations of motion.

On the other hand, space-time does not have to be discrete for the study of discrete equations to be interesting. On several occasions, discrete equations arise naturally. A very simple example is the description of billiard dynamics [71]. A billiard is a system where a particle moves on a flat surface and reflects elastically at the boundary. Since the motion between two reflections is that of a free particle, its description is trivial. The dynamical problem can be reduced to the study of bounces (point of impact, reflection angle) and the time between two impacts is not important. Thus the dynamical variables can be labelled by just a discrete variable (the bounce number) and the equations of motion are, clearly, discrete.

However there exists a domain, one of the most important in modern science, where discrete equations are unavoidable: numerics. The digital simulation of physical phenomena that has become so important with the advent of high-speed computers, is based entirely on discrete equations. In fact all the numerical simulations of mathematical models can be related to some discretisation. Thus, quite often, it is difference equations rather than differential ones that provide us with the results that make possible a better undestanding of nature. The implicit assumption is that there exists a (close) parallel between the properties of the continuous system we wish to study and its discrete analogue. Sometimes this assumption is fully justified [72].

In other instances the parallel between the continuous and the discrete system is only approximate and only valid if the discretisation step is small enough. In fact, this is true most of the time as far as the property of integrability is concerned. Suppose that the task at hand is to study numerically an equation that is known to be integrable. Let us present two simple examples. The Riccati equation that we have already encountered,

$$x' = \alpha x^2 + \beta x + \gamma, \tag{7.1}$$

is linearizable through a Cole-Hopf transformation, $x = P/Q$. How can we discretise it? First we decide to seek its discretisation as a two-point mapping involving only $x(t)$ and $x(t+\Delta t)$, where Δt is the discretisation step. Let us write, for simplicity reasons, $x(t) = x_n$ and $x(t + \Delta t) = x_{n+1}$ (in obvious notations, since $t = t_0 + n\Delta t$). The discretisation of the first derivative is straightforward, $x' \to (x_{n+1} - x_n)/\Delta t$, but how about x^2? This is the main difficulty, namely how one will choose the discretisation of the nonlinear terms. Several possibilities exist: $x^2 \to x_n^2, x_{n+1}^2, x_n x_{n+1}, (x_n + x_{n+1})x_n/2, (x_n + x_{n+1})^2/4, (x_n^2 + x_{n+1}^2)/2$. In fact there is an infinity of possibilities if one considers rational forms. It turns out that the integrable discretisation is $x^2 \to x_n x_{n+1}$. In this case the discrete form of (7.1) is the homographic mapping,

$$x_{n+1} = \frac{bx_n + c}{1 + ax_n}, \tag{7.2}$$

where a, b, c are related to α, β, γ and depend on Δt. Mapping (7.2) is indeeed linearizable by a Cole-Hopf transformation, $x = P/Q$, while other discretisations of (7.1) among which we find the logistic mapping, are known to behave chaotically.

The Painlevé equations offer also a nice example [73]. Let us consider the P_I equation,

$$x'' = 3x^2 + t, \tag{7.3}$$

and look for a discretisation in terms of an explicit three-point mapping. Again, the second derivative is easily discretised, $x'' \to (x_{n+1} - 2x_n + x_{n-1})/(\Delta t)^2$. Several possibilities exist for the nonlinear term but very few lead to an integrable equation. One such example is $3x^2 \to x_n(x_{n+1} + x_n + x_{n-1})$ which yields

$$x_{n+1} - 2x_n + x_{n-1} = \frac{3x_n^2 + z}{1 - x_n}, \tag{7.4}$$

where z is linearly related to n and we have $t = n\Delta t + t_0$. Thus the construction of integrable integrators, i.e. integrable discrete forms of integrable differential equations, is highly nontrivial. The two examples we presented above show that in some cases the discrete analogue of a given system can be found. But how about the remaining integrable equations? Untill recently, very few results were known. The main reason for this was the absence of interest in integrable discrete systems. Then the situation changed, thanks to some recent findings in string theory. While examining a two-dimensional model of quantum gravity (based on closed strings) Brézin and Kazakov [74] were able to reduce the computation of the interesting physical quantities (partition functions) to an integrable recursion relation, the continuous limit of which was precisely P_I. This was not the only instance where (the discrete form of) a Painlevé equation made its appearance in a field theoretical model. The discrete form of P_{II} was also obtained shortly afterwards [75] while higher-order equations of Painlevé type resulted from more complicated models. These results were interesting enough to motivate research in the direction of integrable discrete equations. Integrable discrete systems had already made their appearance in the study of spin systems. Jimbo and Miwa had derived, in 1981 [76], a (very) complicated bilinear equation for the diagonal correlation function of the Ising model, an equation that must be a discrete P_{VI}. However these results did not create much activity as far as integrability research was concerned, and the domain had to wait 10 more years before blossoming. (Quite undestandably, once the integrability community started interesting itself in discrete systems, spin systems and the related integrable discrete equations again attracted the interest of physicists [77]). Once discrete systems began to be studied with integrability in mind, the results evolved from the utmost scarcity to a confortable opulence. The main factor for this spectacular progress was the development of specific techniques suitable for the treatment of discrete systems which were the analogues of the techniques which are currently used in the investigation of continuous integrable systems. Prominent among those is the singularity confinement [78] that, for discrete systems, plays the role of the Painlevé property for continuous ones. Thus it can be used in the detection of discrete integrability and in fact it has been extensively used.

The upshot of all this is that today the discrete forms of most of the well-known (continuous) integrable equations do exist. Let us illustrate this with some examples. We have already encountered the discrete form of the Riccati equation. For the elliptic functions the situation is slightly different. It was already known that the elliptic functions obey addition relations which can be interpreted as discrete equations [79]. In fact the two-point correspondence,

$$\alpha x_{n+1}^2 x_n^2 + \beta x_{n+1} x_n (x_{n+1}+x_n) + \gamma(x_{n+1}^2+x_n^2) + \epsilon x_{n+1}x_n + \zeta(x_{n+1}+x_n)+\mu = 0,$$
$$(7.5)$$

can be parametrized in terms of elliptic functions. This means that the x_n is a sampling of an elliptic function over a one-dimensional mesh of points. This relation has reappeared recently in a slightly different frame. It has been remarked [80] that the three-point mapping,

$$x_{n+1} = \frac{f_1(x_n) - f_2(x_n)x_{n-1}}{f_2(x_n) - f_3(x_n)x_{n-1}},$$
$$(7.6)$$

where the f_i are specific functions of x_n, has (7.5) as its integral. Thus mapping (7.6) is the discrete form of the generic second-order equation (equation 23 in the Gambier list) solvable in terms of elliptic functions. The discrete form of the Painlevé equations will be presented in detail in section 9. We have already given the form of d-P_I. Let us just give here the form of d-P_{III} [81] for the sake of illustration,

$$x_{n+1}x_{n-1} = \frac{ab(x_n - p)(x_n - q)}{(x_n - a)(x_n - b)}, \qquad (7.7)$$

where $p = p_0\lambda^n$, $q = q_0\lambda^n$ and a, b are constants. It goes without saying that the discrete forms of KdV and mKdV are known [75]. For d-KdV,

$$x_n^{k+1} = x_n^{k-1} + \frac{1}{x_{n+1}^k - x_{n-1}^k}, \qquad (7.8)$$

while for the d-mKdV (where μ is a constant),

$$x_n^{k+1} = x_n^{k-1} \frac{x_{n+1}^k - \mu x_{n-1}^k}{\mu x_{n+1}^k - x_{n-1}^k}. \qquad (7.9)$$

The discrete form of KdV was first given by Hirota [81], who derived it using the discrete form of his bilinear formalism. However it turned out that this equation was already known under a different aspect. In fact, equation (7.8) above had already been obtained by Wynn [83]. It is known under the name of ϵ-algorithm and can be used to accelerate the convergence of sequences [84]. This example shows clearly that integrable systems are universal and widely applicable.

In order to fix the ideas let us summarize here what we mean by "integrable" in the discrete case. Discrete integrability in our sense means one of the following things:

a) existence of a sufficient number of rational expressions $\Phi_k(x_1, \ldots x_N) = C_k$, the values of which are invariant under the action of the mapping,

b) linearizability of the mapping through a Cole-Hopf type transformation $x_i = u_i/v_i$ whereupon the mapping reduces to a linear one for the u_i's and v_i's,

c) linearizability through a Lax pair. In this case, the mapping is the compatibility condition of a linear system of differential-difference, q-difference or pure difference equations.

The above are not definitions but rather illustrations of the various types of integrability. It may well occur, as in the case of the mappings (7.6), that the existence of one invariant reduces the mapping to a correspondence of the form (7.5) that can be parametrized in terms of elliptic functions. In other cases, integration using the rational invariants may lead to some transcendental equation like the discrete Painlevé ones. All of the above types of integrability have been encountered in the discrete systems that we have studied. The reason for the above classification is to emphasize the parallel existing between the continuous and discrete cases.

8 Singularity Confinement:
The Discrete Painlevé Property

As we said in the previous section, the appearance of an integrability detector for discrete systems played a crucial role in opening the field for research. This detector took quite some time to be developed because nobody knew precisely how to tackle the problem. Of course, it was everybody's intuition that singularities would play a role.

The singularity confinement criterion did not arise suddenly as a divine inspiration but, rather, resulted from considerations on singularities of integrable discrete systems. We started by considering possible restrictions of the lattice KdV equation [78],

$$x_j^{i+1} = x_{j+1}^{i-1} + \frac{1}{x_j^i} - \frac{1}{x_{j+1}^i}, \tag{8.1}$$

where the motion would be restricted on a region of the lattice through the existence of 'infinite walls'. Then the question arose naturally, "what if a singularity appears spontaneously?" How does it evolve under the mapping (8.1)? Does it create a (semi) infinite wall? Astonishingly enough the result turned out to be the following: a $x = 0$ at (i, j) leads to divergent x's at both $(i + 1, j - 1)$ and $(i + 1, j)$ and a vanishing x at $(i + 2, j - 1)$. Then at both sites $(i + 3, j - 2)$ and $(i + 3, j - 1)$ a fine cancellation occurs and one obtains finite values, $x_{j-1}^{i+3} = x_j^{i-1} + 1/x_{j-1}^i - 1/x_j^{i+2}$, and a similar one for x_{j-2}^{i+3}. Thus the singularity does not propagate beyond a few lattice points and is confined to a small region [78]. This was reminiscent of the continuous Painlevé property: absence of natural boundaries in integrable 2-D Hamiltonians [46]. We wondered whether this was a general property of integrable discrete systems. It turned out that it was indeed!

Although the singularity confinement was discovered in the case of a 2-D lattice, the best way to understand how it operates is in the case of a 1-D system. So let us take the most classical integrable mapping example, the McMillan mapping,

$$x_{n+1} + x_{n-1} = \frac{2\mu x_n}{1 - x_n^2}. \tag{8.2}$$

This mapping is well known for its integrability. In fact, it can be completely integrated in terms of elliptic functions: $x = x_0 \operatorname{cn}(\Omega n, \kappa)$, where $\kappa = x_0 \operatorname{dn}(\Omega)/\operatorname{sn}(\Omega)$ and Ω is related to μ through $\mu = \operatorname{cn}(\Omega)/\operatorname{dn}^2(\Omega)$. A singularity may appear in the recursion (8.2) whenever x passes through the value 1. So let us assume that x_0 is finite and that $x_1 = 1+\epsilon$. (This can be obtained from a perfectly regular x_{-1}.) We find then the following values: $x_2 = -\mu/\epsilon-(x_0+\mu/2)+\mathcal{O}(\epsilon)$, $x_3 = -1 + \epsilon + \mathcal{O}(\epsilon^2)$ and $x_4 = x_0 + \mathcal{O}(\epsilon)$. Thus, not only is the singularity confined at this step but, also, the mapping has recovered a memory of the initial conditions through x_0.

Starting from such simple observations we were led to the formulation of the conjecture that the confinement of singularities is a necessary condition for integrability. As in the case of the continuous ARS-Painlevé conjecture this requirement is so strong that, once a system satisfies it, one expects it to be integrable. However this is too simplistic an attitude. First, as we shall see below, the notion of 'singularity' itself must be refined.

The extension of the notion of singularity was introduced in [85], where we considered mappings of the form

$$x_{n+1} = \frac{f_1(x_n) - x_{n-1}f_2(x_n)}{f_4(x_n) - x_{n-1}f_3(x_n)}, \tag{8.3}$$

where f_i are linear in x_n: $f_i = a_i x_n + b_i$. In this case an infinite value for x_i, $i = n, n \pm 1$, does not play any particular role. In fact, relation (8.3) is 'bi-homographic' and thus infinity can be taken to any finite value by a simple homographic transformation of variables. However (8.3) may pose a subtler problem. It may turn out that, for a certain n, the mapping (apparently) loses one degree of freedom. This occurs when x_{n+1} is defined independenly of x_{n-1} and this happens whenever

$$f_1(x_n)f_3(x_n) - f_2(x_n)f_4(x_n) = 0. \tag{8.4}$$

The fact that f_i are linear is not really important. The argument of the loss of degree of freedom can be repeated for any polynomial f_i. Once x_n is obtained from (8.4) one can compute x_{n+1} simply as $x_{n+1} = f_1(x_n)/f_4(x_n) = f_2(x_n)/f_3(x_n)$, unless x_{n-1} was such that both the numerator and the denominator of the fraction defining x_{n+1} vanished, that is

$$x_{n-1} = f_1(x_n)/f_2(x_n) = f_4(x_n)/f_3(x_n). \tag{8.5}$$

Thus one sees two ways in which the singularity confinement can be preserved: either relation (8.5) is satisfied or it is not, in which case x_{n+1} is determined and is independent of x_{n-1}. In the latter case one degree of freedom will be definitely lost, as x_{n+2} will be determined in terms of x_n only, unless both the numerator and the denominator of the fraction that define it vanish, that is $x_n = f_1(x_{n+1})/f_2(x_{n+1}) = f_4(x_{n+1})/f_3(x_{n+1})$. In the case where (8.5) is satisfied, on the other hand, it would appear that a degree of freedom suddenly appears at step $n + 1$. The only way out is to demand that x_n be determined by x_{n-1} only, independent of x_{n-2}, which means that one already had at the previous step: $x_n = f_1(x_{n-1})/f_4(x_{n-1}) = f_2(x_{n-1})/f_3(x_{n-1})$.

The main idea in the above analysis is that what we call singularity is the loss or appearance of a degree of freedom. How can this singularity be confined, i.e. how can the mapping recover the lost degree of freedom? For rational mappings of the kind we are considering, this can be realized if some of the mapping's variables assume an indeterminate form, $0/0$. In that case new free parameters can be introduced and the mapping recovers its full dimensionality.

Even with this extension the singularity confinement criterion is not sufficiently strong. Another ingredient is needed. In order to illustrate the difficulty let us consider the singularity confinement of the two-point mapping

$x_{n+1} = f(x_n)$, where f is rational. We find that this requirement leads to

$$x_{n+1} = \alpha + \sum_k \frac{1}{(x_n - \beta_k)^{\nu_k}} \qquad (8.6)$$

with integral ν_k, provided that for all k, $\beta_k \neq \alpha$. Indeed, if $x_n = \beta_k$, then x_{n+1} diverges, $x_{n+2} = \alpha$ and x_{n+3} is finite. So the mapping propagates without any further difficulty. However, if we consider the 'backward' evolution, then (8.6) solved for x_n in terms of x_{n+1} leads to multideterminacy and the number of preimages grows exponentially with the number of 'backward' iterations. Indeed, the only mapping of the form (8.6) with no growth is the homographic mapping:

$$x_{n+1} = \frac{ax_n + b}{cx_n + d}, \qquad (8.7)$$

which is the discrete form of the Riccati equation.

So, singularity confinement is still a valid criterion: it must only be complemented by what we call the preimage nonproliferation condition [86]. We claim that a prerequisite for integrability of a mapping is that each point must have a single preimage. The latter is clearly not sufficient for integrability but can be used as a fast screening procedure. Unless a mapping has unique preimages there is no point implementing the singularity confinement algorithm.

9 Applying the Confinement Method: Discrete Painlevé Equations and Other Systems

A method is useful only if it provides results. In this domain the singularity confinement method has been as successful as its continuous predecessor, the Painlevé method, and an impressive amount of results has been obtained in very few years. In this section we can only summarize some selected topics. Let us begin with the most important application: the discrete Painlevé equations.

9.1 The Discrete Painlevé Equations

Discrete Painlevé equations first attracted attention when d-P_I was derived in a field theoretical model [74],

$$x_{n+1} + x_n + x_{n-1} = \frac{z_n}{x_n} + a, \qquad (9.1)$$

where $z_n = \alpha n + \beta$. This was not the oldest known example of d-P. In fact, Jimbo, Miwa and Ueno had already obtained [87] an alternate form of d-P_I,

$$\frac{z_n}{x_{n+1} + x_n} + \frac{z_{n-1}}{x_n + x_{n-1}} = -x_n^2 + a \qquad (9.2)$$

Several d-P's are known to date, being the discrete analogues to the Painlevé equations I to VI. (Note that more than one discrete analogue may exist for

each Painlevé equation [88]). The numbering of the discrete Painlevé equations
is based on their continuous limit.

Several methods have been used for their derivation [89]:

- the direct method, through singularity confinement,
- the orthogonal polynomial method,
- use of continuous auto-Bäcklund and Schlesinger transformations,
- the discrete AKNS method,
- use of discrete analogues of Miura transformations,
- use of discrete auto-Bäcklund and Schlesinger transformations,
- similarity reductions of integrable lattices,
- use of the discrete dressing method,
- limits and degeneracies of other discrete Painlevé equations.

Here are the first five discrete Painlevé equations. They are considered as the
'standard' forms, essentially for historical reasons.

$$x_{n+1} + x_{n-1} = -x_n + \frac{z}{x_n} + a \qquad (9.3a)$$

$$x_{n+1} + x_{n-1} = \frac{zx_n + a}{1 - x_n^2} \qquad (9.3b)$$

$$x_{n+1}x_{n-1} = \frac{ab(x_n - p)(x_n - q)}{(x_n - a)(x_n - b)} \qquad (9.3c)$$

$$(x_{n+1} + x_n)(x_n + x_{n-1}) = \frac{(x_n^2 - a^2)(x_n^2 - b^2)}{(x_n - z)^2 - c^2} \qquad (9.3d)$$

$$(x_{n+1}x_n - 1)(x_nx_{n-1} - 1) = \frac{pq(x_n - a)(x_n - 1/a)(x_n - b)(x_n - 1/b)}{(x_n - p)(x_n - q)}, \qquad (9.3e)$$

where $z = \alpha n + \beta$, $p = p_0\lambda^n$, $q = q_0\lambda^n$ and a, b, c constants. For d-P$_{VI}$ the only
form known to date is that of a system of two-point mappings [90],

$$x_{n+1}x_n = \frac{cd(y_n - p)(y_n - q)}{(y_n - a)(y_n - b)} \qquad (9.3f)$$

$$y_ny_{n-1} = \frac{ab(x_n - r)(x_n - s)}{(x_n - c)(x_n - d)},$$

where a, b, c, d are constants, p, q, r, s are proportional to λ^n and we have the
constraint $pqcd = \lambda rsab$.

The properties of the discrete Painlevé equations are in perfect analogy with
those of the continuous ones. In what follows we present a (non exhaustive) list
of these properties.

a. Coalescence Cascades

The continuous Painlevé equations form a coalescence cascade, i.e. the "lower"
(in order) equations can be obtained from the "higher" ones through adequate

limiting processes involving the dependent variable and the free constants entering the equation. The reduction scheme follows the pattern $P_{VI} \rightarrow P_V \rightarrow \{P_{IV}, P_{III}\} \rightarrow P_{II} \rightarrow P_I$. We have shown in [81] that the discrete P's follow *exactly* the same pattern but one must keep in mind that several coalescence cascades may exist. In particular the d-P_{VI} (9.3f) is not related to the d-P's (9.3a-e) but to some other ones.

In order to illustrate the process, let us work out in full detail the case d-P_{II} $\rightarrow$ d-P_I. We start with the equation

$$X_{n+1} + X_{n-1} = \frac{ZX_n + A}{1 - X_n^2}. \tag{9.4}$$

We put $X = 1 + \delta x$, whereupon the equation becomes

$$4 + 2\delta(x_{n+1} + x_{n-1} + x_n) = -\frac{Z(1 + \delta x_n) + A}{\delta x_n}. \tag{9.5}$$

Now, clearly, Z must cancel A up to order δ and this suggests the ansatz $Z = -A - 2\delta^2 z$. Moreover, the $\mathcal{O}(\delta^0)$ term in the right-hand side must cancel the 4 of the left-hand side and we are thus led to $A = 4 + 2\delta a$. Using these values of Z and A we find (at $\delta \rightarrow 0$),

$$x_{n+1} + x_{n-1} + x_n = \frac{z}{x_n} + a, \tag{9.6}$$

i.e. precisely d-P_I.

Mapping (9.6) is not the only coalescence limit of d-P_{II}. Putting $X = x/\delta$, $Z = -z/\delta^2$, and $c = -\gamma/\delta^3$ we recover an alternate d-P_I at the limit $\delta \rightarrow 0$,

$$x_{n+1} + x_{n-1} = \frac{\gamma}{x_n^2} + \frac{z}{x_n}. \tag{9.7}$$

On the other hand the alternate d-P_I (9.2) does not belong to the same cascade but comes from an alternate d-P_{II},

$$\frac{z_{n+1}}{x_{n+1}x_n + 1} + \frac{z_n}{x_n x_{n-1} + 1} = -x_n + \frac{1}{x_n} + z_n + \mu. \tag{9.8}$$

Similar results hold for the other known discrete Painlevé equations.

b. Lax Pairs

The ultimate proof of the integrablity of the d-P's is their effective linearization, i.e. their transcription as the compatibility condition for a linear isospectral deformation problem. In most cases of known Lax pairs the linear system assumes the form

$$\zeta \Phi_{n,\zeta} = L_n(\zeta)\Phi_n$$

$$\Phi_{n+1} = M_n(\zeta)\Phi_n, \tag{9.9}$$

leading to the compatibility condition,

$$\zeta M_{n,\zeta} = L_{n+1}M_n - M_n L_n. \tag{9.10}$$

The latter yields the discrete Painlevé equation. Thus for d-P_I we have found [91]:

$$L(\zeta) = \begin{pmatrix} \kappa & v_2 & 1 \\ \zeta & \lambda & v_3 \\ \zeta v_1 & \zeta & \mu \end{pmatrix} \qquad M(\zeta) = \begin{pmatrix} d_1 & 1 & 0 \\ 0 & d_2 & 1 \\ \zeta & 0 & 0 \end{pmatrix}. \tag{9.11}$$

Taking $d_2 = 0$ leads to a Lax pair for d-P_I, equation (9.1). A more interesting result is obtained when one does *not* assume $d_2 = 0$. Here one obtains: $d_1 = (\kappa - \mu)/v_1(n+1)$, and $\kappa =$ constant, $\lambda =$ constant, $\mu = z(n) =$ linear in n. Setting $x_n = v_1$ and $y_n = v_3$ we find

$$x_{n+1} + x_n = C - y_n + \frac{z_n - \lambda}{y_n}$$

$$y_{n-1} + y_n = C - x_n + \frac{z_{n-1} - \kappa}{x_n}. \tag{9.12}$$

This is precisely a d-P_I that includes the even-odd dependence, in perfect agreement with the results of the singularity confinement approach. A careful calculation of the continuous limit of (9.12) shows that it is in fact a d-P_{II}.

A most interesting result is the isospectral problem associated to d-P_{III}. Here a q-difference scheme is necessary instead of a differential one,

$$\Phi_n(q\zeta) = L_n(\zeta)\Phi_n(\zeta)$$

$$\Phi_{n+1}(\zeta) = M_n(\zeta)\Phi_n(\zeta), \tag{9.13}$$

leading to

$$M_n(q\zeta)L_n(\zeta) = L_{n+1}(\zeta)M_n(\zeta). \tag{9.14}$$

The resulting Lax pair is written in terms of 4×4 matrices. The recently derived d-P_{VI} equation (9.3f) is also associated to a q-difference scheme.

Although the list of Lax pairs for the known d-P's is far from being complete there is reasonable hope that eventually all of them will be derived.

c. Miura and Bäcklund Relations
Just as in the continuous case [92], the d-P's possess (auto-) Bäcklund and Miura transformations that allow to establish a dense net of relationships among them.

Let us illustrate this point with the example of d-P_{II}, written as

$$x_{n+1} + x_{n-1} = \frac{x_n(z_n + z_{n-1}) + \delta + z_n - z_{n-1}}{1 - x_n^2}. \tag{9.15}$$

We introduce the Miura transformation [93],

$$y_n = (x_n - 1)(x_{n+1} + 1) + z_n. \tag{9.16}$$

and we obtain,

$$(y_n + y_{n+1})(y_n + y_{n-1}) = \frac{-4y_n^2 + \delta^2}{y_n - z_n}. \tag{9.17}$$

Equation (9.17) is d-P$_{34}$, i.e. the discrete form of equation 34 in the Gambier classification, in perfect analogy to what happens in the continuous case.

An example of auto-Bäcklund transformation will be given in the case of d-P$_{\text{IV}}$ [94]. It is written as the pair of equations,

$$y_n = -\frac{x_n x_{n+1} + x_{n+1}(\tilde{z} + \kappa) + x_n(\tilde{z} - \kappa) + \mu}{x_n + x_{n+1}} \tag{9.18a}$$

$$x_n = -\frac{y_n y_{n-1} + y_n(z - \tilde{\kappa}) + y_{n-1}(z + \tilde{\kappa}) + \lambda}{y_n + y_{n-1}}, \tag{9.18b}$$

where $\tilde{z} = z + \alpha/2$, $\tilde{\kappa} = \kappa + \alpha/2$ and α is the lattice spacing in the discrete variable n. The meaning of these equations is that, when one eliminates either x or y between the two he ends up with d-P$_{\text{IV}}$ in the form

$$(x_{n+1} + x)(x + x_{n-1}) = \frac{(x^2 - \mu)^2 - 4\kappa^2 x^2}{(x + z)^2 - \tilde{\kappa}^2 - \lambda} \tag{9.19a}$$

$$(y_{n+1} + y)(y + y_{n-1}) = \frac{(y^2 - \lambda)^2 - 4\tilde{\kappa}^2 y^2}{(y + \tilde{z})^2 - \kappa^2 - \mu}. \tag{9.19b}$$

The important remark here is that (9.19a) and (9.19b) are *not* on the same lattice (since in (9.19b) the quantity $\tilde{z}$ figures in the denominator, instead of z) but, rather, on 'staggered' lattices.

d. Particular Solutions

It is well known [95] that the continuous Painlevé equations P$_{\text{II}}$ to P$_{\text{VI}}$ possess elementary solutions for specific values of their parameters. Some of them are in terms of special functions (Airy, Bessel, Weber-Hermite, Whittaker and hypergeometric), while the others are just rational ones. Quite remarkably the discrete P's have the same property and, in fact, their "special function"-type solutions are solutions of linear difference equations that are discretizations of the corresponding equations for the continuous special functions.

As we have shown in our previous work [94,96], a simple way to obtain particular solutions of a d-P is through factorization whenever this is possible, of course. This allows one to reduce the equation to a discrete Riccati, i.e. a homographic transformation, which is subsequently linearized and reduced to the equation for some special function. We shall present here the case of q-P$_{\text{V}}$ [97],

$$(x_{n+1}x_n - 1)(x_n x_{n-1} - 1) = \frac{pr(x_n - u)(x_n - 1/u)(x_n - v)(x_n - 1/v)}{(x_n - p)(x_n - r)}. \tag{9.20}$$

We propose the following factorization:

$$x_n x_{n+1} - 1 = \frac{p(x_n - u)(x_n - v)}{uv(x_n z - p)} \tag{9.21a}$$

$$x_n x_{n-1} - 1 = \frac{uvr(x_n - 1/u)(x_n - 1/v)}{(x_n z - r)}. \tag{9.21b}$$

The two equations are compatible only when the following condition holds,

$$uv = p/r\lambda. \tag{9.22}$$

In this case, equation (9.21) can be cast in a more symmetric form that is in fact a discrete Riccati,

$$z(x_n x_{n+1} - 1) = p x_{n+1} + \lambda r(x_n - u - v) \tag{9.23}$$

We can easily show that this equation is indeed related to the confluent hypergeometric/Whittaker equation. For the continuous limit we must take $\lambda = 1 + \epsilon$, $p = 1/\epsilon + p_0$, $r = -1/\epsilon + p_0$, $u = 1 + \epsilon u_1$, $v = -1 + \epsilon v_1$, $z = e^{-n\epsilon}$. We thus obtain at the $\epsilon \to 0$ limit the Riccati (where $'$ denotes the z derivative),

$$x' = -x^2 + \frac{2p_0 - 1}{z}x + \frac{\kappa}{z} + 1, \tag{9.24}$$

where κ is related to u_1 and v_1. Next we linearize, introducing the Cole-Hopf transformation $x = a'/a$ and obtain

$$a'' = \frac{2p_0 - 1}{z}a' + \left(\frac{\kappa}{z} + 1\right)a. \tag{9.25}$$

Finally, we transform once more $a = we^z$ and obtain a confluent hypergeometric equation for w,

$$zw'' = (2p_0 - 1 - 2z)w' + (\kappa + 2p_0 - 1)w. \tag{9.26}$$

We can also show that (9.23) can indeed be linearized. Solving for x_{n+1}, we rewrite it as

$$x_{n+1} = \frac{\lambda r(x_n - u - v) + z}{z x_n - p} \tag{9.27}$$

We introduce the discrete equivalent of a Cole-Hopf, $x = B/A$, and obtain the system,

$$B_{n+1} = \lambda r B_n + (z_n - \lambda r(u + v))A_n,$$

$$A_{n+1} = z_n B_n - p A_n. \tag{9.28}$$

Eliminating B we get the linear three-point mapping,

$$A_{n+2} + (p - r)A_{n+1} - (z_n z_{n+1} - z_n r(u + v) + pr)A_n = 0, \tag{9.29}$$

which in view of our analysis above is indeed a discrete form of the confluent hypergeometric equation, up to some straightforward transformations.

The discrete Painlevé equations have another type of solutions, namely rational ones. We shall examine them in the case of q-P_V. One obvious solution of this type is $x = \pm 1$ which exists whenever either u or v takes the value ± 1. Nontrivial solutions also exist. We have, in fact two families of such rational solutions. The first has a most elementary member,

$$x = \pm 1 + (p + r)/z, \tag{9.30}$$

provided u (or $1/u$) $= \mp 1/\lambda$ and v (or $1/v$) $= \mp p/r$ (or $u \leftrightarrow v$). For the second we find

$$x = (p + r)/z, \tag{9.31}$$

which exists for $u = \sqrt{\lambda}$, $v = -\sqrt{\lambda}$. These rational solutions exist only on a codimension-two submanifold and, moreover, they do not contain any free integration constants.

e. Bilinear Forms

As we have shown in [98], the bilinear forms of the discrete Painlevé equations can be obtained in a straightforward way if one uses the information provided by the structure of the singularities of the equation. Thus we have concluded that the number of necessary τ-functions is given by the number of different singularity patterns. In the case of d-P_{II} (9.3b), a singularity appears whenever x_n in the denominator takes the value $+1$ or -1. Thus we have two singularity patterns, which, in this case, turn out to be $\{-1, \infty, +1\}$ and $\{+1, \infty, -1\}$ and we expect two τ-functions, F and G, to appear in the expression for x. In [98] the following simple expression was found for x,

$$x_n = -1 + \frac{F_{n+1}G_{n-1}}{F_n G_n} = 1 - \frac{F_{n-1}G_{n+1}}{F_n G_n}. \tag{9.32}$$

Equation (9.32) provides the first equation of the system. By eliminating the denominator, $F_n G_n$, we obtain

$$F_{n+1}G_{n-1} + F_{n-1}G_{n+1} - 2F_n G_n = 0. \tag{9.33}$$

In order to obtain the second equation we rewrite d-P_{II} as $(x_{n+1} + x_{n-1})(1 - x_n)(1 + x_n) = zx_n + a$. We use the two possible definitions of x_n in terms of F and G in order to simplify the expressions $1 - x_n$ and $1 + x_n$. Next, we obtain two equations by using these two definitions for x_{n+1} combined with the alternate definition for x_{n-1}. We thus obtain:

$$F_{n+2}F_{n-1}G_{n-1} - F_{n-2}F_{n+1}G_{n+1} = F_n^2 G_n(zx_n + a) \tag{9.34a}$$

and

$$G_{n-2}G_{n+1}F_{n+1} - G_{n+2}G_{n-1}F_{n-1} = G_n^2 F_n(zx_n + a). \tag{9.34b}$$

Finally, we add equation (9.34a) multiplied by G_{n+2} and (9.34b) multiplied by F_{n+2}. Up to the use of the upshift of (9.33), a factor $F_{n+1}G_{n+1}$ appears in both sides of the resulting expression. After simplification, the remaining equation is indeed bilinear,

$$F_{n+2}G_{n-2} - F_{n-2}G_{n+2} = z(F_{n+1}G_{n-1} - F_{n-1}G_{n+1}) + 2aF_nG_n. \qquad (9.35)$$

Equations (9.33) and (9.35), taken together, are the bilinear form of d-P_{II}. One interesting result of [98] is that in some cases (in particular for d-P_I) no bilinear form can be obtained and one must introduce a trilinear one.

Although the d-P's have the look and the feel of Painlevé equations they are more general objects than the continuous ones and they present a fundamental difference, they do not possess a single canonical form. Rather, one continuous Painlevé equation may have more than one discrete counterpart. Thus an iteresting classification problem is still open.

9.2 Multidimensional Lattices and Their Similarity Reductions

Discrete systems in several dimensions have been examined from the point of view of integrability using the singularity confinement approach. We have shown thus [99] that the Hirota-Miwa equation [82,100],

$$[Z_1e^{D_1} + Z_2e^{D_2} + Z_3e^{D_3}]f{\cdot}f = 0 \qquad (9.36)$$

where the exponentials of the Hirota bilinear operators D_i introduce finite shifts in the corresponding lattice direction, e.g., $e^{D_1}f{\cdot}f = f(n_1+1,\dots)f(n_1-1,\dots)$, satisfies the integrability requirement. With appropriate restrictions of (9.36) we can obtain the discrete forms of KdV, mKdV, sine-Gordon and so on. An interesting result is that the natural form of d-KdV is *trilinear*. Still the latter can be bilinearized, leading to

$$Z_1f(m+1,n)f(m-1,n-1) + Z_2f(m+1,n-1)f(m-1,n) +$$
$$Z_3f(m,n)f(m,n-1) = 0. \qquad (9.37)$$

The similarity reductions of discrete lattices have been studied in [101]. Contrary to the continuous case, no explicit similarity variable exists for discrete systems. What we have here is a similarity constraint. In practice this means that one complements the autonomous lattice equation with a non-autonomous one *of the same bilinear form*, where the dependence on the lattice variable is linear. In the case of d-KdV it turns out, unsurprisingly enough, that the similarity constraint is trilinear and moreover non reducible to a bilinear form. Singularity confinement can be used in order to obtain the precise form of the nonautonomous equation. The difficult step in this calculation is to show that the two equations of the system are indeed compatible. However we can limit somewhat our scope and use the similarity reduction only to obtain particular solutions. An interesting approach, introduced in [75], considers semi-continuous limits of the system where one of the lattice variables goes to infinity. In the case of 2-D lattices one thus obtains nonautonomous mappings which are nothing but discrete Painlevé equations. Recent developments indicate that this limiting process may be unnecessary: one can obtain a discrete Painlevé equation while staying within the discrete framework.

9.3   Linearizable Mappings

In a series of articles we have derived the analogues of the linearizable equations of the Painlevé-Gambier list. These discrete equations written as a one-component mapping, belong to the class

$$x_{n+1} = \frac{f_1(x_n) - f_2(x_n)x_{n-1}}{f_4(x_n) - f_3(x_n)x_{n-1}} \tag{9.38}$$

where f_i are quadratic in x_n. The general form of the linearizable class is the Gambier mapping [102] that is best given in the two-component form,

$$y_{n+1} = \frac{by_n + c}{y_n + 1} \tag{9.39a}$$

$$x_{n+1} = \frac{dx_n y_n + \sigma}{1 - ax_n}, \tag{9.39b}$$

where $\sigma=0$ or 1. The coefficients, $a(n)$, $b(n)$, $c(n)$ and $d(n)$ are *not* all free. The integrability of this mapping becomes clear under this form: the mapping consists in two Riccati's in cascade. Still, integrability constraints do exist. When y from the first equation assumes a particular value, this may lead to a singularity in the second equation that needs the satisfaction of a condition for its confinement.

Particular forms of the Gambier mapping are also interesting. Among them one can distinguish a particular mapping where the f_i are (special) linear functions. Its canonical form is

$$x_{n-1} = \frac{x_{n+1}(px_n + q) + r}{x_{n+1}x_n}, \tag{9.40}$$

where p, q, r are functions of n. This mapping is linearizable [85] through a generalisation of the Cole-Hopf transformation.

10   Discrete/Continuous Systems: Blending Confinement with Singularity Analysis

Let us start with a continuous/discrete system and consider a system of interacting particles where the particle position is the dependent variable, depending on the continuous time and the discrete particle index. In order to fix the ideas we shall consider the classical paradigmatic integrable system: the Toda lattice,

$$\ddot{x}_n = e^{x_{n+1}-x_n} - e^{x_n-x_{n-1}}. \tag{10.1}$$

It is clear that one does not have to consider the particle index as an independent dynamical variable and, in fact, in the traditional approach this was never the case. Two different approaches to the singularity analysis of such systems can be encountered. First, one can consider a chain of a fixed number of particles and perform the singularity analysis completely through some periodicity argument. The drawback of this method is that, as the number of particles changes new singular behaviour starts appearing. As an alternative, one can study the infinite chain case for some singular behaviours only. These traditional approaches are not satisfactory.

The spirit of the confined-singularity approach is different [103]. The discrete system is considered as a recursion allowing one to compute a given term from a knowledge of the preceding ones. The idea is to look for the possible singularities and their propagation under this recursion. In order to make these considerations more precise, we start by transforming (10.1) into a purely algebraic form through the transformation $a_n = e^{x_{n+1} - x_n}$, $b_n = \dot{x}_n$, leading to

$$\dot{a}_n = a_n(b_{n+1} - b_n) \tag{10.2a}$$

$$\dot{b}_n = a_n - a_{n-1}. \tag{10.2b}$$

We look for the spontaneous appearance of a singularity for some n, when the particle number is interpreted as the number of steps in the recursion. Thus we do not study the solutions that are allowed to be singular for every n but only those that *become* singular at some n. In this context relation (10.2) is to be interpreted as

$$a_n = a_{n-1} + \dot{b}_n \tag{10.3a}$$

$$b_{n+1} = b_n + \frac{\dot{a}_n}{a_n}. \tag{10.3b}$$

We start by assuming that both b_n and a_n are non-divergent and that the singularity appears in step $n + 1$. In fact, due to the presence of the logarithmic derivative in (10.3b), a pole may appear in b_{n+1} if a_n vanishes at some time t_0. Let us start with the simplest case of a single zero, i.e.

$$a_n = \alpha\tau,$$

where $\tau = t - t_0$ and $\alpha = \alpha(t)$ with $\alpha(t_0) \neq 0$. Substituting in (10.3b) we find

$$b_{n+1} = \frac{1}{\tau} + b_n + \frac{\dot{\alpha}}{\alpha}$$

$$a_{n+1} = -\frac{1}{\tau^2} + \dot{b}_n + \frac{\ddot{\alpha}}{\alpha} - \left(\frac{\dot{\alpha}}{\alpha}\right)^2 + \alpha\tau.$$

Iterating further we obtain

$$b_{n+2} = -\frac{1}{\tau} + b_n + \frac{\dot{\alpha}}{\alpha} - 2\tau(\dot{b}_n + \frac{\ddot{\alpha}}{\alpha} - \left(\frac{\dot{\alpha}}{\alpha}\right)^2) - A\tau^2 + \mathcal{O}(\tau^3)$$

$$a_{n+2} = (4A - 7\alpha)\tau + \mathcal{O}(\tau^2),$$

where A is a quantity depending on α and b_n. Iterating further we obtain a finite result for b_{n+3}. Thus the singularity that appeared at b_{n+1} due to the simple root in a_n is confined after two steps.

Clearly the vanishing-a_n behaviour examined above which induces the divergence of b_{n+1} is not the only one. One can imagine higher-order zeros of the type $a_n = \alpha\tau^k$. Depending on the value of k, more and more intermediate steps will be necessary for the confinement of the singularity. In principle, the case $a_n \propto \tau^k$ would necessitate $k + 1$ steps. However the simplest singular behaviour is also the most generic one and its study yields the most important integrability constraints for the system.

What does this analysis teach us? First, the singularities that appear do have the Painlevé property (absence of branching). Second, they do not propagate *ad infinitum* under the recursion (10.3) but are confined to a few iteration steps. The first is the usual Painlevé-property integrability requirement. The second property, had it been discovered in this context, could have introduced the singularity confinement notion.

10.1   Integrodifferential Equations of the Benjamin-Ono Type

The main difficulty for the treatment of discrete systems from the point of view of singularity analysis is their nonlocal character (since the discrete variable takes only integral values). Thus it was natural to try to adapt the discrete techniques to nonlocal continuous systems [104]. An integrable family of such systems is known, the Intermediate-long wave (ILW), Benjamin-Ono and their extensions, i.e., equations involving the Hilbert integral transform and its generalizations. Let us write the ILW in bilinear form,

$$(iD_t + \frac{i}{h}D_x - D_x^2)\overline{F}{\cdot}F = 0, \tag{10.4}$$

where $F = F(x - ih, t)$ and $\overline{F} = F(x + ih, t)$, where h is a real parameter. The operator D is the Hirota bilinear operator defined by $D_x F \cdot F \equiv (\partial_x - \partial_{x'})F(x)F(x')|_{x=x'}$. The essential nonlocality of (10.4) comes from the fact that it relates the function at point $(x + ih)$ to the function at $(x - ih)$. From the point of view of the traditional singularity analysis, the functions F, $\overline{F}$ are considered as *different* objects and, thus, we have only one equation for two unknowns. It is therefore impossible to perform the singularity analysis on this bilinear form. The confined-singularity approach is based on the observation that F is defined (in the complex-x plane) in strips of width h parallel to the x-axis. Thus (10.4) can be considered as a (discrete) recursion relating the F's of two adjacent strips. It can then be treated along the same lines as the Toda system. We are not going to present here the details of this analysis. It suffices to say that the ILW equation does indeed satisfy the singularity confinement criterion.

10.2   Multidimensional Discrete/Continuous Systems

One particular application of this approach is on multidimensional systems. Discrete/continuous multidimensional systems have also been examined in [105] and in particular trilinear equations in determinental form [106]. We have shown there that while the 1-D relativistic Toda Lattice statisfies the singularity confinement this is not the case for its 2-D extension [107]. Thus the latter is not integrable, despite the richness of its particular solutions.

10.3   Delay-Differential Equations

Another very interesting application is that of delay-differential equations. We have examined in [108] a particular class of discrete/continuous systems, differential-delay equations where the dependent function appears at a given time t and also at previous times $t - \tau$, $t - 2\tau, \ldots$, where τ is the delay. Our approach treats hysterodifferential equations as differential-difference systems. The $u(t + k\tau)$ for various k's are treated as different functions of the continuous variable indexed by k, $u(t + k\tau) = u_k(t)$. A detailed analysis of a particular class of such delay systems, i.e. equations of the form

$$F(u_k, u_{k-1}, u'_k, u'_{k-1}) = 0 \tag{10.6}$$

which are bi-Riccati has led to the discovery of a new class of transcendents, the delay-Painlevé equations. We have, for example, with $u = u(t)$ and $\overline{u} = u(t + \tau)$, a particular form of D-P$_I$,

$$u' + \overline{u}' = (u - \overline{u})^2 + k(u + \overline{u}) + \lambda t. \tag{10.7}$$

Our results indicate that the delays P's are objects that may go beyond the Painlevé transcendents.

First-order, three-point delay P's have also been identified although our investigation in this case is still in an initial phase. An example of D-P$_I$ may be written:

$$\frac{u'}{u} = \frac{\overline{u}}{\underline{u}} + \lambda t \tag{10.8}$$

The general form that contains all integrable cases is the mapping,

$$\overline{u} = \frac{f_1(u, u') - f_2(u, u')\underline{u}}{f_4(u, u') - f_3(u, u')\underline{u}}, \tag{10.9}$$

where the f_i are 'Riccati-like' objects, $f_i = \alpha_i u' + \beta_i u^2 + \gamma_i u + \delta_i$, with $\alpha, \beta, \gamma, \delta$ functions of t.

11   Conclusion

We would like to begin this conclusion with our statement of faith. We are deeply convinced that an integrability detector must be used in order to produce new integrable systems. This can be done either through a systematic approach where one examines the equations of a given class or by examining the interesting physical equations of motion for which there exist indications (or mere hunches) that they may be integrable. Thus this course has been based on the more 'practical' aspects of integrability without any attempt at a rigorous formulation. Similarly we have not included a presentation of the (admittedly interesting) Ziglin's approach that may be used in order to prove nonintegrability. A most interesting direction that has hardly been explored and not presented at all here is the application of the techniques of integrable systems to nonintegrable ones.

The emphasis of our presentation has been on discrete systems. This is due not only to the fact that the domain lies now at the centre of interest but also because the world of discrete systems is far richer than that of continuous ones. What is more important, we believe that physical reality can be described in terms of discrete systems based on a discrete space-time in a way as satisfactory as that of continuous formulations. Integrable discrete systems could serve as paradigms in this case, enabling us to undestand and explore the discrete world.

Acknowledgements

The authors wish to express their gratitude to all who have assisted them over the years in exploring the world of integrable systems either through direct collaboration or through discussions and exchange of correspondence and in particular: M. Ablowitz, P. Clarkson, G. Contopoulos, B. Dorizzi, A. Fokas, J. Hietarinta, R. Hirota, A. Its, M. Jimbo, N. Joshi, K. Kajiwara, Y. Kosmann-Schwarzbach, M. Kruskal, R. Miura, F. Nijhoff, Y. Ohta, V. Papageorgiou, V. Ravoson, J. Satsuma, H. Segur, J-M. Strelcyn, K.M. Tamizhmani and P. Winternitz.

They are particularly grateful to J. Hietarinta, Y. Kosmann-Schwarzbach and B.E. Schwarzbach who read and criticized the first draft of the course, spotted mistakes, ommissions and misprints: if more of those remain, this is due to the fact that there were so many to start with, leading to the saturation of the most intrepid readers!

The financial help of the CEFIPRA, through the contract 1201-1, is also gratefully acknowledged.

References

1. Although the present notes on integrability detectors can stand by themselves the reader can find a wealth of information on integrable systems, as well as a detailed bibliography, in the book of M.J. Ablowitz and P.A. Clarkson *Solitons, Nonlinear Evolution Equations and Inverse Scattering*, Cambridge University Press, 1992
2. H. Poincaré, *Les méthodes nouvelles de la Mécanique Céleste*, Gauthier Villars, Paris (1892)
3. S. Novikov, Math. Intel. 14 (1992) 13
4. We refer here to the famous Kolmogorov-Arnold-Moser (KAM) theorem and its applications in the study on near-integrable systems

5. H. Segur, Physica D 51 (1991) 343. We must point out here that the title of this section has been anashamedly stolen from this article of Segur.
6. F. Calogero, in *What is Integrability?*, ed., V. Zakharov, Springer, New York, (1990) 1
7. A.S. Fokas, Physica D 87 (1995) 145
8. E.T. Whittaker, *Analytical Dynamics of Particles*, Cambridge U.P.,Cambridge, (1959)
9. R. Conte, in *An introduction to methods of complex analysis and geometry for classical mechanics and nonlinear waves*, eds., D. Benest and C. Froeschlé, Editions Frontières, Gif-sur-Yvette (1994)
10. H. Yoshida and B. Grammaticos, A. Ramani, Acta Appl. Math. 8 (1987) 75
11. L. Fuchs, Sitz. Akad. Wiss. Berlin, 32 (1884) 669
12. P. Painlevé, C. R. Acad. Sc. Paris, 107 (1888) 221, 320, 724
13. S. Kovalevskaya, Acta Math. 12 (1889) 177
14. D.J. Korteweg and G. de Vries, Philos. Mag. Ser 5, 39 (1895) 422
15. P. Painlevé, Acta Math. 25 (1902) 1
16. B. Gambier, Acta Math. 33 (1909) 1
17. J. Chazy, Acta Math. 34 (1911) 317
18. R. Garnier, Ann. Sci. Ecole Norm. Sup. 29 (1912) 1
19. E. Fermi, J. Pasta and S. Ulam, *Studies of nonlinear problems*, Los Alamos Report LA1940 (1955)
20. N.J. Zabusky and M.D. Kruskal, Phys. Rev. Lett. 15 (1965) 240
21. C.S. Gardner, J.M. Greene, M.D. Kruskal, and R. Miura, Phys. Rev. Lett. 19 (1967) 1095
22. R. Miura, J. Math. Phys. 9 (1968) 1202
23. R. Hirota, Phys. Rev. Lett. 27 (1971) 1192
24. P.D. Lax, Commun. Pure Appl. Math. 21 (1968) 467
25. M. Toda, J. Phys. Soc. Japan, 22 (1967) 431
26. V.E. Zakharov and A.B. Shabat, Sov. Phys. JETP 34 (1972) 62
27. M.J. Ablowitz, D.J. Kaup, A.C. Newell and H. Segur, Phys. Rev. Lett. 30 (1973) 1262
28. M.J. Ablowitz and H. Segur, Phys. Rev. Lett. 38 (1977) 1103
29. M.J. Ablowitz, A. Ramani and H. Segur, Lett. Nuov. Cim. 23 (1978) 333
30. J. Weiss, M. Tabor, and G. Carnevale, J. Math. Phys. 24 (1983) 522
31. M.D. Kruskal and P.A. Clarkson, Stud. Appl. Math. 86 (1992) 87
32. M.D. Kruskal, summary talk at the "Kruskal Symposium", Boulder (1995)
33. This section is directly inspired from our course at les Houches 89, M.D. Kruskal, A. Ramani, and B. Grammaticos, NATO ASI Series C 310, Kluwer 1989, 321
34. A. Ramani, B. Grammaticos, B. Dorizzi, and T. Bountis, J. Math. Phys. 25 (1984) 878
35. B. Dorizzi, B. Grammaticos and A. Ramani, J. Math.Phys. 25 (1984) 481
36. H. Yoshida, Celest. Mech. 31 (1983) 363 and 381
37. B. Grammaticos, J. Moulin-Ollagnier, A. Ramani, J.M Strelcyn and S. Wojciechowski, Physica 163A (1990) 683
38. M. Adler and P. van Moerbecke, *Algebraic completely integrable systems: a systematic approach*, Perspectives in Mathematics, Academic Press, New York, (1988)
39. E. Gutkin, Physica 16D (1985) 235
40. H. Yoshida, A. Ramani, B. Grammaticos and J. Hietarinta, Physica 144A (1987), 310

41. P.J. Richens and M.V. Berry, Physica 2D (1981) 495
42. A. Ramani, A. Kalliterakis, B. Grammaticos, and B. Dorizzi, Phys. Lett. A 115 (1986) 25
43. J. Hietarinta, Phys. Rep. 147 (1987), 87
44. V. Ravoson, A. Ramani, and B. Grammaticos, Phys. Lett. A191 (1994) 91
45. M. Hénon, *Numerical exploration of Hamiltonian systems* in Les Houches 1981, North Holland (1983) 55
46. A. Ramani, B. Grammaticos. and T. Bountis, Phys. Rep. 180 (1989) 159
47. E.L. Ince, *Ordinary Differential Equations*, Dover, London, (1956)
48. M.J. Ablowitz, A. Ramani, and H. Segur, J. Math. Phys. 21 (1980) 715 and 1006.
49. A. Ramani, B. Dorizzi, and B. Grammaticos, Phys. Rev. Lett., 49 (1982) 1539
50. Exceptions to this empirical rule are known to exist, Chazy's equation [17] being the best known example where no generic leading behaviour seems to exist
51. C. Briot and J.C. Bouquet, J. École Imp. Polytech. 21 (1856) 36
52. F. Bureau, A. Garcet, and J. Goffar, Ann. Math. Pura Appl. 92 (1972) 177
53. F. Bureau, Ann. Math. Pura Appl. 91 (1972) 163
54. C.M. Cosgrove and G. Scoufis, Stud. Appl. Math. 88 (1993) 25
55. C.M. Cosgrove, Stud. Appl. Math. 90 (1993) 119
56. R.S. Ward, Phys. Lett. 102A (1984), 279
57. M. Jimbo, M.D. Kruskal, and T. Miwa, Phys. Lett. 92A (1982) 59
58. M. Musette and R. Conte, J. Math. Phys. 32 (1991) 1450.
59. M.D. Kruskal, *Flexibility in applying the Painlevé test*, NATO ASI B278, Plenum (1992) 187
60. R. Conte, A.P. Fordy, and A. Pickering, Physica D 69 (1993) 33
61. V. A. Belinski and I. M. Khalatnikov, Sov. Phys. JETP 29 (1969) 911
62. G. Contopoulos, B. Grammaticos, and A. Ramani, J. Phys. A 26 (1993) 5795
63. G. Contopoulos, B. Grammaticos, and A. Ramani, J. Phys. A 27 (1994) 5357
64. G. Contopoulos, B. Grammaticos, and A. Ramani, J. Phys. A 28 (1995) 5313
65. A. Latifi, M. Musette, and R. Conte, Phys. Lett. A 194 (1994) 83
66. H. Segur, Lectures at International School "Enrico Fermi", Varenna, Italy (1980)
67. M. Tabor and J. Weiss, Phys. Rev. A 24 (1981), 2157
68. T. Bountis, A. Ramani, B. Grammaticos, and B. Dorizzi, Physica 128A (1984) 268
69. B. Grammaticos, B. Dorizzi, and A. Ramani, J. Math. Phys. 24 (1983) 2289
70. R.P. Feynman, Intl. Jour. Theor. Phys. 21 (1982) 467
71. V. Kozlov and D. Treshchëv, *Billiards*, AMS Transl. of Mathematical Monographs Vol. 89, 1991
72. B. Grammaticos and B. Dorizzi, Europhys. Lett. 14 (1991) 169
73. B. Grammaticos and B. Dorizzi, J. Math. Comp. Sim. 37 (1994) 341
74. E. Brézin and V.A. Kazakov, Phys. Lett. 236B (1990) 144
75. F.W. Nijhoff and V.G. Papageorgiou, Phys. Lett. 153A (1991) 337
76. M. Jimbo and T. Miwa, Proc. Jap. Acad. 56, Ser. A (1980) 405
77. Proceedings of the Intl. Conf. *Yang-Baxter equations in Paris*, Int. J. Mod. Phys. B 7 (1993) Nos 21 & 22
78. B. Grammaticos, A. Ramani, and V.G. Papageorgiou, Phys. Rev. Lett. 67 (1991) 1825.
79. R.J. Baxter, *Exactly Solved Models in Statistical Mechanics*, Associated Press, (London) 1982
80. G.R.W. Quispel, J.A.G. Roberts and C.J. Thompson, Physica D34 (1989) 183

81. A. Ramani, B. Grammaticos, and J. Hietarinta, Phys. Rev. Lett. 67 (1991) 1829
82. R. Hirota, J. Phys. Soc. Jpn 50 (1981) 3785
83. P. Wynn, Math. Tables and Aids to Computing 10 (1956) 91
84. V.G. Papageorgiou, B. Grammaticos, and A. Ramani, Phys. Lett. A 179 (1993) 111
85. A. Ramani, B. Grammaticos, and G. Karra, Physica A 181 (1992) 115
86. B. Grammaticos, A. Ramani, and K.M. Tamizhmani, Jour. Phys. A 27 (1994) 559
87. M. Jimbo, T. Miwa, and K. Ueno, Physica D 2 (1981) 306
88. A. Ramani and B. Grammaticos, Physica A 228 (1996) 160
89. A.S. Fokas, B. Grammaticos, and A. Ramani, J. Math. An. and Appl. 180 (1993) 342
90. M. Jimbo and H. Sakai, Lett. Math. Phys. 38 (1996) 145
91. V.G. Papageorgiou, F.W. Nijhoff, B. Grammaticos, and A. Ramani, Phys. Lett. A 164 (1992) 57
92. A.S. Fokas and M.J. Ablowitz, J. Math. Phys. 23 (1982) 2033
93. A. Ramani and B. Grammaticos, Jour. Phys. A 25 (1992) L633
94. K.M. Tamizhmani, B. Grammaticos, and A. Ramani, Lett. Math. Phys. 29 (1993) 49
95. V.A. Gromak and N.A. Lukashevich, *The analytic solutions of the Painlevé equations*, (Universitetskoye Publishers, Minsk 1990), in Russian
96. B. Grammaticos, F.W. Nijhoff, V.G. Papageorgiou, A. Ramani, and J. Satsuma, Phys. Lett. A 185 (1994) 446
97. K.M. Tamizhmani, B. Grammaticos, A. Ramani, and Y. Ohta, Lett. Math. Phys. 38 (1996) 289
98. A. Ramani, B. Grammaticos, and J. Satsuma, J. Phys. A 28 (1995) 4655
99. A. Ramani, B. Grammaticos, and J. Satsuma, Phys. Lett. A169 (1992) 323
100. T. Miwa, Proc. Japan. Acad. 58 (1982) 9
101. J. Satsuma, A. Ramani, and B. Grammaticos, Phys. Lett. A 174 (1993) 387
102. A. Ramani and B. Grammaticos, Physica A 223 (1995) 125
103. A. Ramani, B. Grammaticos, and K.M. Tamizhmani, J. Phys. A 25 (1992) L883
104. A. Ramani, B. Grammaticos, and K.M. Tamizhmani, J. Phys. A 26 (1993) L53
105. J. Satsuma, B. Grammaticos, and A. Ramani, RIMS Kokyuroku 868 (1994) 129
106. J. Satsuma and J. Matsukidaira, J. Phys. Soc. Japan 59 (1990) 3413
107. J. Hietarinta and J. Satsuma, Phys. Lett. A 161 (1991) 267
108. B. Grammaticos, A. Ramani, and I.C. Moreira, Physica A 196 (1993) 574

Introduction to the Hirota Bilinear Method

J. Hietarinta

Department of Physics, University of Turku, 20014 Turku, Finland, `hietarin@utu.fi`

Abstract. We give an elementary introduction to Hirota's direct method of constructing multi-soliton solutions to integrable nonlinear evolution equations. We discuss in detail how this works for equations in the Korteweg–de Vries class, with some comments on the more complicated cases. We also show how Hirota's method can be used to search for new integrable evolution equations and list some equations found this way.

1 Why the Bilinear Form?

In 1971 Hirota introduced a new direct method for constructing multi-soliton solutions to integrable nonlinear evolution equations [1]. The idea was to make a transformation into new variables, so that in these new variables multi-soliton solutions appear in a particularly simple form. The method turned out to be very effective and was quickly shown to give N-soliton solutions to the Korteweg–de Vries (KdV) [1], modified Korteweg–de Vries (mKdV) [2], sine-Gordon (sG) [3] and nonlinear Schrödinger (nlS) [4] equations. It is also useful in constructing their Bäcklund transformations [5]. Later it was observed that the new dependent variables (called "τ-functions") have very good properties and this has become a starting point for further developments [6].

Here our point of view is practical: we want to describe how multi-soliton solutions can be constructed for a given equation using Hirota's method. The approach works well for integrable equations, and even for nonintegrable ones we can sometimes get two-soliton solutions with this approach. However, the method is not entirely algorithmic, in particular the bilinearization of a given equation may be difficult. Multi-soliton solutions can, of course, be derived by many other methods, e.g., by the inverse scattering transform (IST) and various dressing methods. The advantage of Hirota's method over the others is that it is algebraic rather than analytic. The IST method is more powerful in the sense that it can handle more general initial conditions, but at the same time it is restricted to a smaller set of equations. Accordingly, if one just wants to find soliton solutions, Hirota's method is the fastest in producing results.

2 From Nonlinear to Bilinear

The (integrable) PDE's that appear in some particular (physical) problem are not usually in the best form for further analysis. For constructing soliton solu-

J. Hietarinta, Introduction to the Hirota Bilinear Method, Lect. Notes Phys. **638**, 95–105 (2004)
`http://www.springerlink.com/`

tions the best form is Hirota's bilinear form (discussed below) and soliton solutions appear as polynomials of simple exponentials only in the corresponding new dependent variables. The first problem we face is therefore to find the bilinearizing transformation. This is not algorithmic and can require the introduction of new dependent and sometimes even independent variables.

2.1   Bilinearization of the KdV Equation

Let us consider in detail the KdV equation

$$u_{xxx} + 6uu_x + u_t = 0. \tag{1}$$

Since we have not yet defined what bilinear is, let us first concentrate on transforming the equation into a form that is quadratic in the dependent variable and its derivatives. One guideline in searching for the transformation is that the leading derivative should go together with the nonlinear term, and, in particular, have the same number of derivatives. If we count a derivative with respect to x having degree 1, then to balance the first two terms of (1) u should have degree 2. Thus we introduce the transformation to a new dependent variable w (having degree 0) by

$$u = \partial_x^2 w. \tag{2}$$

After this the KdV equation can be written

$$w_{xxxxx} + 6w_{xx}w_{xxx} + w_{xxt} = 0, \tag{3}$$

which can be integrated once with respect to x to give

$$w_{xxxx} + 3w_{xx}^2 + w_{xt} = 0. \tag{4}$$

(In principle this would introduce an integration constant (function of t), but since (2) defines w only up to $w \to w + \rho(t) + x\lambda(t)$, we can use this freedom to absorb it.)

Equations of the above form can usually be bilinearized by introducing a new dependent variable whose natural degree (in the above sense) is zero, e.g., $\log F$ or f/g. In this case the first one works, so let us define

$$w = \alpha \log F, \tag{5}$$

with a free parameter α. This results in an equation that is fourth degree in F, with the structure

$$F^2 \times (\text{something quadratic}) + 3\alpha(2 - \alpha)(2FF'' - F'^2)F'^2 = 0. \tag{6}$$

Thus we get a quadratic equation if we choose $\alpha = 2$, and the result is

$$F_{xxxx}F - 4F_{xxx}F_x + 3F_{xx}^2 + F_{xt}F - F_xF_t = 0. \tag{7}$$

In addition to being quadratic in the dependent variable and its derivatives, an equation in the Hirota bilinear form must also satisfy a condition with respect to the derivatives: they should only appear in combinations that can be expressed using Hirota's D-operator, which is defined by:

$$D_x^n f \cdot g = (\partial_{x_1} - \partial_{x_2})^n f(x_1) g(x_2)\big|_{x_2 = x_1 = x}. \tag{8}$$

Thus D operates on a product of two functions like the Leibniz rule, except for a crucial sign difference. For example

$$D_x f \cdot g = f_x g - f g_x,$$
$$D_x D_t f \cdot g = f_{xt} g - f_x g_t - f_t g_x + f g_{xt}.$$

Using the D-operator we can write (7) in the following condensed form

$$(D_x^4 + D_x D_t) F \cdot F = 0. \tag{9}$$

To summarize: what we needed in order to obtain the bilinear form (9) for (1) is a dependent variable transformation

$$u = 2 \partial_x^2 \log F, \tag{10}$$

but note that we also had to integrate the equation once.

2.2   Another Example: The Sasa-Satsuma Equation

Unfortunately the bilinearization of a given equation can be difficult. It is even difficult to find out beforehand how many new independent and/or dependent variables are needed for the bilinearization. Furthermore, it is not sufficient that the result be bilinear: the functions that appear in it should furthermore be regular functions (τ-functions), which in the soliton case means that all of them should be expressible as polynomials of exponentials. (This problem is illustrated by the Kaup-Kupershmidt equation that has three bilinearizations of which only one is in terms of genuine τ-functions [7].)

As a further example let us consider the Sasa-Satsuma equation

$$q_t + q_{xxx} + 6|q|^2 q_x + 3q|q^2|_x = 0. \tag{11}$$

Here q is a complex field. Note that if we were to take reduction where q is real (or equally well, if it were complex with a *constant* phase) we would obtain mKdV. Since q is complex one could try the substitution used for the prototypical complex case of nlS and try

$$q = G/F, \quad G \text{ complex}, F \text{ real}.$$

A direct substitution then yields

$$F^2[(D_x^3 + D_t)G \cdot F] - 3GF(D_x G \cdot G^*) - 3(D_x G \cdot F)[D_x^2 F \cdot F - 4|G|^2] = 0, \quad (12)$$

which is quartic in F, G. At this point it would be tempting to bilinearize the above by

$$\begin{cases} (D_x^3 + D_t)G \cdot F = 0, \\ D_x G \cdot G^* = 0, \\ D_x^2 F \cdot F = 4|G|^2, \end{cases}$$

but this kind of brute force bilinearization is wrong, because it would introduce 3 equations for 2 functions. (In fact it is easy to see that the middle equation is solved by $G = g(x, t)e^{ik(t)}$ with g, k real, and then the first equation implies that k is in fact a constant, thereby reducing the equation to mKdV.) One correct way is to keep one bilinear and one trilinear equation,

$$\begin{cases} F[(D_x^3 + D_t)G \cdot F] = 3G[D_x G \cdot G^*], \\ D_x^2 F \cdot F = 4|G|^2. \end{cases} \tag{13}$$

(Further comments on the trilinear forms can be found in [8].)

As a matter of fact it is indeed possible to bilinearize (13). The first trilinear equation can be split into two bilinear equations by introducing an additional dependent variable, and this can be done in two different ways resulting in

$$\begin{cases} (D_x^3 + 4D_t) G \cdot F = 3D_x H \cdot F, \\ (D_x^3 + 4D_t) G^* \cdot F = 3D_x H^* \cdot F, \\ D_x^2 G \cdot F = -HF, \\ D_x^2 G^* \cdot F = -H^* F, \\ D_x^2 F \cdot F = 4|G|^2, \end{cases} \quad \text{or} \quad \begin{cases} (D_x^3 + D_t) G \cdot F = 3SG, \\ (D_x^3 + D_t) G^* \cdot F = -3SG^*, \\ D_x G \cdot G^* = SF, \\ D_x^2 F \cdot F = 4|G|^2, \end{cases} \tag{14}$$

where the new dependent variables have been called H and S, respectively. Here S is pure imaginary, H is complex, and they are related by $HG^* - H^*G = D_x FS$. The bilinearizations (14) are both acceptable, because they introduce an equal number of new functions and new equations, and furthermore the new functions are true τ-functions [9].

2.3   Comments

For a further discussion of bilinearization, see, e.g., [10, 11]. As a guide on the bilinearization one can use the form of the one- and two-soliton solutions, if known. Furthermore, singularity analysis has also been used for this purpose, although only for particular cases, see, e.g., [12].

One important property of equations in Hirota's bilinear form is their gauge invariance. One can show [8] that for a quadratic expression homogeneous in the derivatives, i.e., of the form $\sum_{i=0}^{n} c_i \left(\partial_x^i f \right) \left(\partial_x^{n-i} g \right)$, the requirement of gauge invariance under $f \to e^{kx} f, g \to e^{kx} g$ implies that the expression can be written in terms of Hirota derivatives. This gauge invariance can be taken as a starting point for further generalizations [8, 13].

Finally in this section we would like to list some useful properties of the bilinear derivative [5]:

$$P(D)f \cdot g = P(-D)g \cdot f, \tag{15}$$

$$P(D)1 \cdot f = P(-\partial)f, \quad P(D)f \cdot 1 = P(\partial)f, \tag{16}$$

$$P(D)e^{px} \cdot e^{qx} = P(p-q)e^{(p+q)x}, \tag{17}$$

$$\partial_x^2 \log f = (D_x^2 f \cdot f)/(2f^2), \tag{18}$$

$$\partial_x^4 \log f = (D_x^4 f \cdot f)/(2f^2) - 3(D_x^2 f \cdot f)^2/(2f^4). \tag{19}$$

3 Constructing Multi-soliton Solutions

3.1 The Vacuum, and the One-Soliton Solution

Now that we have obtained the KdV equation in the bilinear form (9), let us start constructing soliton solutions for it. In fact, it is equally easy to consider a whole class of bilinear equations of the form

$$P(D_x, D_y, ...)F \cdot F = 0, \tag{20}$$

where P is some polynomial in the Hirota partial derivatives D. We may assume that P is even, because the odd terms cancel due to the antisymmetry of the D-operator. Note that we do not impose any restrictions on the dimensionality of the problem.

Let us start with the zero-soliton solution or the vacuum. We know that the KdV equation has a solution $u \equiv 0$ and now we want to find the corresponding F. From (10) we see that $F = e^{2\phi(t)x+\beta(t)}$ yields a u that solves (1), and in view of the gauge freedom we can choose $F = 1$ as our vacuum solution. It solves (20) provided that

$$P(0, 0, \dots) = 0. \tag{21}$$

This is then the first condition that we have to impose on the polynomial P in (20).

The multi-soliton solutions are obtained by finite perturbation expansions around the vacuum $F = 1$:

$$F = 1 + \epsilon f_1 + \epsilon^2 f_2 + \epsilon^3 f_3 + \cdots \tag{22}$$

Here ϵ is a formal expansion parameter. For the one-soliton solution (1SS) only one term is needed. If we substitute

$$F = 1 + \epsilon f_1 \tag{23}$$

into (20) we obtain

$$P(D_x, \dots)\{1 \cdot 1 + \epsilon 1 \cdot f_1 + \epsilon f_1 \cdot 1 + \epsilon^2 f_1 \cdot f_1\} = 0.$$

The term of order ϵ^0 vanishes because of (21). For the terms of order ϵ^1 we use property (16) so that, since now P is even, we get

$$P(\partial_x, \partial_y, \dots)f_1 = 0. \tag{24}$$

Soliton solutions correspond to the exponential solutions of (24). For a 1SS we take an f_1 with just one exponential

$$f_1 = e^{\eta}, \quad \eta = px + qy + \cdots + \text{const.}, \tag{25}$$

and then (24) becomes the *dispersion relation* on the parameters $p, q, \dots$

$$P(p, q, \dots) = 0. \tag{26}$$

Finally, the order ϵ^2 term vanishes because

$$P(\mathbf{D})e^{\eta} \cdot e^{\eta} = e^{2\eta} P(\mathbf{p} - \mathbf{p}) = 0,$$

by (21).

In summary, the 1SS is given by (23,25) where the parameters are constrained by (26). For KdV, $\eta = px + \omega t + \text{const.}$ and the dispersion relation is $\omega = -p^3$.

3.2   The Two-Soliton Solution

The 2SS is built from two 1SS's, and indeed the important principle is that

- for integrable systems one must be able to combine *any* pair of 1SS's built over the same vacuum.

Thus if we have two 1SS's, $F_1 = 1 + e^{\eta_1}$ and $F_2 = 1 + e^{\eta_2}$, we should be able to combine them into $F = 1 + f_1 + f_2$, where $f_1 = e^{\eta_1} + e^{\eta_1}$. Gauge invariance suggests that we should try the combination

$$F = 1 + e^{\eta_1} + e^{\eta_2} + A_{12}e^{\eta_1 + \eta_2} \tag{27}$$

where there is just one arbitrary constant A_{12}. Substituting this into (20) yields

$$
P(\mathbf{D})\{ \begin{array}{llll}
1 \cdot 1 + & 1 \cdot e^{\eta_1} + & 1 \cdot e^{\eta_2} + & A_{12}\, 1 \cdot e^{\eta_1 + \eta_2} + \\
e^{\eta_1} \cdot 1 + & e^{\eta_1} \cdot e^{\eta_1} + & e^{\eta_1} \cdot e^{\eta_2} + & A_{12}\, e^{\eta_1} \cdot e^{\eta_1 + \eta_2} + \\
e^{\eta_2} \cdot 1 + & e^{\eta_2} \cdot e^{\eta_1} + & e^{\eta_2} \cdot e^{\eta_2} + & A_{12}\, e^{\eta_2} \cdot e^{\eta_1 + \eta_2} + \\
\underline{A_{12}e^{\eta_1 + \eta_2} \cdot 1} + & A_{12}e^{\eta_1 + \eta_2} \cdot e^{\eta_1} + & A_{12}e^{\eta_1 + \eta_2} \cdot e^{\eta_2} + & A_{12}^2 e^{\eta_1 + \eta_2} \cdot e^{\eta_1 + \eta_2}
\end{array} \} = 0.
$$

In this equation all non-underlined terms vanish due to (21,26). Since P is even, the underlined terms combine as $2A_{12}P(\mathbf{p}_1 + \mathbf{p}_2) + 2P(\mathbf{p}_1 - \mathbf{p}_2) = 0$, which can be solved for A_{12} as

$$A_{12} = -\frac{P(\mathbf{p}_1 - \mathbf{p}_2)}{P(\mathbf{p}_1 + \mathbf{p}_2)}. \tag{28}$$

The actual form of this *phase factor* carries information about the hierarchy to which the equation may belong.

Note that we were able to construct a two-soliton solution for a huge class of equations, namely all those whose bilinear form is of type (20), for whatever P. In particular this includes many nonintegrable systems.

3.3   Multi-soliton Solutions

The above shows that for the KdV class (20) the existence of 2SS is not strongly related to integrability, but it turns out that the existence on 3SS is very restrictive.

A 3SS should start with $f_1 = e^{\eta_1} + e^{\eta_2} + e^{\eta_3}$ and, if the above is any guide, contain terms up to f_3. If we now use the natural requirement that the 3SS should reduce to a 2SS when the third soliton goes to infinity (which corresponds to $\eta_k \to \pm\infty$) then one finds that F must have the form

$$F = 1 + e^{\eta_1} + e^{\eta_2} + e^{\eta_3}$$
$$+ A_{12}e^{\eta_1+\eta_2} + A_{13}e^{\eta_1+\eta_3} + A_{23}e^{\eta_2+\eta_3} + A_{12}A_{13}A_{23}e^{\eta_1+\eta_2+\eta_3}. \quad (29)$$

Note in particular that this expression contains no additional freedom. The parameters p_i are only required to satisfy the dispersion relation (26) and the phase factors A were already determined (28). This extends to NSS [14]:

$$F = \sum_{\substack{\mu_i=0,1 \\ 1\leq i\leq N}} \exp\left(\sum_{1\leq i<j\leq N} \varphi(i,j)\mu_i\mu_j + \sum_{i=1}^{N} \mu_i\eta_i\right), \quad (30)$$

where $A_{ij} = e^{\varphi(i,j)}$. Thus the ansatz for a NSS is completely fixed and the requirement that it be a solution of (20) implies conditions on the equation itself. Only for integrable equations can we combine solitons in this simple way. More precisely, let us make the

DEFINITION: *A set of equations written in the Hirota bilinear form is* **Hirota integrable,** *if one can combine any number N of one-soliton solutions into an NSS, and the combination is a finite polynomial in the e^η's involved.*

In all cases known so far, Hirota integrability has turned out to be equivalent to more conventional definitions of integrability.

4   Searching for Integrable Evolution Equations

Since the existence of a 3SS is very restrictive, one can use it as a method for searching for new integrable equations. All search methods contain a definition of the class of equations to be considered. In the present case we assume that the nonlinear PDE can be put into a bilinear form of type (20), but no assumptions are made for example on the number of independent variables. This is in contrast with many other searches, in which the leading (linear) derivative terms are fixed but the nonlinearity is left open.

4.1 KdV

If one substitutes (29) into (20) one obtains the condition

$$\sum_{\sigma_i=\pm 1} P(\sigma_1\mathbf{p}_1 + \sigma_2\mathbf{p}_2 + \sigma_3\mathbf{p}_3)$$
$$\times P(\sigma_1\mathbf{p}_1 - \sigma_2\mathbf{p}_2)P(\sigma_2\mathbf{p}_2 - \sigma_3\mathbf{p}_3)P(\sigma_3\mathbf{p}_3 - \sigma_1\mathbf{p}_1) \doteq 0, \qquad (31)$$

where the symbol $\doteq$ means that the equality is required to hold only when the parameters $\mathbf{p}_i$ satisfy the dispersion relation $P(\mathbf{p}_i) = 0$.

In order to find possible solutions of (31), we made a computer assisted study [15] and the result was that the only genuinely nonlinear equations that solved (31) were

$$(D_x^4 - 4D_xD_t + 3D_y^2)F \cdot F = 0, \qquad (32)$$
$$(D_x^3D_t + aD_x^2 + D_tD_y)F \cdot F = 0, \qquad (33)$$
$$(D_x^4 - D_xD_t^3 + aD_x^2 + bD_xD_t + cD_t^2)F \cdot F = 0, \qquad (34)$$
$$(D_x^6 + 5D_x^3D_t - 5D_t^2 + D_xDy)F \cdot F = 0. \qquad (35)$$

and their reductions. These equations also have 4SS and they all pass the Painlevé test [16]. Among them we recognize the Kadomtsev-Petviashvili (containing KdV and Boussinesq) (32), Hirota-Satsuma-Ito (33) and Sawada-Kotera-Ramani (35) equations; they also appear in the Jimbo-Miwa classification [6]. The only new equation is (34). It is somewhat mysterious. It has not been identified within the Jimbo-Miwa classification because it has no nontrivial scaling invariances, furthermore we do not know its Lax pair or Bäcklund transformation.

4.2 mKdV and sG

As was mentioned before, Hirota's bilinear method has been applied to many other equations beside KdV. Here we would like to mention briefly some of them.

For example the modified Korteweg–de Vries (mKdV) and sine–Gordon (sG) equations have a bilinear form of the type

$$\begin{cases} B(\mathbf{D})\,G \cdot F = 0, \\ A(\mathbf{D})(F \cdot F + G \cdot G) = 0, \end{cases} \qquad (36)$$

where A is even and B either odd (mKdV) or even (sG). For mKdV we have $B = D_x^3 + D_t$, $A = D_x^2$, and for sG, $B = D_xD_y - 1$, $A = D_xD_y$. This class of equations also has 2SS for any choice of A and B. [If B is odd one can make a rotation $F = f + g$, $G = i(f - g)$ after which the pair (36) becomes $B\,g \cdot f = 0$, $A\,g \cdot f = 0$.]

In principle the pair (36) can have two different kinds of solitons,

$$\begin{cases} F = 1 + e^{\eta_A},\ G = 0,\ \text{with dispersion relation } A(\mathbf{p}) = 0, \\ F = 1,\ G = e^{\eta_B},\qquad \text{with dispersion relation } B(\mathbf{p}) = 0. \end{cases} \qquad (37)$$

In mKdV and sG the A polynomial is too trivial to make the first kind of soliton interesting. In [11,17] we searched for polynomials A and B for which any set of three solitons could be combined for a 3SS. The final result contains 5 equations of mKdV type, three of them have a nonlinear B polynomial but a factorizable A part (and hence only one kind of soliton with B acting as the dispersion relation),

$$\begin{cases} (aD_x^7 + bD_x^5 + D_x^2 D_t + D_y)\,G \cdot F = 0, \\ \qquad\qquad\qquad\qquad D_x^2\,G \cdot F = 0, \end{cases} \tag{38}$$

$$\begin{cases} (aD_x^3 + bD_x^3 + D_y)\,G \cdot F = 0, \\ \qquad\qquad\quad D_x D_t\,G \cdot F = 0, \end{cases} \tag{39}$$

$$\begin{cases} (D_x D_y D_t + aD_x + bD_t)\,G \cdot F = 0, \\ \qquad\qquad\quad D_x D_t\,G \cdot F = 0. \end{cases} \tag{40}$$

For a discussion of the nonlinear versions of the last two equations, see [18].

We also found two cases where both A and B are nonlinear enough to support solitons:

$$\begin{cases} (D_x^3 + D_y)\,G \cdot F = 0, \\ (D_x^3 D_t + aD_x^2 + D_t D_y)\,G \cdot F = 0, \end{cases} \tag{41}$$

$$\begin{cases} (D_x^3 + D_y)\,G \cdot F = 0, \\ (D_x^6 + 5D_x^3 D_y - 5D_y^2 + D_t D_x)\,G \cdot F = 0. \end{cases} \tag{42}$$

Note that the B polynomials are the same and that the A parts have already appeared in the KdV list.

Two equations of sine-Gordon type were also found:

$$\begin{cases} (D_x D_t + b)\,G \cdot F = 0, \\ (D_x^3 D_t + 3bD_x^2 + D_t D_y)(F \cdot F + G \cdot G) = 0, \end{cases} \tag{43}$$

$$\begin{cases} (aD_x^3 D_t + D_t D_y + b)\,G \cdot F = 0, \\ \qquad\qquad D_x D_t(F \cdot F + G \cdot G) = 0. \end{cases} \tag{44}$$

4.3   nlS

A similar search was performed [11,19,20] on equations of nonlinear Schrödinger (nlS) type,

$$\begin{cases} B(\mathbf{D})\,G \cdot F = 0, \\ A(\mathbf{D})\,F \cdot F = |G|^2, \end{cases} \tag{45}$$

where F is real and G complex. Again two kinds of solitons exist,

$$\begin{cases} F = 1 + e^{\eta_A},\ G = 0, \qquad\qquad \text{with dispersion relation } A(\mathbf{p}) = 0, \\ F = 1 + Ke^{\eta_B + \eta_B^*},\ G = e^{\eta_B}, \ \text{with dispersion relation } B(\mathbf{p}) = 0. \end{cases} \tag{46}$$

In this case the existence of a 2SS is not automatic because the 1SS already involves terms of order ϵ^2 and a 2SS therefore ϵ^4, whereas in the previous cases ϵ^2 contributions were sufficient for 2SS.

Three equations with 3SS were found in this search:

$$\begin{cases} (D_x^2 + iD_y + c) \, G \cdot F = 0, \\ (a(D_x^4 - 3D_y^2) + D_x D_t) \, F \cdot F = |G|^2, \end{cases} \tag{47}$$

$$\begin{cases} (i\alpha D_x^3 + 3cD_x^2 + i(bD_x - 2dD_t) + g) \, G \cdot F = 0, \\ (\alpha D_x^3 D_t + a D_x^2 + (b + 3c^2) D_x D_t + d D_t^2) \, F \cdot F = |G|^2, \end{cases} \tag{48}$$

$$\begin{cases} (i\alpha D_x^3 + 3D_x D_y - 2iD_t + c) \, G \cdot F = 0, \\ (a(\alpha^2 D_x^4 - 3D_y^2 + 4\alpha D_x D_t) + bD_x^2) \, F \cdot F = |G|^2. \end{cases} \tag{49}$$

Perhaps the most interesting new equation in the above list is the combination in (49) of the two most important $(2+1)$-dimensional equations, Davey-Stewartson and Kadomtsev-Petviashvili equations ([20], for further discussion of this equation, see [21, 22]).

The above lists contain many equations of which nothing is known other than that they have 3SS and 4SS. This alone suggests that they are good candidates to be integrable $(2+1)$-dimensional equations, but the relations to other definitions of integrability, such as Lax pairs, are not yet obvious.

Further applications of Hirota's bilinear approach can be found in the lectures of J. Satsuma in this volume.

Acknowledgments

This work was supported in part by the Academy of Finland, project 31445.

References

1. R. Hirota, Phys. Rev. Lett. **27**, 1192 (1971)
2. R. Hirota, J. Phys. Soc. Japan **33**, 1456 (1972)
3. R. Hirota, J. Phys. Soc. Japan **33**, 1459 (1972)
4. R. Hirota, J. Math. Phys. **14**, 805 (1973)
5. R. Hirota, Progr. Theor. Phys. **52**, 1498 (1974)
6. M. Jimbo and T. Miwa, Publ. RIMS, Kyoto Univ. **19**, 943 (1983)
7. J. Springael, Ph.D. thesis, Vrije Universiteit Brussel (1999), p. 36
8. B. Grammaticos, A. Ramani, and J. Hietarinta, Phys. Lett. A **190**, 65 (1994)
9. C. Gilson, J. Hietarinta, J. Nimmo, and Y. Ohta, Phys. Rev. E **68**, 016614 (2003)
10. R. Hirota in "Solitons", R.K. Bullough and P.J. Caudrey (eds.), Springer (1980), p. 157
11. J. Hietarinta, in "Partially Integrable Evolution Equations in Physics", R. Conte and N. Boccara (eds.), Kluwer Academic (1990), p. 459
12. P. Estévez et al., J. Phys. A **26**, 1915 (1993)
13. J. Hietarinta, B. Grammaticos and A. Ramani, in "NEEDS '94", V. Makhankov et al. (eds.), World Scientific (1995), p. 54

14. R. Hirota, J. Math. Phys. **14**, 810 (1973)
15. J. Hietarinta, J. Math. Phys. **28**, 1732 (1987)
16. B. Grammaticos, A. Ramani and J. Hietarinta, J. Math. Phys. **31**, 2572 (1990)
17. J. Hietarinta, J. Math. Phys. **28**, 2094, 2586 (1987)
18. J.J.C. Nimmo, in "Applications of Analytic and Geometric Methods to Nonlinear Differential Equations", P. Clarkson (ed.), Kluwer Academic (1992), p. 183
19. J. Hietarinta, J. Math. Phys. **29**, 628 (1988)
20. J. Hietarinta, in "Nonlinear Evolution Equations: integrability and spectral methods", A. Degasperis, A.P. Fordy, and M. Lakshmanan (eds.), Manchester University Press (1990), p. 307
21. R. Hirota and Y. Ohta, J. Phys. Soc. Japan **60** 798 (1991)
22. S. Isojima, R. Willox, and J. Satsuma, J. Phys. A: Math. Gen. **35** 6893 (2002)

Lie Bialgebras, Poisson Lie Groups, and Dressing Transformations

Y. Kosmann-Schwarzbach

Centre de Mathématiques, UMR 7640 du CNRS, École Polytechnique,
91128 Palaiseau, France, yks@math.polytechnique.fr

Abstract. In this course, we present an elementary introduction, including the proofs of the main theorems, to the theory of Lie bialgebras and Poisson Lie groups and its applications to the theory of integrable systems. We discuss r-matrices, the classical and modified Yang-Baxter equations, and the tensor notation. We study the dual and double of Poisson Lie groups, and the infinitesimal and global dressing transformations.

Introduction

What we shall study in these lectures are classical objects, not in the sense that they date back to the nineteenth century, but in the sense that they admit a quantum counterpart. In fact, the theory of Lie bialgebras and Poisson Lie groups, due for the most part to V.G. Drinfeld and M.A. Semenov-Tian-Shansky, dates back to the early 80's, while the concept of a classical r-matrix was introduced a few years earlier by E.K. Sklyanin. It is somewhat surprising that these structures first appeared as the classical limit (the expression "semi-classical limit" is sometimes used instead) of the mathematical structures underlying the quantum inverse scattering method (QISM) developed by L. Faddeev and his school in St. Petersburg (then Leningrad). The commutativity property of the row-to-row transfer matrices for solvable lattice models was found to be a consequence of the existence of the so-called quantum R-matrix, which figures in the now famous equation,

$$RT_1T_2 = T_2T_1R \ .$$

Such an R-matrix satisifies the equation,

$$R_{12}R_{13}R_{23} = R_{23}R_{13}R_{12} \ ,$$

called the *quantum Yang-Baxter equation (QYBE)*. The notion of a *quantum group*, a deformation of either the algebra of functions on a Lie group or the universal enveloping algebra of the associated Lie algebra, evolved from these considerations.

Just as quantum R-matrices and quantum groups play an important role in QISM, their classical limits, classiical r-matrices and Poisson Lie groups enter into the theory of classical integrable systems.

Poisson Lie groups are Lie groups equipped with an additional structure, a Poisson bracket satisfying a compatibility condition with the group multiplication. The infinitesimal object associated with a Poisson Lie group is the tangent

Y. Kosmann-Schwarzbach, Lie Bialgebras, Poisson Lie Groups, and Dressing Transformations, Lect. Notes Phys. **638**, 107–173 (2004)
http://www.springerlink.com/

vector space at the origin of the group, which is, in a natural way, a Lie algebra, $\mathfrak{g}$. The Poisson structure on the group induces on the Lie algebra an additional structure, which is nothing but a Lie algebra structure on the dual vector space $\mathfrak{g}^*$ satisfying a compatibility condition with the Lie bracket on $\mathfrak{g}$ itself. Such a Lie algebra together with its additional structure is called a *Lie bialgebra*. In most applications, the group is a group of matrices, while its Lie algebra is also an algebra of matrices of the same size, say $p \times p$. The classical r-matrices are then matrices of size $p^2 \times p^2$. What is the relationship between such a matrix and the notion of bialgebra? The answer involves taking the Lie-algebra coboundary of the r-matrix (see Sect. 2.1). Another way to explain this relationship is as follows: Assume that we can identify the Lie algebra with its dual vector space by means of an invariant scalar product. Considering a Lie algebra structure on the dual vector space then amounts to considering a second Lie algebra structure on $\mathfrak{g}$ itself. When the $p^2 \times p^2$ r-matrix is identified with a linear map from $\mathfrak{g}$ to itself, which we denote by R, the second Lie bracket is given by

$$[x, y]_R = [Rx, y] + [x, Ry] \ .$$

The *modified Yang-Baxter equation (MYBE)* is a sufficient condition for R to define a second Lie bracket on $\mathfrak{g}$ by this formula, while the *classical Yang-Baxter equation (CYBE)*,

$$[r_{12}, r_{13}] + [r_{12}, r_{23}] + [r_{13}, r_{23}] = 0 \ ,$$

is the r-matrix version of this condition and is in fact obtained as a limit of the quantum Yang-Baxter equation.

In these lectures, we have tried to give a self-contained account of the theory of Lie bialgebras and Poisson Lie groups, including the basic definitions concerning the adjoint and coadjoint representations, Lie-algebra cohomology, Poisson manifolds and the Lie-Poisson structure of the dual of a Lie algebra, to explain all notations, including the 'tensor notation' which is ubiquitous in the physics literature, and to present the proofs of all the results. We have included the definitions of Manin triples, coboundary Lie bialgebras (triangular, quasi-triangular and factorizable r-matrices), as well as the corresponding notions for Poisson Lie groups.

The examples that we discuss in detail are elementary, so we refer to the literature for a wealth of further examples.

Among the properties of Lie bialgebras and Lie groups, the existence of the dual and of the double of a Lie bialgebra, the integration theorem of a Lie bialgebra into a Poisson Lie group, whence the existence of the dual and of the double of a Poisson Lie group, are the most important. In the proofs, we use the Schouten bracket, whose importance in this theory was first pointed out by Gelfand and Dorfman [17] and emphasized by Magri and myself [38–40].

Semenov-Tian-Shansky's theorem, generalizing the theorem of Adler-Kostant-Symes, has important applications to the theory of integrable Hamiltonian systems. It establishes that, when a Lie algebra with an invariant scalar product is equipped with an R-matrix, the dynamical systems defined by an invariant function are in Lax form, and possess conserved quantities in involution (see Sect. 3.6). Whenever the original Lie algebra splits as a direct sum of two Lie subalgebras, the difference of the projections onto the Lie subalgebras (or a scalar multiple of this difference) is an R-matrix. (This occurs in many cases. The infinite-dimensional examples are the most interesting for the applications, but require some extension of the theory presented here. See [27,28].) In this situation, such Lax equations can be solved by factorization, a method which replaces an initial-value problem with a problem of factorization in the associated Poisson Lie group. This is the reason for the name "factorizable R-matrix" given to the solutions of the modified Yang-Baxter equation. Thus, the R-matrix formalism can be considered to be an infinitesimal version of the Riemann-Hilbert factorization problem.

An essential ingredient of this theory and its applications to integrable systems is the notion of a Poisson action (of a Poisson Lie group on a Poisson manifold). It is a new concept which reduces to that of a Hamiltonian action when the Poisson structure on the Lie group vanishes. It was necessary to introduce such a generalization of Hamiltonian actions in order to account for the properties of the dressing transformations, under the "hidden symmetry group", of fields satisfying a zero-curvature equation. There are naturally defined actions of any Poisson Lie group on the dual Lie group, and conversely, and these are Poisson actions. (We give a one-line proof of this fact in Appendix 2, using the Poisson calculus.) In the case of a Poisson Lie group defined by a factorizable R-matrix, the explicit formulæ for these *dressing actions* coincide with the dressing of fields that are solutions of zero-curvature equations. There is a notion of momentum mapping for Poisson actions, and in this case it coincides with the monodromy matrix of the linear system. This establishes the connection between soliton equations which admit a zero-curvature representation (the compatibility condition for an auxiliary linear problem) in which the wave function takes values in a group and the theory of Poisson Lie groups.

In the bibliography, we have given references to

A. a variety of books from which all the prerequisites for a study of these lectures, and much more, can be learnt,

B. some of the articles that founded the subject, Drinfeld [15,16], Gelfand and Dorfman [17], Semenov-Tian-Shansky [18,19].

C. later expositions in books and surveys, *e.g.*, Reyman and Semenov-Tian-Shansky [27], Chari and Pressley [22], Vaisman [30], Reyman [26] and, for a survey of the Lie-algebraic approach to integrable systems, Perelomov [25].

D. a few of the most relevant publications that have further developed the subject, foremost among which is the article by Lu and Weinstein [44].

These lecture notes are meant as the necessary background for the survey of "Quantum and classical integrable systems" by Semenov-Tian-Shansky in this volume,[1] where he briefly recalls the notions and results that we explain here, and some of their generalizations, and uses them extensively in the study of classical integrable systems, before studying their quantum counterpart.

These notes have been revised for clarity, and misprints have been corrected for the second edition of this book.

1 Lie Bialgebras

We shall study Lie algebras $\mathfrak{g}$ whose dual vector space $\mathfrak{g}^*$ carries a Lie-algebra structure satisfying a compatibility condition, to be described in Sect. 1.3, with that of $\mathfrak{g}$ itself. Such objects are called *Lie bialgebras*. The corresponding Lie groups carry a Poisson structure compatible with the group multiplication (see Sects. 3 and 4). They constitute the semi-classical limit of quantum groups. In this section we shall study the general, abstract framework, and in Sect. 2 we shall describe the Lie-bialgebra structures defined by r-matrices, *i.e.*, solutions of the classical Yang-Baxter equation.

1.1 An Example: $\mathfrak{sl}(2, \mathbb{C})$

Let us consider the Lie algebra $\mathfrak{g} = \mathfrak{sl}(2, \mathbb{C})$, with basis

$$H = \begin{pmatrix} 1 & 0 \\ 0 & -1 \end{pmatrix}, \ X = \begin{pmatrix} 0 & 1 \\ 0 & 0 \end{pmatrix}, \ Y = \begin{pmatrix} 0 & 0 \\ 1 & 0 \end{pmatrix}$$

and commutation relations,

$$[H, X] = 2X, \ \ [H, Y] = -2Y, \ \ [X, Y] = H \ .$$

The dual vector space $\mathfrak{g}^*$ has the dual basis H^*, X^*, Y^*, where, by definition, $\langle H^*, H \rangle = 1$, $\langle X^*, X \rangle = 1$, $\langle Y^*, Y \rangle = 1$, and all other duality brackets are 0. We shall consider the following commutation relations on $\mathfrak{g}^*$,

$$[H^*, X^*]_{\mathfrak{g}^*} = \frac{1}{4} X^*, \ \ [H^*, Y^*]_{\mathfrak{g}^*} = \frac{1}{4} Y^*, \ \ [X^*, Y^*]_{\mathfrak{g}^*} = 0 \ .$$

---

[1] *Remark on notations and conventions.* Throughout these lectures, we reserve the term r-matrix on $\mathfrak{g}$ for elements of $\mathfrak{g} \otimes \mathfrak{g}$, and the term R-matrix for endomorphisms of $\mathfrak{g}$, while Semenov-Tian-Shansky uses the same letter r, and the term r-matrix, in both cases. In any case, these classical r-matrices and R-matrices should not be confused with the quantum R-matrices satisfying the quantum Yang-Baxter equation.

The bracket that we have associated here with a given R-matrix is twice the one which is defined in Semenov-Tian-Shansky's lectures. However, in the important special case where the second bracket is obtained as a result of the splitting of a Lie algebra into complementary Lie subalgebras, this bracket coincides with the bracket in his lectures, because the R-matrix that we consider is one-half of the difference of the projections, while he considers the difference of the projections.

Now consider $\mathfrak{g} \oplus \mathfrak{g}^*$. We can turn $\mathfrak{g} \oplus \mathfrak{g}^*$ into a Lie algebra, denoted by $\mathfrak{g} \bowtie \mathfrak{g}^*$ or $\mathfrak{d}$, and called the *double* of $\mathfrak{g}$, such that both $\mathfrak{g}$ and $\mathfrak{g}^*$ are Lie subalgebras of $\mathfrak{d}$, by setting

$$[H, H^*]_\mathfrak{d} = 0, \quad [H, X^*]_\mathfrak{d} = -2X^*, \quad [H, Y^*]_\mathfrak{d} = 2Y^*$$
$$[X, H^*]_\mathfrak{d} = \tfrac{1}{4}X - Y^*, \quad [X, X^*]_\mathfrak{d} = -\tfrac{1}{4}H + 2H^*, \quad [X, Y^*]_\mathfrak{d} = 0$$
$$[Y, H^*]_\mathfrak{d} = \tfrac{1}{4}Y + X^*, \quad [Y, X^*]_\mathfrak{d} = 0, \quad [Y, Y^*]_\mathfrak{d} = -\tfrac{1}{4}H - 2H^*.$$

We can prove the following facts :

(i) This is a Lie-algebra bracket on $\mathfrak{d}$, since we can show that it satisfies the Jacobi identity.

(ii) There is a natural scalar product $(\,|\,)$ on $\mathfrak{d}$, defined by

$$(H|H^*) = (X|X^*) = (Y|Y^*) = 1 \ ,$$

while all other scalar products of elements in the basis vanish. With respect to this scalar product, $\mathfrak{g}$ and $\mathfrak{g}^*$ are isotropic, because the definition of an *isotropic subspace* is that the scalar product vanishes on it.

(iii) The scalar product is invariant for the Lie-algebra structure of $\mathfrak{d}$ defined above. Recall that a scalar product $(\,|\,)$ on a Lie algebra $\mathfrak{a}$ with bracket $[\,,\,]$ is called *invariant* if, for any $u, v, w \in \mathfrak{a}$,

$$([u,v]|w) = (u|[v,w]) \ .$$

1.2   Lie-Algebra Cohomology

In order to formulate the definition and properties of Lie bialgebras in general, we shall need a few definitions from the theory of Lie-algebra cohomology.

Let $\mathfrak{g}$ be a Lie algebra over the field of complex or real numbers. When M is the vector space of a representation ρ of $\mathfrak{g}$, we say that $\mathfrak{g}$ *acts on* M, or that M is a $\mathfrak{g}$-*module*. For $x \in \mathfrak{g}$, $a \in M$, we often denote $(\rho(x))(a)$ simply by $x.a$.

Examples. Any Lie algebra $\mathfrak{g}$ acts on itself by the adjoint representation, ad : $x \in \mathfrak{g} \mapsto ad_x \in \mathrm{End}\ \mathfrak{g}$, defined, for $y \in \mathfrak{g}$, by $ad_x(y) = [x, y]$.

More generally, $\mathfrak{g}$ acts on any tensor product of $\mathfrak{g}$ with itself in the following way. For decomposable elements, $y_1 \otimes \cdots \otimes y_p$ in $\overset{p}{\otimes} \mathfrak{g} = \mathfrak{g} \otimes \cdots \otimes \mathfrak{g}$ (p times),

$$x \cdot (y_1 \otimes \cdots \otimes y_p) = ad_x^{(p)}(y_1 \otimes \cdots \otimes y_p)$$
$$= ad_x y_1 \otimes y_2 \otimes \cdots \otimes y_p + y_1 \otimes ad_x y_2 \otimes y_3 \otimes \cdots \otimes y_p + \cdots$$
$$+ y_1 \otimes y_2 \otimes \cdots \otimes y_{p-1} \otimes ad_x y_p \ .$$

For example, for $p = 2$,

$$ad_x^{(2)}(y_1 \otimes y_2) = ad_x y_1 \otimes y_2 + y_1 \otimes ad_x y_2 = [x, y_1] \otimes y_2 + y_1 \otimes [x, y_2].$$

Thus, denoting the identity map from $\mathfrak{g}$ to $\mathfrak{g}$ by 1,

$$ad_x^{(2)} = ad_x \otimes 1 + 1 \otimes ad_x \ .$$

Since, when $y \in \mathfrak{g}$, $ad_x y = [x, y]$, one often writes

$$(ad_x \otimes 1 + 1 \otimes ad_x)(u) = [x \otimes 1 + 1 \otimes x, u] \qquad (1.1)$$

for $u \in \mathfrak{g} \otimes \mathfrak{g}$.

Now, let $\mathfrak{g}$ be a finite-dimensional Lie algebra, and let $(e_1, \cdots, e_n)$ be a basis of $\mathfrak{g}$. Using the Einstein summation convention, any element b in $\mathfrak{g} \otimes \mathfrak{g}$ can be written, $b = b^{ij} e_i \otimes e_j$, and then

$$ad_x^{(2)} b = b^{ij} ([x, e_i] \otimes e_j + e_i \otimes [x, e_j]) .$$

We could expand this quantity further in terms of the structure constants of the Lie algebra $\mathfrak{g}$ and the components of x.

Similarly, $\mathfrak{g}$ acts on the p-th exterior power of $\mathfrak{g}$, $\bigwedge^p \mathfrak{g}$, for any p. For example, for $p = 2$,

$$x.(y_1 \wedge y_2) = [x, y_1] \wedge y_2 + y_1 \wedge [x, y_2] ,$$

$$x.(\sum_{i<j} a^{ij} e_i \wedge e_j) = \sum_{i<j} a^{ij} ([x, e_i] \wedge e_j + e_i \wedge [x, e_j]) .$$

Definition. *For each nonnegative integer k, the vector space of skew-symmetric k-linear mappings on $\mathfrak{g}$ with values in M, where M is the vector space of a representation of $\mathfrak{g}$, is called the space of k-cochains on $\mathfrak{g}$ with values in M.*

A 1-cochain on $\mathfrak{g}$ with values in M is just a linear map from $\mathfrak{g}$ to M, while a 0-cochain on $\mathfrak{g}$ with values in M is an element of M.

We can now define the coboundary of a k-cochain u on $\mathfrak{g}$ with values in M, denoted by δu. Since we shall need only the cases where $k = 0$ or 1, we shall first write the definition in these two cases,

$$k = 0, \ u \in M, \ x \in \mathfrak{g} , \ \delta u(x) = x.u ,$$

$$k = 1, \ v : \mathfrak{g} \to M, \ x, y \in \mathfrak{g}, \ \delta v(x, y) = x.v(y) - y.v(x) - v([x, y]).$$

We immediately observe that for any 0-cochain u on $\mathfrak{g}$ with values in M,

$$\delta(\delta u) = 0 .$$

In fact, for $x, y \in \mathfrak{g}$,

$$(\delta(\delta u))(x, y) = x.(y.u) - y.(x.u)) - [x, y].u ,$$

and this quantity vanishes identically because $x \mapsto \rho(x)$ is a representation of $\mathfrak{g}$ in M. More generally,

Definition. *The coboundary of a k-cochain u on $\mathfrak{g}$ with values in M is the $(k+1)$-cochain, δu, with values in M defined by*

$$\delta u(x_0, x_1, \cdots, x_k) = \sum_{i=0}^{k} (-1)^i x_i.(u(x_0, \cdots, \hat{x}_i, \cdots, x_k))$$

$$+ \sum_{i<j} (-1)^{i+j} u([x_i, x_j], x_0, \cdots, \hat{x}_i, \cdots, \hat{x}_j, \cdots, x_k),$$

for $x_0, x_1, \cdots, x_k \in \mathfrak{g}$, where $\hat{x}_i$ indicates that the element x_i is omitted.

Proposition. *The property $\delta(\delta u) = 0$ is valid for any k-cochain u, $k \geq 0$.*

This is a standard result, which generalizes the property proved above for $k = 0$.

Definition. *A k-cochain u is called a k-cocycle if $\delta u = 0$. A k-cochain u $(k \geq 1)$ is called a k-coboundary if there exists a $(k-1)$-cochain, v, such that $u = \delta v$.*

By the proposition, any k-coboundary is a k-cocycle. By definition, the quotient of the vector space of k-cocycles by the vector space of k-coboundaries is called the k-th *cohomology vector space* of $\mathfrak{g}$, with values in M.

Remark. The 0-cocyles of $\mathfrak{g}$ with values in M are the invariant elements in M, *i.e.*, the elements $u \in M$ such that $x.u = 0$, for each $x \in \mathfrak{g}$.

1.3   Definition of Lie Bialgebras

Let us now assume that $\mathfrak{g}$ is a Lie algebra and that γ is a linear map from $\mathfrak{g}$ to $\mathfrak{g} \otimes \mathfrak{g}$ whose transpose we denote by ${}^t\gamma : \mathfrak{g}^* \otimes \mathfrak{g}^* \to \mathfrak{g}^*$. (If $\mathfrak{g}$ is infinite-dimensional, $\mathfrak{g}^* \otimes \mathfrak{g}^*$ is a subspace of $(\mathfrak{g} \otimes \mathfrak{g})^*$, and what we are considering is in fact the restriction of the transpose of γ.) Recall that a linear map on $\mathfrak{g}^* \otimes \mathfrak{g}^*$ can be identified with a bilinear map on $\mathfrak{g}^*$.

Definition. *A Lie bialgebra is a Lie algebra $\mathfrak{g}$ with a linear map $\gamma : \mathfrak{g} \to \mathfrak{g} \otimes \mathfrak{g}$ such that*

(i) ${}^t\gamma : \mathfrak{g}^* \otimes \mathfrak{g}^* \to \mathfrak{g}^*$ defines a Lie bracket on $\mathfrak{g}^*$, i.e., is a skew-symmetric bilinear map on $\mathfrak{g}^*$ satisfying the Jacobi identity, and

(ii) γ is a 1-cocycle on $\mathfrak{g}$ with values in $\mathfrak{g} \otimes \mathfrak{g}$, where $\mathfrak{g}$ acts on $\mathfrak{g} \otimes \mathfrak{g}$ by the adjoint representation $ad^{(2)}$.

Condition *(ii)* means that the 2-cochain $\delta\gamma$ vanishes, *i.e.*, for $x, y \in \mathfrak{g}$,

$$ad_x^{(2)}(\gamma(y)) - ad_y^{(2)}(\gamma(x)) - \gamma([x,y]) = 0 . \tag{ii'}$$

Let us introduce the notation

$$[\xi, \eta]_{\mathfrak{g}^*} = {}^t\gamma(\xi \otimes \eta) ,$$

for $\xi, \eta \in \mathfrak{g}^*$. Thus, by definition, for $x \in \mathfrak{g}$,

$$\langle [\xi, \eta]_{\mathfrak{g}^*}, x \rangle = \langle \gamma(x), \xi \otimes \eta \rangle .$$

Condition *(i)* is equivalent to the following,

$$\begin{cases} [\xi, \eta]_{\mathfrak{g}^*} = -[\eta, \xi]_{\mathfrak{g}^*}, \\ [\xi, [\eta, \zeta]_{\mathfrak{g}^*}]_{\mathfrak{g}^*} + [\eta, [\zeta, \xi]_{\mathfrak{g}^*}]_{\mathfrak{g}^*} + [\zeta, [\xi, \eta]_{\mathfrak{g}^*}]_{\mathfrak{g}^*} = 0 , \end{cases}$$

An alternate way of writing condition $(ii\,')$ is

$$\langle [\xi,\eta]_{\mathfrak{g}^*}, [x,y] \rangle = \langle \xi \otimes \eta, (ad_x \otimes 1 + 1 \otimes ad_x)(\gamma(y)) \rangle$$
$$- \langle \xi \otimes \eta, (ad_y \otimes 1 + 1 \otimes ad_y)(\gamma(x)) \rangle \; .$$

(Recall that, by definition, $(ad_x \otimes 1)(y_1 \otimes y_2) = [x, y_1] \otimes y_2$.)

1.4   The Coadjoint Representation

We now introduce the important definition of the coadjoint representation of a Lie algebra on the dual vector space.

Let $\mathfrak{g}$ be a Lie algebra and let $\mathfrak{g}^*$ be its dual vector space. For simplicity, we shall assume that $\mathfrak{g}$ is finite-dimensional. For $x \in \mathfrak{g}$, we set

$$ad_x^* = -{}^t(ad_x) \; .$$

Thus ad_x^* is the endomorphism of $\mathfrak{g}^*$ satisfying

$$\langle \xi, ad_x y \rangle = -\langle ad_x^* \xi, y \rangle \; ,$$

for $y \in \mathfrak{g}$, $\xi \in \mathfrak{g}^*$. Then, it is easy to prove that the map $x \in \mathfrak{g} \mapsto ad_x^* \in \mathrm{End}\,\mathfrak{g}^*$ is a representation of $\mathfrak{g}$ in $\mathfrak{g}^*$.

Definition. *The representation* $x \mapsto ad_x^*$ *of* $\mathfrak{g}$ *in* $\mathfrak{g}^*$ *is called the* coadjoint re-presentation *of* $\mathfrak{g}$.

1.5   The Dual of a Lie Bialgebra

In the notation of the preceding section, $(ii\,')$ of Sect. 1.3 can be written

$$\langle [\xi,\eta]_{\mathfrak{g}^*}, [x,y] \rangle + \langle [ad_x^*\xi, \eta]_{\mathfrak{g}^*}, y \rangle + \langle [\xi, ad_x^*\eta]_{\mathfrak{g}^*}, y \rangle$$

$$-\langle [ad_y^*\xi, \eta]_{\mathfrak{g}^*}, x \rangle - \langle [\xi, ad_y^*\eta]_{\mathfrak{g}^*}, x \rangle = 0 \; .$$

We now see that there is a symmetry between $\mathfrak{g}$ with its Lie bracket $[\,,\,]$ and $\mathfrak{g}^*$ with its Lie bracket $[\,,\,]_{\mathfrak{g}^*}$ defined by γ. In the same fashion as above, let us set

$$ad_\xi \eta = [\xi, \eta]_{\mathfrak{g}^*}$$

and

$$\langle ad_\xi \eta, x \rangle = -\langle \eta, ad_\xi^* x \rangle \; ,$$

for $\xi, \eta \in \mathfrak{g}^*, x \in \mathfrak{g}$. Then $\xi \in \mathfrak{g}^* \mapsto ad_\xi^* \in \mathrm{End}\,\mathfrak{g}$ is the coadjoint representation of $\mathfrak{g}^*$ in the dual of $\mathfrak{g}^*$ which is isomorphic to $\mathfrak{g}$.

Now relation (ii) of Sect. 1.3 is equivalent to

$$\langle [\xi,\eta]_{\mathfrak{g}^*}, [x,y] \rangle + \langle ad_x^*\xi, ad_\eta^* y \rangle - \langle ad_x^*\eta, ad_\xi^* y \rangle \qquad (ii'')$$

$$-\langle ad_y^*\xi, ad_\eta^* x \rangle + \langle ad_y^*\eta, ad_\xi^* x \rangle = 0 \; .$$

It is now obvious that $\mathfrak{g}$ and $\mathfrak{g}^*$ play symmetric roles. Let us call $\mu : \mathfrak{g} \otimes \mathfrak{g} \to \mathfrak{g}$ the skew-symmetric bilinear mapping on $\mathfrak{g}$ defining the Lie bracket of $\mathfrak{g}$. Transforming relation (ii'') again, it is easy to see that it is equivalent to the condition that ${}^t\mu : \mathfrak{g}^* \to \mathfrak{g}^* \otimes \mathfrak{g}^*$ be a 1-cocycle on $\mathfrak{g}^*$ with values in $\mathfrak{g}^* \otimes \mathfrak{g}^*$, where $\mathfrak{g}^*$ acts on $\mathfrak{g}^* \otimes \mathfrak{g}^*$ by the adjoint action. In fact, since the left-hand side of (ii'') is

$$\langle {}^t\mu[\xi,\eta]_{\mathfrak{g}^*}, x \otimes y\rangle - \langle \xi, [x, ad_\eta^* y]\rangle + \langle \eta, [x, ad_\xi^* y]\rangle$$

$$+ \langle \xi, [y, ad_\eta^* x]\rangle - \langle \eta, [y, ad_\xi^* x]\rangle ,$$

condition (ii) is equivalent to

$$\langle {}^t\mu[\xi,\eta]_{\mathfrak{g}^*}, x \otimes y\rangle + \langle (ad_\eta \otimes 1 + 1 \otimes ad_\eta)({}^t\mu(\xi)), x \otimes y\rangle$$

$$- \langle (ad_\xi \otimes 1 + 1 \otimes ad_\xi)({}^t\mu(\eta)), x \otimes y\rangle = 0$$

or

$$ad_\xi^{(2)}(({}^t\mu)(\eta)) - ad_\eta^{(2)}(({}^t\mu)(\xi)) - ({}^t\mu)([\xi,\eta]_{\mathfrak{g}^*}) = 0 .$$

Therefore

Proposition. *If $(\mathfrak{g}, \gamma)$ is a Lie bialgebra, and μ is the Lie bracket of $\mathfrak{g}$, then $(\mathfrak{g}^*, {}^t\mu)$ is a Lie bialgebra, where ${}^t\gamma$ is the Lie bracket of $\mathfrak{g}^*$.*

By definition $(\mathfrak{g}^*, {}^t\mu)$ is called the *dual* of Lie bialgebra $(\mathfrak{g}, \gamma)$. Thus each Lie bialgebra has a dual Lie bialgebra whose dual is the Lie bialgebra itself.

1.6   The Double of a Lie Bialgebra. Manin Triples

Proposition. *Let $(\mathfrak{g}, \gamma)$ be a Lie bialgebra with dual $(\mathfrak{g}^*, {}^t\mu)$. There exists a unique Lie-algebra structure on the vector space $\mathfrak{g} \oplus \mathfrak{g}^*$ such that $\mathfrak{g}$ and $\mathfrak{g}^*$ are Lie subalgebras and that the natural scalar product on $\mathfrak{g} \oplus \mathfrak{g}^*$ is invariant.*

Proof. The natural scalar product on $\mathfrak{g} \oplus \mathfrak{g}^*$ is defined by

$$(x|y) = 0 \;,\; (\xi|\eta) = 0 \;,\; (x|\xi) = \langle \xi, x\rangle \;,\text{ for } x, y \in \mathfrak{g},\; \xi, \eta \in \mathfrak{g}^* .$$

Let us denote by $[u, v]_{\mathfrak{d}}$ the Lie bracket of two elements u, v in $\mathfrak{d} = \mathfrak{g} \oplus \mathfrak{g}^*$. By the invariance condition on the natural scalar product and by the fact that $\mathfrak{g}$ is a Lie subalgebra, we obtain

$$(y|[x, \xi]_{\mathfrak{d}}) = ([y, x]_{\mathfrak{d}}|\xi) = ([y, x]|\xi)$$

$$= \langle \xi, [y, x]\rangle = \langle ad_x^* \xi, y\rangle = (y|ad_x^* \xi) ,$$

and similarly $(\eta|[x, \xi]_{\mathfrak{d}}) = -(\eta|ad_\xi^* x)$, which proves that $[x, \xi]_{\mathfrak{d}} = -ad_\xi^* x + ad_x^* \xi$. One must now prove that the formulæ

$$\begin{cases} [x, y]_{\mathfrak{d}} = [x, y] \\ [x, \xi]_{\mathfrak{d}} = -ad_\xi^* x + ad_x^* \xi \\ [\xi, \eta]_{\mathfrak{d}} = [\xi, \eta]_{\mathfrak{g}^*} \end{cases} \tag{1.2}$$

define a Lie-algebra structure on $\mathfrak{g} \oplus \mathfrak{g}^*$. The proof of the Jacobi identity uses conditions (i) and (ii) of the definition of a Lie bialgebra.

Definition. *When $\mathfrak{g}$ is a Lie bialgebra, $\mathfrak{g} \oplus \mathfrak{g}^*$ equipped with the Lie bracket $[\, , \,]_\mathfrak{d}$ defined by (1.2) is called the* double *of $\mathfrak{g}$, and denoted $\mathfrak{g} \bowtie \mathfrak{g}^*$ or $\mathfrak{d}$.*

We described an example in Sect. 1.1.

Note that $\mathfrak{d} = \mathfrak{g} \bowtie \mathfrak{g}^*$ is also the double of $\mathfrak{g}^*$. In the Lie algebra $\mathfrak{d}$, the subspaces $\mathfrak{g}$ and $\mathfrak{g}^*$ are complementary Lie subalgebras, and both are isotropic, *i.e.*, the scalar product vanishes on $\mathfrak{g}$ and on $\mathfrak{g}^*$. Thus we see that, for any Lie bialgebra $\mathfrak{g}$, $(\mathfrak{d}, \mathfrak{g}, \mathfrak{g}^*)$ is an example of a Manin triple, defined as follows:

Definition. *A* Manin triple *is a triple $(\mathfrak{p}, \mathfrak{a}, \mathfrak{b})$, where $\mathfrak{p}$ is a Lie algebra with an invariant, non-degenerate, symmetric bilinear form, and $\mathfrak{a}$ and $\mathfrak{b}$ are complementary isotropic Lie subalgebras.*

In the finite-dimensional case, we can show that, conversely, when $(\mathfrak{p}, \mathfrak{a}, \mathfrak{b})$ is a Manin triple, $\mathfrak{a}$ has a Lie-bialgebra structure. Since $\mathfrak{a}$ and $\mathfrak{b}$ play symmetric roles, $\mathfrak{b}$ also has a Lie-bialgebra structure, and the Lie bialgebra $\mathfrak{b}$ can be identified with the dual of the Lie bialgebra $\mathfrak{a}$. Let $(\, | \,)$ be the given scalar product on $\mathfrak{p}$. To $b \in \mathfrak{b}$ we associate the 1-form $\iota(b)$ on $\mathfrak{a}$ defined by $\iota(b)(a) = (a|b)$. The linear map $b \mapsto \iota(b)$ from $\mathfrak{b}$ to $\mathfrak{a}^*$ is injective. In fact, if $\iota(b) = 0$, then $(a|b) = 0$ for all $a \in \mathfrak{a}$, and therefore for all $a \in \mathfrak{p}$, since $\mathfrak{b}$ is isotropic and $\mathfrak{p} = \mathfrak{a} \oplus \mathfrak{b}$. By the non-degeneracy of the scalar product, we find that $b = 0$. Counting dimensions, we see that $\mathfrak{b}$ is isomorphic to $\mathfrak{a}^*$. The Lie bracket on $\mathfrak{b}$ therefore defines a Lie bracket on $\mathfrak{a}^*$. To see that it defines a Lie-bialgebra structure on $\mathfrak{a}$, we use the Jacobi identity in $\mathfrak{p}$, and the invariance of the scalar product. Thus

Theorem. *There is a one-to-one correspondence between finite-dimensional Lie bialgebras and finite-dimensional Manin triples.*

For a short, conceptual proof of this theorem, see Appendix 1.

1.7 Examples

Simple Lie Algebras over $\mathbb{C}$. Let $\mathfrak{g}$ be a simple Lie algebra over $\mathbb{C}$, of rank r, with Cartan subalgebra $\mathfrak{h}$, and with positive (resp., negative) Borel subalgebra $\mathfrak{b}_+$ (resp., $\mathfrak{b}_-$), generated by $\mathfrak{h}$ and positive (resp., negative) root vectors.

Set $\mathfrak{p} = \mathfrak{g} \oplus \mathfrak{g}$ (direct sum of Lie algebras); let $\mathfrak{p}_1$ be the diagonal subalgebra, and $\mathfrak{p}_2 = \{(x, y) \in \mathfrak{b}_- \oplus \mathfrak{b}_+ | \mathfrak{h}\text{-components of } x \text{ and } y \text{ are opposite}\}$. Define the scalar product of (x, y) and (x', y') to be $-\frac{1}{2}((x|x')_\mathfrak{g} - (y|y')_\mathfrak{g})$ where $(\, | \,)_\mathfrak{g}$ is the Killing form of $\mathfrak{g}$. (The factor $-\frac{1}{2}$ is conventional.) Then $(\mathfrak{p}, \mathfrak{p}_1, \mathfrak{p}_2)$ is a Manin triple, and the Lie brackets thus defined in $\mathfrak{g}^*$ can be explicitly written in terms of Weyl generators, (H_j, X_j, Y_j), $j = 1, \ldots, r$.

Let us illustrate this fact for $\mathfrak{g} = \mathfrak{sl}(2, \mathbb{C})$. The Killing form of $\mathfrak{g}$ is such that

$$(H|H)_\mathfrak{g} = 8 \, , \quad (X|Y)_\mathfrak{g} = 4 \, ,$$

and all other scalar products vanish. In this case $\mathfrak{b}_+$ (resp., $\mathfrak{b}_-$) is generated by H and X (resp., H and Y). A basis for $\mathfrak{p}_1$ is $e_1 = (H, H), e_2 = (X, X), e_3 = (Y, Y)$. A basis for $\mathfrak{p}_2$ is $f_1 = (H, -H), f_2 = (0, X), f_3 = (Y, 0)$. We find that

$$(e_i|e_j) = 0, \ (f_i|f_j) = 0, \ (e_1|f_1) = -8, \ (e_3|f_2) = 2, \ (e_2|f_3) = -2 \ ,$$

and that all other scalar products vanish. We now identitfy $\mathfrak{p}_2$ with $\mathfrak{p}_1^* \simeq \mathfrak{g}^*$ by means of the scalar product, and we denote this identification map by ι. Then $\iota(f_1) = -8H^*, \iota(f_2) = 2Y^*, \iota(f_3) = -2X^*$. Now

$$\begin{aligned}
[f_1, f_2] &= (0, -[H, X]) = -2(0, X) = -2f_2, \\
[f_1, f_3] &= ([H, Y], 0) = -2(Y, 0) = -2f_3, \\
[f_2, f_3] &= 0,
\end{aligned}$$

and therefore we recover the commutation relations for $\mathfrak{g}^*$ given in Sect. 1.1,

$$[H^*, X^*] = \frac{1}{4}X^*, \ [H^*, Y^*] = \frac{1}{4}Y^*, \ [X^*, Y^*] = 0 \ .$$

As a consequence, we see that the double of the Lie bialgebra $\mathfrak{sl}(2, \mathbb{C})$ is isomorphic to $\mathfrak{sl}(2, \mathbb{C}) \oplus \mathfrak{sl}(2, \mathbb{C})$.

Compact Lie Algebras. Let $\mathfrak{g}$ be a simple complex Lie algebra of rank r, with Cartan subalgebra $\mathfrak{h}$ and Weyl basis $(H_j, X_\alpha, Y_\alpha)$, $j = 1, \cdots, r$, and $\alpha \in \Delta_+$, where Δ_+ is the set of positive roots. Then the real linear span of

$$iH_j \ , \ X_\alpha - Y_\alpha, \ i(X_\alpha + Y_\alpha)$$

is a real subalgebra of $\mathfrak{g}^{\mathbb{R}}$, i.e., of $\mathfrak{g}$ considered as a real Lie algebra, denoted by $\mathfrak{k}$ and called the *compact form* of $\mathfrak{g}$. The real linear span of iH_j, $j = 1, \cdots, r$, is a Cartan subalgebra $\mathfrak{t}$ of $\mathfrak{k}$, and $\mathfrak{h} = \mathfrak{t} \oplus i\mathfrak{t}$. Let $\mathfrak{b} = i\mathfrak{t} \oplus \mathfrak{n}_+$, where $\mathfrak{n}_+$ is the Lie subalgebra generated by X_α, $\alpha \in \Delta_+$. Then $\mathfrak{b}$ is a solvable, real Lie subalgebra of $\mathfrak{g}^{\mathbb{R}}$, and

$$\mathfrak{g}^{\mathbb{R}} = \mathfrak{k} \oplus \mathfrak{b} \ .$$

Define the scalar product on $\mathfrak{g}^{\mathbb{R}}$, $(\ |\ )_{\mathfrak{g}^{\mathbb{R}}} = \text{Im}(\ |\ )_{\mathfrak{g}}$, where $(\ |\ )_{\mathfrak{g}}$ is the Killing form of $\mathfrak{g}$, and Im denotes the imaginary part of a complex number. Then $(\mathfrak{g}^{\mathbb{R}}, \mathfrak{k}, \mathfrak{b})$ is a Manin triple. Therefore $\mathfrak{k}$ (resp., $\mathfrak{b}$) is a Lie bialgebra with dual $\mathfrak{b}$ (resp., $\mathfrak{k}$).

We derive this Lie-bialgebra structure explicitly for $\mathfrak{g} = \mathfrak{sl}(2, \mathbb{C})$, in which case $\mathfrak{k} = \mathfrak{su}(2)$. Let

$$e_1 = iH, e_2 = X - Y, e_3 = i(X + Y), f_1 = H, f_2 = X, f_3 = iX \ ,$$

where H, X, Y are as in Sect. 1.1. Then (e_1, e_2, e_3) is a basis of $\mathfrak{su}(2)$, while (f_1, f_2, f_3) is a basis of the Lie subalgebra $\mathfrak{b}$ of $\mathfrak{g}^{\mathbb{R}}$ of complex, upper triangular 2×2 matrices, with real diagonal and vanishing trace. Thus $\mathfrak{sl}(2, \mathbb{C}) = \mathfrak{su}(2) \oplus \mathfrak{b}$. We observe that

$$e_1 = \begin{pmatrix} i & 0 \\ 0 & -i \end{pmatrix}, \quad e_2 = \begin{pmatrix} 0 & 1 \\ -1 & 0 \end{pmatrix}, \quad e_3 = \begin{pmatrix} 0 & i \\ i & 0 \end{pmatrix}.$$

We now identify $\mathfrak{b}$ with $\mathfrak{su}(2)^*$ by the mapping ι, where

$$\iota(Z)(T) = \mathrm{Im}(Z|T)_{\mathfrak{g}} .$$

Then $\iota(f_1) = 8e_1^*$, $\iota(f_2) = 4e_3^*$, $\iota(f_3) = -4e_2^*$, where (e_1^*, e_2^*, e_3^*) is the basis dual to the basis (e_1, e_2, e_3) of $\mathfrak{su}(2)$. Since $[f_1, f_2] = 2f_2$, $[f_1, f_3] = 2f_3$, $[f_2, f_3] = 0$, we find that

$$[e_1^*, e_2^*] = \frac{1}{4}e_2^*, \quad [e_1^*, e_3^*] = \frac{1}{4}e_3^*, \quad [e_2^*, e_3^*] = 0 .$$

These are the commutation relations of $\mathfrak{su}(2)^*$, which is a solvable Lie algebra. Had we chosen the scalar product $tr X_1 X_2$ on $\mathfrak{g}$ instead of the Killing form, we would have obtained, as the commutation relations of $\mathfrak{su}(2)^*$,

$$[e_1^*, e_2^*] = e_2^*, \quad [e_1^*, e_3^*] = e_3^* , \quad [e_2^*, e_3^*] = 0 .$$

Remark. Let γ_0 be the 1-cocycle on $\mathfrak{k}$ with values in $\bigwedge^2 \mathfrak{k}$ defining the above Lie-bialgebra structure of $\mathfrak{k}$. It can be shown (see Soibelman [49]) that the most general Lie-bialgebra structure on the compact Lie algebra $\mathfrak{k}$ is

$$\gamma = \lambda\gamma_0 + \delta u ,$$

where λ is a real constant and u is an arbitrary element of $\bigwedge^2 \mathfrak{k}$. (Recall that δu was defined in Sect. 1.2.)

Infinite-Dimensional Lie Bialgebras. The construction given for simple Lie algebras is also valid for Kac-Moody algebras.

Let $\mathfrak{a}$ be a finite-dimensional simple Lie algebra over $\mathbb{C}$, and let $\mathfrak{p} = \mathfrak{a}((u^{-1}))$ be the Lie algebra of the $\mathfrak{a}$-valued Laurent series in u^{-1}, $\mathfrak{p}_1 = \mathfrak{a}[u]$ the Lie subalgebra of $\mathfrak{a}$-valued polynomials in u, and $\mathfrak{p}_2 = u^{-1}\mathfrak{a}[[u^{-1}]]$, the Lie algebra of $\mathfrak{a}$-valued formal series in u^{-1}, with no constant term. Given $f, g \in \mathfrak{p}$, we define their scalar product to be the coefficient of u^{-1} in the scalar-valued Laurent series in u^{-1} obtained by taking the scalar product of the coefficients by means of the Killing form of $\mathfrak{a}$.

Then $(\mathfrak{p}, \mathfrak{p}_1, \mathfrak{p}_2)$ is a Manin triple. The corresponding 1-cocycle γ on $\mathfrak{p}_1 = \mathfrak{a}[u]$ can be written as follows. Since $(\mathfrak{a} \otimes \mathfrak{a})[u, v] \simeq \mathfrak{a}[u] \otimes \mathfrak{a}[v]$, for any $f \in \mathfrak{p}_1$, $\gamma(f)$ is an $(\mathfrak{a} \otimes \mathfrak{a})$-valued polynomial in two variables that can be expressed in terms of t, the Killing form of $\mathfrak{a}$, viewed as an element of $\mathfrak{a} \otimes \mathfrak{a}$. (A priori, t is an element of $\mathfrak{a}^* \otimes \mathfrak{a}^*$, but, by means of the Killing form itself, this twice covariant tensor can be mapped to a twice contravariant tensor.) In fact,

$$(\gamma(f))(u, v) = (ad_{f(u)} \otimes 1 + 1 \otimes ad_{f(v)})\frac{t}{u - v},$$

where 1 is the identity map of $\mathfrak{a}$ onto itself. If (I^k) is an orthonormal basis of $\mathfrak{a}$ with respect to the Killing form, then $t = \sum_k I^k \otimes I^k$, and

$$(\gamma(f))(u,v) = \frac{1}{u-v} \sum_k ([f(u), I^k] \otimes I^k + I^k \otimes [f(v), I^k]).$$

1.8 Bibliographical Note

For this section, see Drinfeld [15,16], Chari and Pressley [22], Chapter 1 or Vaisman [30], Chapter 10. (See also Kosmann-Schwarzbach [38], Verdier [31].) There are summaries of results on simple Lie algebras in Perelomov [25] and in Chari and Pressley [22, Appendix A] (page 562, line 4 of A2, read $\pm a_{ij}$, and page 564. line 9 of A6, read $T_i(x_j^-)$). For Lie-bialgebra structures on compact Lie algebras, see Lu and Weinstein [45] and Soibelman [49]. For the infinite-dimensional example in Sect. 1.7, see Drinfeld [16], Chari and Pressley [22].

2 Classical Yang-Baxter Equation and r-Matrices

In this section, we shall study the Lie-bialgebra structures on a Lie algebra $\mathfrak{g}$ defined by a cocycle δr which is the coboundary of an element $r \in \mathfrak{g} \otimes \mathfrak{g}$. Such elements $r \in \mathfrak{g} \otimes \mathfrak{g}$ are called r-matrices. We shall show that the classical Yang-Baxter equation $(CYBE)$ is a sufficient condition for δr to define a Lie bracket on $\mathfrak{g}^*$. We shall also define triangular, quasi-triangular and factorizable Lie bialgebras, show that the double of any Lie bialgebra is a factorizable Lie bialgebra, and we shall study examples.

2.1 When Does δr Define a Lie-Bialgebra Structure on $\mathfrak{g}$?

We already noted that a 1-cochain on $\mathfrak{g}$ with values in $\mathfrak{g} \otimes \mathfrak{g}$ which is the coboundary of a 0-cochain on $\mathfrak{g}$ with values in $\mathfrak{g} \otimes \mathfrak{g}$, i.e., of an element $r \in \mathfrak{g} \otimes \mathfrak{g}$, is necessarily a 1-cocycle. So, in order for $\gamma = \delta r$ to define a Lie-bialgebra structure, there remain two conditions:

(i) δr must take values in $\bigwedge^2 \mathfrak{g}$ (skew-symmetry of the bracket on $\mathfrak{g}^*$ defined by δr),

(ii) the Jacobi identity for the bracket on $\mathfrak{g}^*$ defined by δr must be satisfied.

Let us denote by a (resp., s) the skew-symmetric (resp., symmetric) part of r. Thus $r = a + s$, where $a \in \bigwedge^2 \mathfrak{g}$, $s \in S^2\mathfrak{g}$.

Let us assume for simplicity that $\mathfrak{g}$ is finite-dimensional. To any element r in $\mathfrak{g} \otimes \mathfrak{g}$, we associate the map $\underline{r} : \mathfrak{g}^* \to \mathfrak{g}$ defined by

$$\underline{r}(\xi)(\eta) = r(\xi, \eta),$$

for $\xi, \eta \in \mathfrak{g}^*$. Here an element r in $\mathfrak{g} \otimes \mathfrak{g}$ is viewed as a bilinear form on $\mathfrak{g}^*$, and an element $\underline{r}(\xi)$ in $\mathfrak{g}$ is viewed as a linear form on $\mathfrak{g}^*$. Another notation for

$\underline{r}(\xi)(\eta)$ is $\langle \eta, \underline{r}\xi \rangle$. Let $^t\underline{r} : \mathfrak{g}^* \to \mathfrak{g}$ denote the transpose of the map $\underline{r}$. Then, by definition,

$$\underline{a} = \frac{1}{2}(\underline{r} - {}^t\underline{r}) \ , \quad \underline{s} = \frac{1}{2}(\underline{r} + {}^t\underline{r}) \ .$$

Now let $\gamma = \delta r$. Then, by definition,

$$\gamma(x) = ad_x^{(2)} r = (ad_x \otimes 1 + 1 \otimes ad_x)(r) \ ,$$

where 1 is the identity map of $\mathfrak{g}$ into itself. The right-hand side stands for

$$r^{ij}(ad_x e_i \otimes e_j + e_i \otimes ad_x e_j),$$

when (e_i) is a basis of $\mathfrak{g}$ and $r = r^{ij} e_i \otimes e_j$. As we explained in Sect. 1.2, the following notation is also used,

$$ad_x^{(2)}(r) = [x \otimes 1 + 1 \otimes x, r] \ .$$

For $\xi, \eta \in \mathfrak{g}^*$, we have set $[\xi, \eta]_{\mathfrak{g}^*} = {}^t\gamma(\xi, \eta)$. When $\gamma = \delta r$, we shall write $[\xi, \eta]^r$ instead of $[\xi, \eta]_{\mathfrak{g}^*}$.

Condition (i) above is satisfied if and only if $\delta s = 0$, that is, s is invariant under the adjoint action,

$$ad_x^{(2)} s = 0 \ ,$$

for all $x \in \mathfrak{g}$. This condition is often written $[x \otimes 1 + 1 \otimes x, s] = 0$. We shall often make use of the equivalent form of the ad-invariance condition for s,

$$ad_x \circ \underline{s} = \underline{s} \circ ad_x^*, \text{ for all } x \in \mathfrak{g} \ . \tag{2.1}$$

Whenever s is ad-invariant, $\delta r = \delta a$, and conversely. These equivalent conditions are obviously satisfied when $s = 0$, i.e., when r is skew-symmetric ($r = a$).

Proposition. *When r is skew-symmetric, then*

$$[\xi, \eta]^r = ad_{\underline{r}\xi}^* \eta - ad_{\underline{r}\eta}^* \xi \ . \tag{2.2}$$

Proof. By the definition of δr and that of the coadjoint action,

$$\langle {}^t(\delta r)(\xi, \eta), x \rangle = ((ad_x \otimes 1 + 1 \otimes ad_x)(r))(\xi, \eta) = -r(ad_x^* \xi, \eta) - r(\xi, ad_x^* \eta) \ .$$

By the skew-symmetry of r and the definition of $\underline{r}$, we find that

$$-r(ad_x^* \xi, \eta) - r(\xi, ad_x^* \eta) = r(\eta, ad_x^* \xi) - r(\xi, ad_x^* \eta)$$

$$= \underline{r}(\eta)(ad_x^* \xi) - \underline{r}(\xi)(ad_x^* \eta) = \langle ad_x^* \xi, \underline{r}\eta \rangle - \langle ad_x^* \eta, \underline{r}\xi \rangle \ .$$

Since for $x, y \in \mathfrak{g}, \alpha \in \mathfrak{g}^*, \langle ad_x^* \alpha, y \rangle = -\langle \alpha, [x, y] \rangle = \langle \alpha, [y, x] \rangle$, we obtain the general and useful relation,

$$\langle ad_x^* \alpha, y \rangle = -\langle ad_y^* \alpha, x \rangle \ . \tag{2.3}$$

Whence,

$$^t(\delta r)(\xi, \eta) = ad^*_{\underline{r}\xi}\eta - ad^*_{\underline{r}\eta}\xi \ ,$$

and the proposition is proved.

We shall now study condition (ii). We introduce the algebraic Schouten bracket of an element $r \in \bigwedge^2 \mathfrak{g}$ with itself, denoted by $[\![r, r]\!]$. It is the element in $\bigwedge^3 \mathfrak{g}$ defined by

$$[\![r, r]\!](\xi, \eta, \zeta) = -2 \circlearrowleft \langle \zeta, [\underline{r}\xi, \underline{r}\eta] \rangle \ , \tag{2.4}$$

where $\circlearrowleft$ denotes the summation over the circular permutations of ξ, η, ζ. (The factor -2 is conventional.)

Proposition. *A necessary and sufficient condition for $\gamma = \delta r$, $r \in \bigwedge^2 \mathfrak{g}$, to define a Lie bracket on $\mathfrak{g}^*$ is that $[\![r, r]\!] \in \bigwedge^3 \mathfrak{g}$ be ad-invariant.*

Proof. Here we give a computational proof. See Appendix 1 for a shorter, more conceptual proof. Note that the element $[\![r, r]\!]$ is a 0-cochain on $\mathfrak{g}$ with values in $\bigwedge^3 \mathfrak{g}$. It is *ad*-invariant if and only if $\delta([\![r, r]\!]) = 0$. The proposition will then follow from the identity

$$\circlearrowleft \langle [[\xi, \eta]^r, \zeta]^r, x \rangle = \frac{1}{2}\delta([\![r, r]\!])(x)(\xi, \eta, \zeta) \ , \tag{2.5}$$

for $\xi, \eta, \zeta \in \mathfrak{g}^*, x \in \mathfrak{g}$.

By (2.2),

$$\langle [[\xi, \eta]^r, \zeta]^r, x \rangle = \langle [ad^*_{\underline{r}\xi}\eta - ad^*_{\underline{r}\eta}\xi, \zeta]^r, x \rangle$$

$$= \langle ad^*_{\underline{r}(ad^*_{\underline{r}\xi}\eta - ad^*_{\underline{r}\eta}\xi)}\zeta, x \rangle - \langle ad^*_{\underline{r}\zeta}(ad^*_{\underline{r}\xi}\eta - ad^*_{\underline{r}\eta}\xi), x \rangle \ .$$

By (2.3), this expression is equal to

$$-\langle ad^*_x\zeta, \underline{r}(ad^*_{\underline{r}\xi}\eta - ad^*_{\underline{r}\eta}\xi) \rangle + \langle ad^*_x(ad^*_{\underline{r}\xi}\eta - ad^*_{\underline{r}\eta}\xi), \underline{r}\zeta \rangle \ .$$

Using the skew-symmetry of r and the relation

$$ad^*_x ad^*_y - ad^*_y ad^*_x = ad^*_{[x,y]} \ ,$$

valid for any $x, y \in \mathfrak{g}$, applied to $y = \underline{r}\xi$, we obtain

$$\langle [[\xi, \eta]^r, \zeta]^r, x \rangle = \langle \underline{r} \, ad^*_x\zeta, ad^*_{\underline{r}\xi}\eta \rangle - \langle \underline{r} \, ad^*_x\zeta, ad^*_{\underline{r}\eta}\xi \rangle + \langle ad^*_{\underline{r}\xi} ad^*_x\eta, \underline{r}\zeta \rangle$$

$$+ \langle ad^*_{[x,\underline{r}\xi]}\eta, \underline{r}\zeta \rangle - \langle ad^*_x ad^*_{\underline{r}\eta}\xi, \underline{r}\zeta \rangle \ .$$

Using (2.3), we obtain $\langle ad^*_{[x,\underline{r}\xi]}\eta, \underline{r}\zeta \rangle = -\langle ad^*_{\underline{r}\zeta}\eta, [x, \underline{r}\xi] \rangle = \langle ad^*_x ad^*_{\underline{r}\zeta}\eta, \underline{r}\xi \rangle$. Therefore,

$$\circlearrowleft \langle [[\xi, \eta]^r, \zeta]^r, x \rangle = \circlearrowleft (\langle \eta, [\underline{r} \, ad^*_x\zeta, \underline{r}\xi] \rangle + \langle \xi, [\underline{r}\eta, \underline{r} \, ad^*_x\zeta] \rangle + \langle ad^*_x\eta, [\underline{r}\zeta, \underline{r}\xi] \rangle)$$

$$+ \circlearrowleft (\langle ad^*_x ad^*_{\underline{r}\zeta}\eta, \underline{r}\xi \rangle - \langle ad^*_x ad^*_{\underline{r}\eta}\xi, \underline{r}\zeta \rangle) \ .$$

The last summation obviously vanishes. Now, by (2.4),

$$\frac{1}{2}\delta([\![r,r]\!])(x)(\xi,\eta,\zeta) = \frac{1}{2}ad_x^{(3)}[\![r,r]\!](\xi,\eta,\zeta) = -\frac{1}{2}\circlearrowleft[\![r,r]\!](\xi,\eta,ad_x^*\zeta)$$

$$= \circlearrowleft(\langle ad_x^*\zeta,[\underline{r}\xi,\underline{r}\eta]\rangle + \langle\eta,[\underline{r}\ ad_x^*\zeta,\underline{r}\xi]\rangle + \langle\xi,[\underline{r}\eta,\underline{r}\ ad_x^*\zeta]\rangle)\ .$$

Comparing the two expressions we have just found, we see that (2.5) is proved.

Let r be a skew-symmetric element of $\mathfrak{g}\otimes\mathfrak{g}$. The condition that $[\![r,r]\!]$ be *ad*-invariant is sometimes called the *generalized Yang-Baxter equation*. Obviously, a sufficient condition for $[\![r,r]\!]$ to be *ad*-invariant is

$$[\![r,r]\!] = 0\ . \tag{2.6}$$

We shall see that condition (2.6) is a particular case of the classical Yang-Baxter equation. (See Sect. 2.2.)

Definition. *Let r be an element in $\mathfrak{g}\otimes\mathfrak{g}$, with symmetric part s, and skew-symmetric part a. If s and $[\![a,a]\!]$ are ad-invariant, then r is called a* classical r-matrix *or, if no confusion is possible, an r-matrix. If r is skew-symmetric ($r=a$) and if $[\![r,r]\!]=0$, then r is called a* triangular r-matrix.

Remark 1. It follows from the preceding discussion that any r-matrix in $\mathfrak{g}\otimes\mathfrak{g}$ defines a Lie-bialgebra structure on $\mathfrak{g}$ for which the Lie bracket on $\mathfrak{g}^*$ is given by formula (2.2). This bracket is called the *Sklyanin bracket* defined by r.

Remark 2. Some authors (see Babelon and Viallet [35], Li and Parmentier [43], Reiman [26]) define an r-matrix to be an element r in $\mathfrak{g}\otimes\mathfrak{g}$ such that $ad_{r\xi}^*\eta-ad_{r\eta}^*\xi$ is a Lie bracket. In this definition, s is not necessarily *ad*-invariant, and such a Lie bracket is not, in general, a Lie-bialgebra bracket.

In Sect. 2.4 we shall study the modified Yang-Baxter equation and its solutions, which are called classical R-matrices, or when no confusion can arise, R-matrices.

2.2   The Classical Yang-Baxter Equation

Let r be an element in $\mathfrak{g}\otimes\mathfrak{g}$, and let us introduce $\langle r,r\rangle : \bigwedge^2\mathfrak{g}^* \to \mathfrak{g}$, defined by

$$\langle r,r\rangle(\xi,\eta) = [\underline{r}\xi,\underline{r}\eta] - \underline{r}[\xi,\eta]^r\ .$$

Setting

$$\langle r,r\rangle(\xi,\eta,\zeta) = \langle\zeta,\langle r,r\rangle(\xi,\eta)\rangle,$$

the map $\langle r,r\rangle$ is identified with an element $\langle r,r\rangle \in \bigwedge^2\mathfrak{g}\otimes\mathfrak{g}$. We shall show that, whenever the symmetric part of r is *ad*-invariant, the element $\langle r,r\rangle$ is in fact in $\bigwedge^3\mathfrak{g}$.

Theorem. (*i*) *Let a be in* $\mathfrak{g} \otimes \mathfrak{g}$ *and skew-symmetric. Then* $\langle a, a \rangle$ *is in* $\bigwedge^3 \mathfrak{g}$, *and*

$$\langle a, a \rangle = -\frac{1}{2} [\![a, a]\!] \ ,$$

(*ii*) *Let s be in* $\mathfrak{g} \otimes \mathfrak{g}$, *symmetric and ad-invariant. Then* $\langle s, s \rangle$ *is an ad-invariant element in* $\bigwedge^3 \mathfrak{g}$, *and*

$$\langle \underline{s}, s \rangle(\xi, \eta) = [\underline{s}\xi, \underline{s}\eta] \ ,$$

(*iii*) *For* $r = a + s$, *where a is skew-symmetric, and s is symmetric and ad-invariant,* $\langle r, r \rangle$ *is in* $\bigwedge^3 \mathfrak{g}$, *and*

$$\langle r, r \rangle = \langle a, a \rangle + \langle s, s \rangle \ ,$$

Proof.

(*i*) By definition, $\langle a, a \rangle(\xi, \eta, \zeta) = \langle \zeta, [\underline{a}\xi, \underline{a}\eta] \rangle - \langle \zeta, \underline{a}[\xi, \eta]^a \rangle$. By (2.2) and the skew-symmetry of $\underline{a}$,

$$\langle a, a \rangle(\xi, \eta, \zeta) = \langle \zeta, [\underline{a}\xi, \underline{a}\eta] \rangle - \langle \zeta, \underline{a} \, ad^*_{\underline{a}\xi}\eta \rangle + \langle \zeta, \underline{a} \, ad^*_{\underline{a}\eta}\xi \rangle$$

$$= \langle \zeta, [\underline{a}\xi, \underline{a}\eta] \rangle + \langle ad^*_{\underline{a}\xi}\eta, \underline{a}\zeta \rangle - \langle ad^*_{\underline{a}\eta}\xi, \underline{a}\zeta \rangle$$

$$= \langle \zeta, [\underline{a}\xi, \underline{a}\eta] \rangle + \langle \eta, [\underline{a}\zeta, \underline{a}\xi] \rangle + \langle \xi, [\underline{a}\eta, \underline{a}\zeta] \rangle \ .$$

By (2.4), we see that

$$\langle a, a \rangle(\xi, \eta, \zeta) = -\frac{1}{2} [\![a, a]\!](\xi, \eta, \zeta) \ ,$$

and therefore (*i*) is proved.

(*ii*) If s is symmetric and ad-invariant, then clearly $\langle s, s \rangle(\xi, \eta) = [\underline{s}\xi, \underline{s}\eta]$. To see that $\langle s, s \rangle$ is in $\bigwedge^3 \mathfrak{g}$, we use the ad-invariance of s again and the symmetry of $\underline{s}$. Since

$$\langle s, s \rangle(\xi, \eta, \zeta) = \langle \zeta, [\underline{s}\xi, \underline{s}\eta] \rangle \ ,$$

we find that

$$\langle s, s \rangle(\xi, \eta, \zeta) = \langle \zeta, ad_{\underline{s}\xi}\underline{s}\eta \rangle = \langle \zeta, \underline{s}ad^*_{\underline{s}\xi}\eta \rangle$$

$$= \langle ad^*_{\underline{s}\xi}\eta, \underline{s}\zeta \rangle = -\langle \eta, [\underline{s}\xi \, , \, \underline{s}\zeta] \rangle = -\langle s, s \rangle(\xi, \zeta, \eta) \ .$$

Thus $\langle s, s \rangle$ is skew-symmetric in the last two variables, therefore $\langle s, s \rangle \in \bigwedge^3 \mathfrak{g}$.
    To prove that $\langle s, s \rangle$ is ad-invariant, we must prove that

$$\langle \underline{s}, s \rangle(ad^*_x\xi, \eta) + \langle \underline{s}, s \rangle(\xi, ad^*_x\eta) = ad_x(\langle \underline{s}, s \rangle(\xi, \eta)) \ .$$

Now

$$\langle s, s \rangle (ad_x^* \xi, \eta) + \langle s, s \rangle (\xi, ad_x^* \eta)$$

$$= [ad_x \underline{s} \xi, \underline{s} \eta] + [\xi, ad_x \underline{s} \eta] = ad_x [\underline{s} \xi, \underline{s} \eta] = ad_x (\langle s, s \rangle (\xi, \eta)) .$$

(*iii*) From the *ad*-invariance of s, we know that

$$[\xi, \eta]^r = [\xi, \eta]^a .$$

Thus, since $r = a + s$, we find that

$$\langle r, r \rangle (\xi, \eta) = [\underline{a} \xi, \underline{a} \eta] - \underline{a} [\xi, \eta]^a$$

$$+ [\underline{a} \xi, \underline{s} \eta] + [\underline{s} \xi, \underline{a} \eta] - \underline{s} [\xi, \eta]^a + [\underline{s} \xi, \underline{s} \eta] .$$

By (2.2) and the *ad*-invariance of $\underline{s}$,

$$\underline{s} [\xi, \eta]^a = \underline{s} \, ad_{\underline{a} \xi}^* \eta - \underline{s} \, ad_{\underline{a} \eta}^* \xi = ad_{\underline{a} \xi} \underline{s} \eta - ad_{\underline{a} \eta} \underline{s} \xi ,$$

and therefore $[\underline{a} \xi, \underline{s} \eta] + [\underline{s} \xi, \underline{a} \eta] - \underline{s} [\xi, \eta]^a = 0$. Therefore $\langle r, r \rangle = \langle a, a \rangle + \langle s, s \rangle$, thus proving (*iii*), since we already know that $\langle a, a \rangle$ and $\langle s, s \rangle$ are in $\bigwedge^3 \mathfrak{g}$.

It follows from the proof of (*i*) that,

$$\langle a, a \rangle (\xi, \eta, \zeta) = \circlearrowleft \langle \zeta, [\underline{a} \xi, \underline{a} \eta] \rangle ,$$

while, if s is *ad*-invariant, it follows from (*ii*) that

$$\langle s, s \rangle (\xi, \eta, \zeta) = \langle \zeta, [\underline{s} \xi, \underline{s} \eta] \rangle.$$

(In this case there is no summation over the circular permutations of ξ, η, ζ.)

From this theorem, we obtain immediately,

Corollary. *Let $r \in \mathfrak{g} \otimes \mathfrak{g}$, $r = a + s$, where s is symmetric and ad-invariant, and a is skew-symmetric. A sufficient condition for $[\![a, a]\!]$ to be ad-invariant is*

$$\langle r, r \rangle = 0 . \tag{2.7}$$

Thus an element $r \in \mathfrak{g} \otimes \mathfrak{g}$ with *ad*-invariant symmetric part, satisfying $\langle r, r \rangle = 0$ is an *r*-matrix.

Definition. *Condition (2.7), $\langle r, r \rangle = 0$, is called the* classical Yang-Baxter equation. *An r-matrix satisfying the classical Yang-Baxter equation is called* quasi-triangular. *If, moreover, the symmetric part of r is invertible, then r is called* factorizable.

Remark 1. When r is skew-symmetric, (2.7) reduces to (2.6). Thus a triangular *r*-matrix is quasi-triangular but not factorizable.

Remark 2. Sometimes condition (2.7), written in the form $\langle a, a \rangle = -\langle s, s \rangle$ is called the *modified Yang-Baxter equation* (for a), and the term classical Yang-Baxter equation is reserved for the case where r is skew-symmetric, *i.e.*, for condition (2.6). The abbreviations *CYBE* and *MYBE* are commonly used.

Remark 3. It is clear from part *(iii)* of the theorem that if $r = a + s$, then $^t r = -a + s$ satisfies

$$\langle {}^t r, {}^t r \rangle = \langle r, r \rangle .$$

Thus, whenever r is a solution of the classical Yang-Baxter equation, so are $^t r$ and $-{}^t r = a - s$. The notations

$$r_+ = r, \quad r_- = -{}^t r$$

will be used in Sects. 2.4, 2.5 and 4.9 below.

Observe also that if r is a solution of the *CYBE*, then so is any scalar multiple of r. However, if $a \in \bigwedge^2 \mathfrak{g}$ is a solution of the *MYBE* in the sense of Remark 2 above, $\langle a, a \rangle = -\langle s, s \rangle$, for a given $s \in \mathfrak{g} \otimes \mathfrak{g}$, then $-a$ satisfies the same equation, but an arbitrary scalar multiple of a does not.

The following proposition is immediate from the definition of $\underline{\langle r, r \rangle}$.

Proposition. *An r-matrix, r, is quasi-triangular if and only if $\underline{r_+} = \underline{a} + \underline{s}$ and $\underline{r_-} = \underline{a} - \underline{s}$ are Lie-algebra morphisms from $(\mathfrak{g}^*, [\ ,\ ]^r)$ to $\mathfrak{g}$.*

We now prove

Proposition. *On a simple Lie algebra over $\mathbb{C}$, any Lie-bialgebra structure is defined by a quasi-triangular r-matrix.*

Proof. First, the Lie-bialgebra structure of any semi-simple Lie algebra, $\mathfrak{g}$, is necessarily defined by an r-matrix. In fact, when $\mathfrak{g}$ is semi-simple, by Whitehead's lemma, any 1-cocycle is a coboundary, and any 1-cocycle γ with values in $\bigwedge^2 \mathfrak{g}$ is the coboundary of an element $a \in \bigwedge^2 \mathfrak{g}$,

$$\gamma = \delta a .$$

We know that $\langle a, a \rangle = -\frac{1}{2} [\![a, a]\!]$ is an *ad*-invariant element of $\bigwedge^3 \mathfrak{g}$, because $^t \gamma$ is a Lie bracket on $\mathfrak{g}^*$, so a is an r-matrix.

Let t be the Killing form of $\mathfrak{g}$. It is an invariant, non-degenerate, symmetric, bilinear form on $\mathfrak{g}$, and $\langle t, t \rangle$ is an *ad*-invariant element of $\bigwedge^3 \mathfrak{g}$, by *(ii)* of the theorem. We now use the result (see Koszul [11]) that, in a simple Lie algebra over $\mathbb{C}$, the space of *ad*-invariant elements of $\bigwedge^3 \mathfrak{g}$ is 1-dimensional. Therefore there exists a complex number μ such that

$$\langle a, a \rangle = -\mu^2 \langle t, t \rangle .$$

Now, $r = a + \mu t$ is in fact a quasi-triangular r-matrix and $\delta r = \delta a = \gamma$.

2.3 Tensor Notation

We have already observed in Sect. 1.2 that, for $r \in \mathfrak{g} \otimes \mathfrak{g}$, there are various notations for $\delta r(x), x \in \mathfrak{g}$,

$$\delta r(x) = (ad_x \otimes 1 + 1 \otimes ad_x)(r) = [x \otimes 1 + 1 \otimes x, r] .$$

We now introduce a new notation, not to be confused with the usual indicial notation of the tensor calculus. If $r \in \mathfrak{g} \otimes \mathfrak{g}$, we define r_{12}, r_{13}, r_{23} as elements in the third tensor power of the enveloping algebra of $\mathfrak{g}$ (an associative algebra with unit such that $[x, y] = x.y - y.x$),

$$r_{12} = r \otimes 1 ,$$
$$r_{23} = 1 \otimes r ,$$

and, if $r = \Sigma_i u_i \otimes v_i$, then $r_{13} = \Sigma_i u_i \otimes 1 \otimes v_i$, where 1 is the unit of the enveloping algebra of $\mathfrak{g}$. This notation is called the *tensor notation*.

In $\mathfrak{g} \otimes \mathfrak{g} \otimes \mathfrak{g}$, we now define

$$[r_{12}, r_{13}] = [\Sigma_i u_i \otimes v_i \otimes 1, \Sigma_j u_j \otimes 1 \otimes v_j] = \Sigma_{i,j}[u_i, u_j] \otimes v_i \otimes v_j ,$$

and, similarly,

$$[r_{12}, r_{23}] = [\Sigma_i u_i \otimes v_i \otimes 1, \Sigma_j 1 \otimes u_j \otimes v_j] = \Sigma_{i,j} u_i \otimes [v_i, u_j] \otimes v_j ,$$

$$[r_{13}, r_{23}] = [\Sigma_i u_i \otimes 1 \otimes v_i, \Sigma_j 1 \otimes u_j \otimes v_j] = \Sigma_{i,j} u_i \otimes u_j \otimes [v_i, v_j] .$$

In these notations, if the symmetric part s of r is ad-invariant, then

$$\langle r, r \rangle = [r_{12}, r_{13}] + [r_{12}, r_{23}] + [r_{13}, r_{23}] , \tag{2.8}$$

and

$$\langle s, s \rangle = [s_{13}, s_{23}] = [s_{23}, s_{12}] = [s_{12}, s_{13}] . \tag{2.9}$$

In fact,

$$[r_{12}, r_{13}](\xi, \eta, \zeta) = \langle \xi, [{}^t\underline{r}\eta, {}^t\underline{r}\zeta] \rangle$$
$$[r_{12}, r_{13}](\xi, \eta, \zeta) = \langle \eta, [\underline{r}\xi, {}^t\underline{r}\zeta] \rangle$$
$$[r_{13}, r_{23}](\xi, \eta, \zeta) = \langle \zeta, [\underline{r}\xi, \underline{r}\eta] \rangle ,$$

while

$$\langle r, r \rangle(\xi, \eta, \zeta) = \langle \zeta, [\underline{r}\xi, \underline{r}\eta] \rangle - \langle \zeta, \underline{r}(ad^*_{\underline{r}\xi}\eta + ad^*_{\underline{t}\underline{r}\eta}\xi) \rangle$$
$$= \langle \zeta, [\underline{r}\xi, \underline{r}\eta] \rangle + \langle \eta, [\underline{r}\xi, {}^t\underline{r}\zeta] \rangle + \langle \xi, [{}^t\underline{r}\eta, {}^t\underline{r}\zeta] \rangle .$$

(We have used the fact that $ad^*_{\underline{s}\xi}\eta + ad^*_{\underline{s}\eta}\xi = 0$, because of the ad-invariance of s, and therefore

$$ad^*_{\underline{r}\xi}\eta + ad^*_{\underline{t}\underline{r}\eta}\xi = ad^*_{\underline{a}\xi}\eta - ad^*_{\underline{a}\eta}\xi = [\xi, \eta]^r .)$$

Thus (2.8) is proved. Another way to state the ad-invariance of s is

$$\langle \xi, [x, \underline{s}\eta] \rangle + \langle \eta, [x, \underline{s}\xi] \rangle = 0 \, ,$$

and (2.9) follows.

So, in tensor notation, the classical Yang-Baxter equation (2.7) reads

$$[r_{12}, r_{13}] + [r_{12}, r_{23}] + [r_{13}, r_{23}] = 0 \, . \tag{2.10}$$

Example 1. Let $\mathfrak{g}$ be the Lie algebra of dimension 2 with basis X, Y and commutation relation

$$[X, Y] = X \, .$$

Then $r = X \wedge Y = X \otimes Y - Y \otimes X$ is a skew-symmetric solution of $CYBE$, i.e., a triangular r-matrix. This fact can be proved using definition (2.4), or one can prove (2.10). Here, for example,

$$[r_{12}, r_{13}] = [X \otimes Y \otimes 1 - Y \otimes X \otimes 1, X \otimes 1 \otimes Y - Y \otimes 1 \otimes X]$$

$$= -[X, Y] \otimes Y \otimes X - [Y, X] \otimes X \otimes Y = -X \otimes Y \otimes X + X \otimes X \otimes Y \, ,$$

and similarly

$$[r_{12}, r_{23}] = -X \otimes X \otimes Y + Y \otimes X \otimes X,$$
$$[r_{13}, r_{23}] = X \otimes Y \otimes X - Y \otimes X \otimes X,$$

so that $\langle r, r \rangle = 0$.

Then $\delta r(X) = 0$, $\delta r(Y) = -X \wedge Y$. In terms of the dual basis X^*, Y^* of $\mathfrak{g}^*$, $[X^*, Y^*]^r = -Y^*$.

Example 2. On $\mathfrak{sl}(2, \mathbb{C})$, we consider the Casimir element, t (i.e., the Killing form seen as an element in $\mathfrak{g} \otimes \mathfrak{g}$),

$$t = \frac{1}{8} H \otimes H + \frac{1}{4}(X \otimes Y + Y \otimes X),$$

and we set

$$t_0 = \frac{1}{8} H \otimes H \, ,$$

$$t_{+-} = \frac{1}{4} X \otimes Y.$$

If we define

$$r = t_0 + 2t_{+-} = \frac{1}{8}(H \otimes H + 4X \otimes Y) \, ,$$

then the symmetric part of r is t, and the skew-symmetric part is $a = \frac{1}{4} X \wedge Y$, and r is a factorizable r-matrix. Then

$$\delta a(H) = 0, \quad \delta a(X) = \frac{1}{4} X \wedge H, \quad \delta a(Y) = \frac{1}{4} Y \wedge H,$$

or, in terms of the dual basis H^*, X^*, Y^* of $\mathfrak{g}^*$,

$$[H^*, X^*]^r = \frac{1}{4}X^*, \quad [H^*, Y^*]^r = \frac{1}{4}Y^*, \quad [X^*, Y^*]^r = 0.$$

Thus the Lie-bialgebra structure of $\mathfrak{sl}(2, \mathbb{C})$ of Sect. 1.1 is defined by the factorizable r-matrix given above.

Example 3. On $\mathfrak{sl}(2, \mathbb{C})$, we consider $r = X \otimes H - H \otimes X$, which is a triangular r-matrix. Then $\delta r(X) = 0, \delta r(Y) = 2Y \wedge X, \delta r(H) = X \wedge H$.

2.4   R-Matrices and Double Lie Algebras

Let now R be any linear map from $\mathfrak{g}$ to $\mathfrak{g}$. We define

$$[x, y]_R = [Rx, y] + [x, Ry] . \tag{2.11}$$

We consider the skew-symmetric bilinear form $\langle R, R \rangle$ on $\mathfrak{g}$ with values in $\mathfrak{g}$ defined by

$$\langle R, R \rangle(x, y) = [Rx, Ry] - R([Rx, y] + [x, Ry]) + [x, y]. \tag{2.12}$$

for $x, y \in \mathfrak{g}$, and, more generally, we define $\langle R, R \rangle_k$, by

$$\langle R, R \rangle_k(x, y) = [Rx, Ry] - R([Rx, y] + [x, Ry]) + k^2[x, y], \tag{2.13}$$

where k is any scalar.

Let $c \in \bigwedge^3 \mathfrak{g}$, and define $\underline{c} : \bigwedge^2 \mathfrak{g}^* \to \mathfrak{g}$ by

$$\langle \zeta, \underline{c}(\xi, \eta) \rangle = c(\xi, \eta, \zeta) .$$

Then, for $u \in \mathfrak{g}$,

$$\begin{aligned}(ad_u c)(\xi, \eta, \zeta) &= -(c(ad_u^*\xi, \eta, \zeta) + c(\xi, ad_u^*\eta, \zeta) + c(\xi, \eta, ad_u^*\zeta)) \\ &= -\circlearrowleft c(\xi, \eta, ad_u^*\zeta) = -\circlearrowleft \langle ad_u^*\zeta, \underline{c}(\xi, \eta) \rangle,\end{aligned}$$

where $\circlearrowleft$ denotes the sum over cyclic permutations in ξ, η, ζ. We have thus proved

Lemma. *The ad-invariance of c is equivalent to the condition on $\underline{c}$,*

$$\circlearrowleft \langle \zeta, [u, \underline{c}(\xi, \eta)] \rangle = 0 , \tag{2.14}$$

for all $\xi, \eta, \zeta \in \mathfrak{g}^*$, $u \in \mathfrak{g}$.

Proposition. *The Jacobi identity for the bracket $[\ , \ ]_R$ defined by (2.11) is satisfied if and only if*

$$\underset{x,y,z}{\circlearrowleft} [z, \langle R, R \rangle(x, y)] = 0, \tag{2.15}$$

for all x, y, z in $\mathfrak{g}$.

Proof. We can show by a direct computation, using the Jacobi identity for the bracket $[\, , \,]$ repeatedly, that

$$\circlearrowleft [[x,y]_R, z]_R - \circlearrowleft [z, \langle R, R \rangle (x,y)]$$

$$= \circlearrowleft ([R[Rx,y],z]+[[Rx,y],Rz]+[R[x,Ry],z]+[[x,Ry],Rz]) - \circlearrowleft [z, \langle R, R \rangle (x,y)]$$

vanishes identically.

Definition. *Condition*

$$\langle R, R \rangle_k = 0 \tag{2.16}$$

is called the modified Yang-Baxter equation *(MYBE) with coefficient* k^2.

An endomorphism R of $\mathfrak{g}$ satisfying $\langle R, R \rangle_k = 0$ for some scalar k is called a classical R-matrix, *or simply an R-matrix. It is called* factorizable *if k is not equal to 0.*

Thus any R-matrix on $\mathfrak{g}$ defines a second Lie-algebra structure $[\, , \,]_R$ on $\mathfrak{g}$. For this reason, a Lie algebra with an R-matrix is called a *double Lie algebra*. (This definition is not to be confused with that of the double of a Lie bialgebra, given in Sect. 1.6.)

Let $\mathfrak{g}$ be a Lie algebra with an *ad*-invariant, non-degenerate, symmetric bilinear form defining a bijective linear map $\underline{s}$ from $\mathfrak{g}^*$ to $\mathfrak{g}$. A skew-symmetric endomorphism of $(\mathfrak{g}, s)$ is a linear map R from $\mathfrak{g}$ to $\mathfrak{g}$ such that $R \circ \underline{s} : \mathfrak{g}^* \to \mathfrak{g}$ is skew-symmetric.

Proposition. *Let R be a skew-symmetric endomorphism of $(\mathfrak{g}, s)$, and set $r = a + s$, where $\underline{a} = R \circ \underline{s}$. Then*

$$[x,y]_R = \underline{s}[\underline{s}^{-1}x, \underline{s}^{-1}y]^r = \underline{s}[\underline{s}^{-1}x, \underline{s}^{-1}y]^a, \tag{2.17}$$

and

$$\langle R, R \rangle (x,y) = \langle \underline{r}, \underline{r} \rangle (\underline{s}^{-1}x, \underline{s}^{-1}y) . \tag{2.18}$$

Proof. The relations between $r = a + s$ and R are

$$\begin{cases} R &= \underline{a} \circ \underline{s}^{-1} , \\ \underline{r} &= (R + 1) \circ \underline{s} , \end{cases}$$

where 1 is the identity map from $\mathfrak{g}$ to $\mathfrak{g}$. By the *ad*-invariance of $\underline{s}$,

$$[Rx,y] + [x,Ry] = ad_{(R \circ \underline{s})\underline{s}^{-1}x}y - ad_{(R \circ \underline{s})\underline{s}^{-1}y}x$$

$$= \underline{s}(ad^*_{\underline{a}(\underline{s}^{-1}x)}\underline{s}^{-1}y - ad^*_{\underline{a}(\underline{s}^{-1}y)}\underline{s}^{-1}x).$$

This relation and (2.2) prove (2.17), and

$$R([Rx,y] + [x,Ry]) = \underline{a}[\underline{s}^{-1}x, \underline{s}^{-1}y]^a .$$

Thus, by the theorem of Sect. 2.2,

$$\langle R, R \rangle (x, y) = [\underline{a}(\underline{s}^{-1}x), \underline{a}(\underline{s}^{-1}y)] - \underline{a}[\underline{s}^{-1}x, \underline{s}^{-1}y]^a + [\underline{s}(\underline{s}^{-1}x), \underline{s}(\underline{s}^{-1}y)]$$
$$= \langle a, a \rangle (\underline{s}^{-1}x, \underline{s}^{-1}y) + \langle \underline{s}, \underline{s} \rangle (\underline{s}^{-1}x, \underline{s}^{-1}y)$$
$$= \langle a + s, a + s \rangle (\underline{s}^{-1}x, \underline{s}^{-1}y) ,$$

and this is equality (2.18).

More generally, setting $r_k = a + k\ s$, we find that

$$\langle R, R \rangle_k (x, y) = \langle r_k, r_k \rangle (\underline{s}^{-1}x, \underline{s}^{-1}y) .$$

Proposition. *Let R be a skew-symmetric endomorphism of $(\mathfrak{g}, s)$. Then R is an R-matrix (resp., a factorizable R -matrix) if and only if $\underline{r_k} = (R + k1) \circ \underline{s}$ is a quasi-triangular r-matrix (resp., a factorizable r-matrix), for some scalar k.*

Proof. This follows from the preceding proposition.

In other words, $R = \underline{a} \circ \underline{s}^{-1}$ satisfies the *MYBE* with coefficient k^2 if and only if $r = a + ks$ satisfies the *CYBE*. In the terminology of Remark 2 of Sect. 2.2, R is a solution of the *MYBE* with coefficient 1 if and only if a, defined by $\underline{a} = R \circ \underline{s}$ satisfies the *MYBE*.

Applying the lemma to $c = \langle r, r \rangle$, and using the preceding proposition, we see that $r = a + s$ is an r-matrix with invertible symmetric part s if and only if $R = \underline{a} \circ \underline{s}^{-1}$ satisfies the condition

$$\underset{x,y,z}{\circlearrowleft} \langle \underline{s}^{-1}z , [u, \langle R, R \rangle (x, y)] \rangle = 0 ,$$

for each $x, y, z, u \in \mathfrak{g}$. Using the ad-invariance of s, we find that this condition is equivalent to

$$\underset{x,y,z}{\circlearrowleft} [z, \langle R, R \rangle (x, y)] = 0 .$$

This fact furnishes an alternate proof of the first proposition of this section.

Example. If $\mathfrak{g} = \mathfrak{a} \oplus \mathfrak{b}$, where both $\mathfrak{a}$ and $\mathfrak{b}$ are Lie subalgebras of $\mathfrak{g}$, then $R = \frac{1}{2}(p_\mathfrak{a} - p_\mathfrak{b})$, where $p_\mathfrak{a}$ (resp., $p_\mathfrak{b}$) is the projection onto $\mathfrak{a}$ (resp., $\mathfrak{b}$) parallel to $\mathfrak{b}$ (resp., $\mathfrak{a}$), is a solution of the modified Yang-Baxter equation with coefficient $\frac{1}{4}$. The proof is straightforward. In this case, we find that

$$[x, y]_R = [x_\mathfrak{a}, y_\mathfrak{a}] - [x_\mathfrak{b}, y_\mathfrak{b}] . \tag{2.19}$$

As an example, we can consider $\mathfrak{g} = \mathfrak{sl}(n, \mathbb{R})$, $\mathfrak{a}$ the subalgebra of upper triangular matrices, and $\mathfrak{b} = \mathfrak{so}(n)$.

Proposition. *If R is a solution of the modified Yang-Baxter equation with coefficient k^2, then*

$$R_\pm = R \pm k\ 1$$

are Lie-algebra morphisms from $\mathfrak{g}_R$ to $\mathfrak{g}$, where $\mathfrak{g}_R$ denotes $\mathfrak{g}$ equipped with the Lie bracket $[\ ,\ ]_R$ defined by (2.11).

Proof. In fact,

$$R_\pm[x,y]_R - [R_\pm x, R_\pm y] = (R \pm k\,1)([Rx,y]+[x,Ry]) - [(R \pm k\,1)x, (R \pm k\,1)y]$$

$$= R([Rx,y]+[x,Ry]) - [Rx,Ry] - k^2[x,y] = -\langle R,R\rangle_k(x,y)\ .$$

Remark. When R is a solution of the *MYBE* with coefficient 1, and $\underline{s}$ is invertible, then $R_\pm = \underline{r}_\pm \circ \underline{s}^{-1}$, and it follows from (2.11) and (2.17) that

$$R_\pm[x,y]_R - [R_\pm x, R_\pm y] = \underline{r}_\pm[\underline{s}^{-1}x, \underline{s}^{-1}y]^r - [\underline{r}_\pm(\underline{s}^{-1}x), \underline{r}_\pm(\underline{s}^{-1}y)].$$

Therefore the morphism properties of $R_\pm$ can be deduced from those of $r_\pm$ proved in Sect. 2.2.

As a consequence of the preceding proposition we obtain the following

Proposition. *Let R be an R-matrix satisfying $\langle R,R\rangle_k = 0$. Then*

$$J : x \in \mathfrak{g}_R \mapsto (R_+x, R_-x) \in \mathfrak{g} \oplus \mathfrak{g}$$

is an injective map which identifies $\mathfrak{g}_R$ with a Lie subalgebra of the direct sum of Lie algebras $\mathfrak{g} \oplus \mathfrak{g}$.

Proof. The linear map J is clearly injective since $R_+x = R_-x = 0$ implies $(R_+ - R_-)x = 0$, and thus $x = 0$. Moreover,

$$J([x,y]_R) = (R_+[x,y]_R, R_-[x,y]_R) = ([R_+x, R_+y], [R_-x, R_-y]) \in \mathfrak{g} \oplus \mathfrak{g},$$

proving the proposition.

2.5   The Double of a Lie Bialgebra Is a Factorizable Lie Bialgebra

Let $(\mathfrak{g}, \gamma)$ be any Lie bialgebra, and let $\mathfrak{d}$ be its double. Recall that, as a vector space, $\mathfrak{d}$ is just $\mathfrak{g} \oplus \mathfrak{g}^*$. There is a canonical R-matrix defined on the Lie algebra $(\mathfrak{d}, [\ ,\ ]_\mathfrak{d})$, namely

$$R = \frac{1}{2}(p_{\mathfrak{g}^*} - p_\mathfrak{g})\ .$$

Since $\mathfrak{g}$ and $\mathfrak{g}^*$ are Lie subalgebras of $\mathfrak{d}$, this endomorphism of $\mathfrak{d}$ is indeed a factorizable R-matrix, which satisfies the modified Yang-Baxter equation with coefficient $k^2 = \frac{1}{4}$ (with respect to the Lie bracket of the double). So $\mathfrak{d}$ is a double Lie algebra, with

$$[x + \xi, y + \eta]_R = -[x,y] + [\xi,\eta]\ .$$

Moreover $\mathfrak{d} = \mathfrak{g} \oplus \mathfrak{g}^*$ has a natural scalar product, $(\ |\ )$, which defines a linear map $\underline{s}_\mathfrak{d}$ from $\mathfrak{d}^* = \mathfrak{g}^* \oplus \mathfrak{g}$ to $\mathfrak{d}$. It is easily seen that

$$\underline{s}_\mathfrak{d}(\xi, x) = (x, \xi)\ .$$

Note that R is skew-symmetric with respect to this scalar product. Therefore, by the third proposition in Sect. 2.4, $\underline{r}_{\mathfrak{d}} = R \circ \underline{s}_{\mathfrak{d}} + \frac{1}{2}\underline{s}_{\mathfrak{d}}$ defines a factorizable r-matrix on $\mathfrak{d}$.

Explicitly, $\underline{r}_{\mathfrak{d}}$ is the linear map from $\mathfrak{d}^*$ to $\mathfrak{d}$ defined by

$$\underline{r}_{\mathfrak{d}}(\xi, x) = (0, \xi) ,$$

with symmetric part $\frac{1}{2}\underline{s}_{\mathfrak{d}}$ and skew-symmetric part

$$\underline{a}_{\mathfrak{d}}(\xi, x) = \frac{1}{2}(-x, \xi) .$$

The bracket on $\mathfrak{d}^*$ defined by the r-matrix $r_{\mathfrak{d}}$ is

$$[\xi + x, \eta + y]_{\mathfrak{d}^*} = [\xi, \eta] - [x, y] .$$

(We use the notations (ξ, x) or $\xi + x$ for elements of $\mathfrak{g}^* \oplus \mathfrak{g}$.) Thus in the Lie bialgebra $\mathfrak{d}$, the dual $\mathfrak{d}^*$ of $\mathfrak{d}$ is the direct sum of $\mathfrak{g}^*$ and the opposite of $\mathfrak{g}$.

Moreover, if the Lie bialgebra $\mathfrak{g}$ itself is quasi-triangular (the cocycle γ is the coboundary of a quasi-triangular r-matrix), then the Lie algebra $\mathfrak{d}$ is a direct sum that is isomorphic to $\mathfrak{g} \oplus \mathfrak{g}$.

Proposition. *If the Lie-bialgebra structure of $\mathfrak{g}$ is defined by a factorizable r-matrix, then its double is isomorphic to the direct sum of Lie algebras $\mathfrak{g} \oplus \mathfrak{g}$.*

Proof. We first embed $\mathfrak{g}$ in $\mathfrak{g} \oplus \mathfrak{g}$ by $x \mapsto (x, x)$. Then we embed $\mathfrak{g}^*$ in $\mathfrak{g} \oplus \mathfrak{g}$ by $j : \xi \mapsto (r_+\xi, r_-\xi)$. This map j is the map J of the last proposition of Sect. 2.4 composed with $\underline{s} : \mathfrak{g}^* \to \mathfrak{g}_R$, and formula (2.17) shows that it is a morphism of Lie algebras from $\mathfrak{g}^*$ to $\mathfrak{g} \oplus \mathfrak{g}$.

We obtain a Manin triple $(\mathfrak{g} \oplus \mathfrak{g}, \mathfrak{g}^{diag}, j(\mathfrak{g}^*))$, where $\mathfrak{g}^{diag}$ is the diagonal subalgebra of $\mathfrak{g} \oplus \mathfrak{g}$, and $\mathfrak{g} \oplus \mathfrak{g}$ is equipped with the invariant scalar product

$$((x, y), (x', y')) = \langle \underline{s}x, x' \rangle - \langle \underline{s}y, y' \rangle.$$

This Manin triple is isomorphic to $(\mathfrak{d}, \mathfrak{g}, \mathfrak{g}^*)$.

Example. The double of $\mathfrak{g} = \mathfrak{sl}(2, \mathbb{R})$ is $\mathfrak{sl}(2, \mathbb{R}) \oplus \mathfrak{sl}(2, \mathbb{R})$. (This fact was proved directly in Sect. 1.7.) This property extends to $\mathfrak{sl}(n, \mathbb{R})$.

We shall see in Sect. 3.6 that Hamiltonian systems on double Lie algebras give rise to equations in Lax form, and in Sect. 4.9 that the double plays a fundamental role in the theory of dressing transformations.

2.6   Bibliographical Note

This section is based on the articles by Drinfeld [15,16] and Semenov-Tian-Shansky [18,19,28]. See also Kosmann-Schwarzbach and Magri [40]. A survey and examples can be found in Chari and Pressley [22], Chap. 2. (Chap. 3 of [22] deals with the classification of the solutions of the classical Yang-Baxter equation, due to Belavin and Drinfeld, which we have not discussed in these lectures.) The first proposition in Sect. 2.1 is in [15] and, in invariant form, in [38,40]. The proposition in Sect. 2.5 was proved independently by Aminou and Kosmann-Schwarzbach [33] and by Reshetikhin and Semenov-Tian-Shansky [47].

3 Poisson Manifolds. The Dual of a Lie Algebra. Lax Equations

We shall now introduce Poisson manifolds and show that the dual of a finite-dimensional Lie algebra $\mathfrak{g}$ is always a Poisson manifold.

In particular, when $\mathfrak{g}$ is a Lie bialgebra, we obtain a Poisson structure on $\mathfrak{g}$ itself from the Lie-algebra structure on $\mathfrak{g}^*$. We shall first express the Poisson brackets of functions on the dual $\mathfrak{g}^*$ of a Lie algebra $\mathfrak{g}$ in tensor notation and then show that, for a connected Lie group, the co-adjoint orbits in $\mathfrak{g}^*$ are the leaves of the symplectic foliation of $\mathfrak{g}^*$. Finally, we shall show that Hamiltonian systems on double Lie algebras give rise to equations in Lax form.

3.1 Poisson Manifolds

On smooth manifolds, one can define Poisson structures, which give rise to Poisson brackets with the usual properties, on the space of smooth functions on the manifold in the following way. We shall henceforth write manifold for smooth manifold, tensor for smooth field of tensors, etc. Let us denote the space of functions on a manifold M by $C^\infty(M)$. By a *bivector* on a manifold M we mean a skew-symmetric, contravariant 2-tensor, *i.e.*, if P is a bivector, at each point $x \in M$, P_x has skew-symmetric components in local coordinates, $(P^{ij}(x))$, $i, j = 1, 2, \cdots, \dim M$. At each point x, we can view P_x as a skew-symmetric bilinear form on $T_x^* M$, the dual of the tangent space $T_x M$, or as the skew-symmetric linear map $\underline{P}_x$ from $T_x^* M$ to $T_x M$, such that

$$\langle \eta_x, \underline{P}_x(\xi_x) \rangle = P_x(\xi_x, \eta_x), \text{ for } \xi_x, \eta_x \in T_x^* M \ . \tag{3.1}$$

If ξ, η are differential 1-forms on M, we define $P(\xi, \eta)$ to be the function in $C^\infty(M)$ whose value at $x \in M$ is $P_x(\xi_x, \eta_x)$.

If f, g are functions on M, and df, dg denote their differentials, we set

$$\{f, g\} = P(df, dg) \ . \tag{3.2}$$

Note that $\underline{P}(df)$ is a vector field, denoted by X_f, and that

$$\{f, g\} = X_f . g \ . \tag{3.3}$$

It is clear that $\{f, gh\} = \{f, g\}h + g\{f, h\}$, for any functions f, g, h on M, so that the bracket $\{ \, , \, \}$ satisfies the Leibniz rule.

Definition. *A Poisson manifold (M, P) is a manifold M with a Poisson bivector P such that the bracket defined by (3.2) satisfies the Jacobi identity.*

When (M, P) is a Poisson manifold, $\{f, g\}$ is called the Poisson bracket of f and $g \in C^\infty(M)$, and $X_f = \underline{P}(df)$ is called the Hamiltonian vector field *with Hamiltonian f. Functions f and g are said to be* in involution *if $\{f, g\} = 0$.*

Example. If $M = \mathbb{R}^{2n}$, with coordinates (q^i, p_i), $i = 1, \cdots, n$, and if

$$\underline{P}(dq^i) = -\frac{\partial}{\partial p_i}, \quad \underline{P}(dp_i) = \frac{\partial}{\partial q^i},$$

then

$$X_f = \frac{\partial f}{\partial p_i}\frac{\partial}{\partial q^i} - \frac{\partial f}{\partial q^i}\frac{\partial}{\partial p_i}$$

and

$$\{f, g\} = \frac{\partial f}{\partial p_i}\frac{\partial g}{\partial q^i} - \frac{\partial f}{\partial q^i}\frac{\partial g}{\partial p_i},$$

the usual Poisson bracket of functions on phase-space. The corresponding bivector is $P = \frac{\partial}{\partial p_i} \wedge \frac{\partial}{\partial q^i}$.

In local coordinates, a necessary and sufficient condition for a bivector P to be a Poisson bivector is

$$\circlearrowleft P^{i\ell}\frac{\partial P^{jk}}{\partial x^\ell} = 0,$$

where $\circlearrowleft$ denotes the sum over the circular permutations of i, j, k (and the summation over $\ell = 1, 2, \ldots, \dim M$ is understood). In an invariant formalism, we define $[\![P, P]\!]$ to be the contravariant 3-tensor with local components

$$[\![P, P]\!]^{ijk} = -2 \circlearrowleft P^{i\ell}\frac{\partial P^{jk}}{\partial x^\ell}. \tag{3.4}$$

One can show that this defines a tri-vector (skew-symmetric, contravariant 3-tensor), called the *Schouten bracket of P* (with itself), and that

$$[\![P, P]\!](df_1, df_2, df_3) = -2 \circlearrowleft \{f_1, \{f_2, f_3\}\}. \tag{3.5}$$

From (3.5), it follows that P is a Poisson bivector if and only if $[\![P, P]\!] = 0$.

Alternatively, the bivector P is a Poisson bivector if and only if

$$X_{\{f,g\}} = [X_f, X_g], \quad \text{for all } f, g \in C^\infty(M). \tag{3.6}$$

Now let (M, ω) be an arbitrary symplectic manifold, *i.e.*, ω is a closed, non-degenerate differential 2-form on M. To say that ω is non-degenerate means that for each $x \in M$, $\underline{\omega}_x : T_x M \to T_x^* M$ defined by

$$\langle \underline{\omega}_x(X_x), Y_x \rangle = \omega_x(Y_x, X_x),$$

for $X_x, Y_x \in T_x M$, is a bijective linear map.

Let us show that every symplectic manifold has a Poisson structure. Since the linear mapping $\underline{\omega}_x$ is bijective, it has an inverse, which we denote by $\underline{P}_x$. As above, we set $X_f = \underline{P}(df) = \underline{\omega}^{-1}(df)$. Then

$$\{f, g\} = X_f \cdot g = \langle \underline{\omega}(X_g), X_f \rangle = \omega(X_f, X_g).$$

First, one can show, using the classical properties of the Lie derivative, that $d\omega = 0$ implies

$$[X_f, X_g] = X_{\{f,g\}},$$

for all f and g in $C^\infty(M)$. Now

$$\circlearrowleft \{f_1, \{f_2, f_3\}\} = X_{f_1} \cdot X_{f_2} \cdot f_3 - X_{f_2} \cdot X_{f_1} \cdot f_3 - X_{\{f_1, f_2\}} \cdot f_3,$$

which vanishes by the preceding observation. Thus, when ω is a closed, non-degenerate 2-form, the bracket $\{\,,\,\}$ satisfies the Jacobi identity. Therefore, if ω is a symplectic structure on M, then $\underline{P}$ defines a Poisson structure on M. Conversely, if P is a Poisson bivector such that $\underline{P}$ is invertible, it defines a symplectic structure.

Whereas every symplectic manifold has a Poisson structure, the converse does not hold: a symplectic manifold is a Poisson manifold such that the linear map $\underline{P}_x$ is a bijection for each x in M.

3.2   The Dual of a Lie Algebra

We shall show that, given a finite-dimensional Lie algebra $\mathfrak{g}$, its dual $\mathfrak{g}^*$ is a Poisson manifold in a natural way. The Poisson structure on $\mathfrak{g}^*$ is defined by

$$\underline{P}_\xi(x) = -ad_x^* \xi, \tag{3.7}$$

where ξ is a point of $\mathfrak{g}^*$, and $x \in \mathfrak{g}$ is considered as a 1-form on $\mathfrak{g}^*$ at point ξ. This Poisson structure is called the *linear Poisson structure* (because $\underline{P}_\xi$ depends linearly on $\xi \in \mathfrak{g}^*$), or the *Berezin-Kirillov-Kostant-Souriau Poisson structure* (or a subset of those names), or the *Lie-Poisson structure* (not to be confused with the Poisson Lie structures on groups to be defined in Sect. 4.2).

For $f \in C^\infty(\mathfrak{g}^*)$, $d_\xi f$ is in $T_\xi^* \mathfrak{g}^* \simeq (\mathfrak{g}^*)^* \simeq \mathfrak{g}$, and

$$\underline{P}_\xi(d_\xi f) = -ad_{d_\xi f}^* \xi \in \mathfrak{g}^* \simeq T_\xi \mathfrak{g}^*,$$

so that

$$X_f(\xi) = -ad_{d_\xi f}^* \xi,$$

while

$$\{f, g\}(\xi) = X_f(\xi) \cdot g = -\langle ad_{d_\xi f}^* \xi, d_\xi g \rangle,$$

whence

$$\{f, g\}(\xi) = \langle \xi, [d_\xi f, d_\xi g] \rangle. \tag{3.8}$$

If, in particular, $f = x$, $g = y$, where $x, y \in \mathfrak{g}$ are seen as linear forms on $\mathfrak{g}^*$, we obtain

$$\{x, y\}(\xi) = \langle \xi, [x, y] \rangle . \tag{3.9}$$

It is clear that we have thus defined a Poisson structure on $\mathfrak{g}^*$ because, by (3.8), the Jacobi identity for $\{ , \}$ follows from the Jacobi identity for the Lie bracket on $\mathfrak{g}$. In fact

$$\circlearrowright \{x_1, \{x_2, x_3\}\}(\xi) = \langle \xi, \circlearrowright [x_1, [x_2, x_3]] \rangle = 0 .$$

From the Jacobi identity for linear functions on $\mathfrak{g}^*$, we deduce the Jacobi identity for polynomials, and hence for all smooth functions on $\mathfrak{g}^*$ (using a density theorem).

3.3   The First Russian Formula

We now consider the case where $\mathfrak{g}$ is not only a Lie algebra, but has a Lie-bialgebra structure defined by an r-matrix, $r \in \mathfrak{g} \otimes \mathfrak{g}$. In this case $\mathfrak{g}^*$ is a Lie algebra, with Lie bracket $[,]^r$ such that

$$\langle [\xi, \eta]^r, x \rangle = \delta r(x)(\xi, \eta) = ((ad_x \otimes 1 + 1 \otimes ad_x)(r))(\xi, \eta)$$
$$= [x \otimes 1 + 1 \otimes x, r](\xi, \eta) ,$$

in the notations of Sect. 1.2, 2.1 and 2.3. Since the vector space $\mathfrak{g}$ can be identified with the dual of the Lie algebra $\mathfrak{g}^*$, it has the linear Poisson structure, here denoted by $\{ , \}^r$,

$$\{\xi, \eta\}^r(x) = \langle x, [\xi, \eta]^r \rangle ,$$

whence

$$\{\xi, \eta\}^r(x) = [x \otimes 1 + 1 \otimes x, r](\xi, \eta) . \tag{3.10}$$

Let us examine the case where $\mathfrak{g}$ is a Lie algebra of matrices. It is customary to write L for a generic element in $\mathfrak{g}$ (because, in the theory of integrable systems, L denotes the Lax matrix; see, e.g., Perelomov [25]). If L is a $p \times p$ matrix, and 1 is the identity $p \times p$ matrix, then $L \otimes 1$ and $1 \otimes L$ are $p^2 \times p^2$ matrices. If $L = (a_i^j)$, then

$$L \otimes 1 = \begin{pmatrix} a_1^1 & 0 & \cdots & 0 & \cdots & a_p^1 & 0 & \cdots & 0 \\ 0 & a_1^1 & \cdots & 0 & \cdots & 0 & a_p^1 & \cdots & 0 \\ \vdots & \vdots & \ddots & \vdots & \ddots & \vdots & \vdots & \ddots & \vdots \\ 0 & 0 & \cdots & a_1^1 & \cdots & 0 & 0 & \cdots & a_p^1 \\ \vdots & \vdots & \vdots & \vdots & \vdots & \vdots & \vdots & \vdots & \vdots \\ a_1^p & 0 & \cdots & 0 & \cdots & a_p^p & 0 & \cdots & 0 \\ 0 & a_1^p & \cdots & 0 & \cdots & 0 & a_p^p & \cdots & 0 \\ \vdots & \vdots & \ddots & \vdots & \ddots & \vdots & \vdots & \ddots & \vdots \\ 0 & 0 & \cdots & a_1^p & \cdots & 0 & 0 & \cdots & a_p^p \end{pmatrix} , \quad 1 \otimes L = \begin{pmatrix} L & 0 & \cdots & 0 \\ 0 & L & \cdots & 0 \\ & & \ddots & \\ 0 & 0 & \cdots & L \end{pmatrix} .$$

Since r is an element in $\mathfrak{g} \otimes \mathfrak{g}$, it is also a $p^2 \times p^2$ matrix. Then, what we have denoted by $[L \otimes 1 + 1 \otimes L, r]$ is the usual commutator of $p^2 \times p^2$ matrices.

The Poisson structure of $\mathfrak{g} \simeq (\mathfrak{g}^*)^*$ is entirely specified if we know the Poisson brackets of the coordinate functions on $\mathfrak{g}$. Since $\mathfrak{g} \subset \mathfrak{gl}(p)$, it is enough to know the pairwise Poisson brackets of coefficients $\{a_i^j, a_k^\ell\}$, where a_i^j, for fixed indices $i, j \in \{1, \cdots, p\}$, is considered as the linear function on $\mathfrak{g}$ which associates its coefficient in the ith column and j^{th} row to a matrix $L \in \mathfrak{g}$. These Poisson brackets can be arranged in a $p^2 \times p^2$ matrix, which we denote by $\{L \overset{\otimes}{,} L\}$. By definition,

$$\{L \overset{\otimes}{,} L\} = \begin{pmatrix} \{a_1^1, a_1^1\} \cdots \{a_1^1, a_p^1\} \{a_2^1, a_1^1\} \cdots \{a_p^1, a_p^1\} \\ \{a_1^1, a_1^2\} \cdots \qquad \vdots \qquad \vdots \qquad \vdots \\ \vdots \qquad \vdots \qquad \vdots \qquad \vdots \\ \{a_1^1, a_1^p\} \cdots \{a_1^1, a_p^p\} \quad \cdots \qquad \vdots \\ \{a_1^2, a_1^1\} \cdots \{a_1^2, a_p^1\} \quad \cdots \\ \vdots \qquad \vdots \\ \{a_1^2, a_1^p\} \cdots \{a_1^2, a_p^p\} \quad \cdots \\ \vdots \qquad \vdots \\ \{a_1^p, a_1^p\} \cdots \{a_1^p, a_p^p\} \quad \cdots \qquad \{a_p^p, a_p^p\} \end{pmatrix}.$$

Now, evaluating the element $[L \otimes 1 + 1 \otimes L, r] \in \mathfrak{g} \otimes \mathfrak{g}$ on the pair $a_i^j, a_k^\ell \in \mathfrak{g}^*$, amounts to taking the $\begin{smallmatrix} j & \ell \\ i & k \end{smallmatrix}$ coefficient in the $p^2 \times p^2$ matrix $[L \otimes 1 + 1 \otimes L, r]$. Thus formula (3.10) becomes the equality of matrices,

$$\{L \overset{\otimes}{,} L\} = [L \otimes 1 + 1 \otimes L, r] . \tag{3.11}$$

Sometimes the notations $L_1 = L \otimes 1, L_2 = 1 \otimes L, \{L \overset{\otimes}{,} L\} = \{L_1, L_2\}$ are used. Then (3.11) becomes

$$\{L_1, L_2\} = [L_1 + L_2, r] . \tag{3.11'}$$

Formula (3.11) is what I call the *first Russian formula*. The second Russian formula to be explained in Sect. 4.3 will express the Poisson bracket of the coordinate functions on a Poisson Lie group when the elements of that group are matrices.

Remark. In formula (3.11), it is clear that r can be assumed to be a skew-symmetric tensor, since its symmetric part is necessarily *ad*-invariant. This does not mean that the $p^2 \times p^2$ matrix of r, nor that matrix $[L_1 + L_2, r]$ is skew-symmetric.

Examples. We denote an element in $\mathfrak{sl}(2, \mathbb{C})$ by $L = \begin{pmatrix} \alpha & \beta \\ \gamma & \delta \end{pmatrix}$, with $\alpha + \delta = 0$. Let $a = \frac{1}{4}(X \otimes Y - Y \otimes X) \in \bigwedge^2(\mathfrak{sl}(2, \mathbb{C}))$, as in Example 2 of Sect. 2.3. Then we obtain from formula (3.11),

$$\{\alpha, \beta\} = \frac{1}{4}\beta, \ \{\alpha, \gamma\} = \frac{1}{4}\gamma, \ \{\beta, \gamma\} = 0.$$

If we choose $r = X \otimes H - H \otimes X$, as in Example 3 of Sect. 2.3, then

$$r = \begin{pmatrix} 0 & -1 & 1 & 0 \\ 0 & 0 & 0 & -1 \\ 0 & 0 & 0 & 1 \\ 0 & 0 & 0 & 0 \end{pmatrix} , \ \text{and } \{\alpha, \beta\} = -2\alpha, \{\alpha, \gamma\} = 0, \{\beta, \gamma\} = 2\gamma.$$

3.4 The Traces of Powers of Lax Matrices Are in Involution

We shall now show that whenever there exists an r-matrix r such that (3.11) holds, the traces of powers of L are pairwise in involution, *i.e.*,

$$\{trL^k, trL^\ell\} = 0 , \tag{3.12}$$

for all $k, \ell \geq 1$. It will then follow that the eigenvalues of L (as functions on $\mathfrak{g}$) are in involution.

For $p \times p$ matrices A and B, we similarly define

$$\{A \overset{\otimes}{,} B\} = \{A_1, B_2\} .$$

First, we observe that

$$tr\{A_1, B_2\} = \{trA, trB\} . \tag{3.13}$$

For matrices A and B, it is clear that $(AB)_1 = A_1B_1$ and $(AB)_2 = A_2B_2$. It follows therefore from the Leibniz rule that, for matrices A, B and C,

$$\{A_1, (BC)_2\} = \{A_1, B_2C_2\} = B_2\{A_1, C_2\} + \{A_1, B_2\}C_2 .$$

Induction then shows that for any matrix L,

$$\{L_1^k, L_2^\ell\} = \sum_{a=0}^{k-1}\sum_{b=0}^{\ell-1} L_1^{k-a-1}L_2^{\ell-b-1}\{L_1, L_2\}L_1^a L_2^b .$$

From this relation, we obtain

$$\{trL^k, trL^\ell\} = tr\{L_1^k, L_2^\ell\} = k\ell \ trL_1^{k-1}L_2^{\ell-1}\{L_1, L_2\},$$

since $A_1B_2 = (A \otimes 1)(1 \otimes B) = A \otimes B = (1 \otimes B)(A \otimes 1) = B_2A_1$ for any matrices A and B. Now, we use (3.11′) and we obtain

$$\{trL^k, trL^\ell\} = k\ell \ trL_1^{k-1}L_2^{\ell-1}[L_1 + L_2, r] .$$

Now $tr(L_1^{k-1}L_2^{\ell-1}L_1r - L_1^{k-1}L_2^{\ell-1}rL_1) = 0$, and similarly $tr(L_1^{k-1}L_2^{\ell-1}L_2r - L_1^{k-1}L_2^{\ell-1}rL_2) = 0$, and relation (3.12) is proved. (This result can be obtained as a corollary of Semenov-Tian-Shansky's theorem on double Lie algebras to be stated in Sect. 3.6.)

We know that when a dynamical system is in *Lax form*, *i.e.*, can be written as

$$\dot{L} = [L, B] \,, \tag{3.14}$$

where L and B are matrices depending on the phase-space coordinates, the eigenvalues of L are conserved (isospectral evolution), and hence the traces of powers of L are conserved. What we have proved is the following

Proposition. *When the Poisson brackets of the coefficients of the Lax matrix are given by an r-matrix, the traces of powers of the Lax matrix are conserved quantities in involution.*

If sufficiently many of these conserved quantites in involution are functionally independent, the complete integrability of the system in the sense of Liouville and Arnold (see Arnold [2]) follows.

Remark. Relation (3.12) is valid under the weaker assumption that r is an r-matrix in the sense of Remark 2 of Sect. 2.1. In this more general case, we obtain from $[\xi, \eta]^r = ad_{r\xi}^* \eta - ad_{r\eta}^* \xi$,

$$\{L_1, L_2\} = [L_1, -\,{}^t r] + [L_2, r] = [\,{}^t r, L_1] - [r, L_2]. \tag{3.15}$$

(This is formula (11) of Babelon and Viallet [35], where $d = {}^t r$.)

3.5 Symplectic Leaves and Coadjoint Orbits

We have remarked in Sect. 3.1 that any symplectic manifold is a Poisson manifold but that the converse does not hold. In fact, any Poisson manifold is a union of symplectic manifolds, generally of varying dimensions, called the *leaves of the symplectic foliation* of the Poisson manifold. (In the case of a symplectic manifold, there is only one such leaf, the manifold itself.) On a Poisson manifold (M, P), let us consider an open set where the rank of $\underline{P}$ is constant. At each point x, the image of $T_x^* M$ under $\underline{P}_x$ is a linear subspace of $T_x M$ of dimension equal to the rank of $\underline{P}_x$. That the distribution $x \mapsto \mathrm{Im}\,\underline{P}_x$ is integrable follows from the Poisson property of P. (Here, 'distribution' means a vector subbundle of the tangent bundle. See, *e.g.*, Vaisman [30].) By the Frobenius theorem, this distribution defines a foliation, whose leaves are the maximal integral manifolds. If the rank of $\underline{P}_x$ is not constant on the manifold, this distribution defines a *generalized foliation*, the leaves of which are of varying dimension equal to the rank of the Poisson map. This will be illustrated in the case of the dual of a Lie algebra.

First, let us review the adjoint and coadjoint actions of Lie groups.

Let G be a Lie group with Lie algebra $\mathfrak{g}$. Then G acts on its Lie algebra by the adjoint action, denoted by Ad. For $g \in G$, $x \in \mathfrak{g}$,

$$Ad_g x = \frac{d}{dt}(g.\exp tx.g^{-1})|_{t=0} \ . \tag{3.15}$$

If G is a matrix Lie group, then g and x are matrices and we recover the usual formula,

$$Ad_g x = gxg^{-1} \ .$$

By definition, the adjoint orbit of $x \in \mathfrak{g}$ is the set of all $Ad_g x$, for $g \in G$. In the case of a matrix Lie group, it is the set of matrices which are conjugate to x by matrices belonging to G.

If we define

$$ad_x y = \frac{d}{dt}(Ad_{\exp tx} y)|_{t=0} \ ,$$

then

$$ad_x y = \frac{d}{dt}\frac{d}{ds}\exp tx.\exp sy.\exp(-tx)|_{t=s=0} = [x, y] \ .$$

Thus ad does coincide with the adjoint representation of $\mathfrak{g}$ on itself introduced in Sect. 1.2.

We can define an action of G on functions on $\mathfrak{g}$ by

$$(g.f)(x) = f(Ad_{g^{-1}}x) \ , \ \text{ for } x \in \mathfrak{g}, f \in C^\infty(\mathfrak{g}), g \in G \ . \tag{3.16}$$

A function f on $\mathfrak{g}$ is called *Ad-invariant* (or *G-invariant*) if $g.f = f$, for all $g \in G$. This means that f is constant on the orbits of the adjoint action. In the case of a group of matrices, it means that the value of f at x depends only on the equivalence class of x, modulo conjugation by elements of G. For example, we can take $f(x) = \frac{1}{k}tr(x^k)$, for k a positive integer.

We now consider the dual $\mathfrak{g}^*$ of $\mathfrak{g}$, and we define the coadjoint action of G on $\mathfrak{g}^*$, denoted by Ad^*, as

$$\langle Ad_g^* \xi, x \rangle = \langle \xi, Ad_{g^{-1}}x \rangle \ , \tag{3.17}$$

for $g \in G, \xi \in \mathfrak{g}^*, x \in \mathfrak{g}$. By definition, the coadjoint orbit of $\xi \in \mathfrak{g}^*$ is the set of all $Ad_g^*\xi$, for $g \in G$. If we define

$$ad_x^* \xi = \frac{d}{dt}Ad_{\exp tx}^* \xi|_{t=0} \ ,$$

we see that ad^* coincides with the coadjoint representation of $\mathfrak{g}$ on $\mathfrak{g}^*$ which we introduced in Sect. 1.3. The tangent space at $\xi \in \mathfrak{g}^*$ to the coadjoint orbit $\mathcal{O}_\xi$ of ξ is the linear subspace of $T_\xi \mathfrak{g}^* \simeq \mathfrak{g}^*$,

$$T_\xi \mathcal{O}_\xi = \{ad_x^* \xi | x \in \mathfrak{g}\} \ . \tag{3.18}$$

The Lie group G acts on functions on $\mathfrak{g}^*$ by

$$(g.f)(\xi) = f(Ad^*_{g^{-1}}\xi), \text{ for } \xi \in \mathfrak{g}^*, f \in C^\infty(\mathfrak{g}^*), g \in G . \qquad (3.19)$$

A function f on $\mathfrak{g}^*$ is called Ad^*-*invariant* (or G-*invariant*) if $g.f = f$, for all $g \in G$.

Now we give infinitesimal characterizations of Ad-invariant functions on $\mathfrak{g}$ and Ad^*-invariant functions on $\mathfrak{g}^*$.

Proposition. *Assume that the Lie group G is connected. A function $f \in C^\infty(\mathfrak{g})$ is Ad-invariant if and only if*

$$ad^*_x(d_x f) = 0 , \quad \text{for all } x \in \mathfrak{g} . \qquad (3.20)$$

A function $f \in C^\infty(\mathfrak{g}^*)$ is Ad^*-invariant if and only if

$$ad^*_{d_\xi f}\xi = 0 , \quad \text{for all } \xi \in \mathfrak{g}^* . \qquad (3.21)$$

Proof. Let f be a function on $\mathfrak{g}$. If G is connected, by (3.16), relation $g.f = f$ is satisfied for all $g \in G$ if and only if

$$0 = \frac{d}{dt} f(Ad_{\exp(-ty)}x)|_{t=0} = -\langle d_x f, [y, x]\rangle = -\langle ad^*_x(d_x f), y\rangle ,$$

for all $x, y \in \mathfrak{g}$, proving (3.20). Similarly, if $f \in C^\infty(\mathfrak{g}^*)$, then by (3.19), relation $g.f = f$ is satisfied for all $g \in G$ if and only if

$$0 = \frac{d}{dt} f(Ad^*_{\exp(-tx)}\xi)|_{t=0} = -\langle d_\xi f, ad^*_x\xi\rangle = \langle x, ad^*_{d_\xi f}\xi\rangle$$

for all $x \in \mathfrak{g}$, $\xi \in \mathfrak{g}^*$, proving (3.21).

Note that alternate ways of writing (3.20) and (3.21) are

$$d_x f \circ ad_x = 0, \text{ for all } x \in \mathfrak{g} \qquad (3.20')$$

and

$$\xi \circ ad_{d_\xi f} = 0, \text{ for all } \xi \in \mathfrak{g}^* . \qquad (3.21')$$

For example, if $f(x) = \frac{1}{k}tr(x^k)$, then $d_x f(h) = tr(x^{k-1}h)$, for $x, h \in \mathfrak{g}$, and it is clear that, for each $y \in \mathfrak{g}$,

$$tr(x^{k-1}[x, y]) = 0 .$$

Let us examine the case where $\mathfrak{g}^*$ can be identified with $\mathfrak{g}$ by means of an ad-invariant, non-degenerate, symmetric, bilinear form s. This is the case if $\mathfrak{g}$ is semi-simple and we choose s to be the Killing form. Then $\underline{s} : \mathfrak{g}^* \to \mathfrak{g}$ satisfies

$$ad_x \circ \underline{s} = \underline{s} \circ ad^*_x ,$$

for all $x \in \mathfrak{g}$ (see Sect. 2.1), so that in this identification the coadjoint action becomes the adjoint action. If f is a function on $\mathfrak{g}^* \simeq \mathfrak{g}$, its differential at a point L is an element in $\mathfrak{g}$. By (3.20), the function f is G-invariant if and only if

$$[L, d_L f] = 0 .$$

We can now prove

Proposition. *Let G be a Lie group with Lie algebra $\mathfrak{g}$. The symplectic leaves of the linear Poisson structure of $\mathfrak{g}^*$ are the connected components of the coadjoint orbits in $\mathfrak{g}^*$.*

Proof. By (3.7), the linear Poisson bivector P on $\mathfrak{g}^*$ is such that

$$\operatorname{Im} \underline{P}_\xi = \{ad_x^* \xi | x \in \mathfrak{g}\} .$$

Thus, by (3.18), the image of $\underline{P}_\xi$ coincides with the tangent space at $\xi \in \mathfrak{g}^*$ to the coadjoint orbit of ξ. The result follows.

Example. If $G = SO(3), \mathfrak{g}^*$ is a 3-dimensional vector space identified with $\mathbb{R}^3$ in which the coadjoint orbits are the point $\{0\}$ (0-dimensional orbit) and all spheres centered at the origin (2-dimensional orbits).

Note that the coadjoint orbit of the origin 0 in the dual of any Lie algebra is always $\{0\}$. In particular, the dual of a Lie algebra is not a symplectic manifold.

When a Poisson manifold is not symplectic, there are nonconstant functions which are in involution with all functions on the manifold.

Definition. *In a Poisson manifold, those functions whose Poisson brackets with all functions vanish are called* Casimir functions.

Proposition. *In the dual of the Lie algebra of a connected Lie group G, the Casimir functions are the Ad^*-invariant functions.*

Proof. In fact, if $f, g \in C^\infty(\mathfrak{g}^*)$ and $\xi \in \mathfrak{g}^*$, then

$$\{f, g\}(\xi) = \langle \xi, [d_\xi f, d_\xi g] \rangle = -\langle ad_{d_\xi f}^* \xi, d_\xi g \rangle .$$

This quantity vanishes for all $g \in C^\infty(\mathfrak{g}^*)$ if and only if $ad_{d_\xi f}^* \xi = 0$, and therefore this proposition is a consequence of (3.21).

The symplectic leaves of a Poisson manifold are contained in the connected components of the level sets of the Casimir functions. (On a symplectic leaf, each Casimir function is constant.)

3.6   Double Lie Algebras and Lax Equations

Let $(\mathfrak{g}, R)$ be a double Lie algebra in the sense of Sect. 2.4. Then $\mathfrak{g}$ has two Lie algebra structures, $[\,,\,]$ and $[\,,\,]_R$ defined by (2.11). We denote the corresponding adjoint (resp., coadjoint) actions by ad and ad^R (resp., ad^* and ad^{R*}). Thus $\mathfrak{g}^*$ has two linear Poisson structures, P and P^R. By definition, for $x, y \in \mathfrak{g}$, $\xi \in \mathfrak{g}^*$,

$$\langle \underline{P}^R_\xi(x), y \rangle = -\langle ad^{R*}_x \xi, y \rangle = \langle \xi, [x, y]_R \rangle$$

$$= \langle \xi, [Rx, y] + [x, Ry] \rangle \ ,$$

whence

$$\langle \underline{P}^R_\xi(x), y \rangle = -\langle ad^*_{Rx} \xi, y \rangle - \langle ad^*_x \xi, Ry \rangle \ . \tag{3.22}$$

We denote the Poisson bracket defined by P^R by $\{\ ,\ \}_R$. For $f_1, f_2 \in C^\infty(\mathfrak{g}^*)$,

$$\{f_1, f_2\}_R(\xi) = -\langle ad^*_{R(d_\xi f_1)} \xi, d_\xi f_2 \rangle - \langle ad^*_{d_\xi f_1} \xi, R(d_\xi f_2) \rangle$$

$$= \langle ad^*_{d_\xi f_2} \xi, R(d_\xi f_1) \rangle - \langle ad^*_{d_\xi f_1} \xi, R(d_\xi f_2) \rangle \ .$$

From this formula and from the proof of the last proposition of Sect. 3.5, we obtain

Theorem. *Let $\mathfrak{g}^*$ be the dual of a double Lie algebra, with Poisson brackets $\{\ ,\ \}$ and $\{\ ,\ \}_R$. If f_1 and f_2 are Casimir functions on $(\mathfrak{g}^*, \{\ ,\ \})$, they are in involution with respect to $\{\ ,\ \}_R$.*

Now, let f be a Casimir function on $(\mathfrak{g}^*, \{\ ,\ \})$. Its Hamiltonian vector field X_f clearly vanishes. Let us denote the Hamiltonian vector field with Hamiltonian f with respect to $\{\ ,\ \}_R$ by X^R_f. Then

$$X^R_f(\xi) = \underline{P}^R_\xi(d_\xi f) \ .$$

If f is a Casimir function on $(\mathfrak{g}^*, \{\ ,\ \})$, it follows from (3.22) and (3.21) that

$$X^R_f(\xi) = -ad^*_{R(d_\xi f)} \xi \ . \tag{3.23}$$

The corresponding Hamiltonian equation is $\dot{\xi} = -ad^*_{R(d_\xi f)} \xi$. In addition, let us assume that $\mathfrak{g}^*$ is identified with $\mathfrak{g}$ by an ad-invariant, nondegenerate, symmetric $s \in \mathfrak{g} \otimes \mathfrak{g}$. Then, denoting a generic element in $\mathfrak{g}$ by L, the element $-ad^*_{R(d_\xi f)} \xi$ in $\mathfrak{g}^*$ is identified with $-ad_{R(d_L f)} L = [L, R(d_L f)]$ in $\mathfrak{g}$. Thus the Hamiltonian vector field satisfies

$$X^R_f(L) = [L, R(d_L f)] \ ,$$

and the corresponding Hamiltonian equation is

$$\dot{L} = [L, R(d_L f)] \tag{3.24}$$

which is in fact an equation of Lax type, $\dot{L} = [L, B]$.

The theorem and these formulæ constitute *Semenov-Tian-Shansky's theorem*.

If, in particular, $\mathfrak{g} = \mathfrak{g}_+ \oplus \mathfrak{g}_-$, where $\mathfrak{g}_+$ and $\mathfrak{g}_-$ are Lie subalgebras, and $R = \frac{1}{2}(p_+ - p_-)$, where p_+ (resp., p_-) is the projection onto $\mathfrak{g}_+$ (resp., $\mathfrak{g}_-$)

parallel to $\mathfrak{g}_-$ (resp., $\mathfrak{g}_+$) (see the example in Sect. 2.4), then for a Casimir function f on $(\mathfrak{g}^*, \{\ ,\ \})$, from (3.23) and (3.21), we obtain

$$X_f^R(\xi) = -\frac{1}{2}ad^*_{(d_\xi f)_+ - (d_\xi f)_-}\xi = -ad^*_{(d_\xi f)_+}\xi = ad^*_{(d_\xi f)_-}\xi \ ,$$

where $x_\pm = p_\pm(x)$, or, in the Lax form,

$$\dot{L} = [L, (d_L f)_+] \ , \tag{3.25}$$

or, equivalently,

$$\dot{L} = -[L, (d_L f)_-] \ .$$

Example. Let $\mathfrak{g} = \mathfrak{sl}(n, \mathbb{R})$, and let $\mathfrak{g}_+ = \mathfrak{so}(n)$, $\mathfrak{g}_-$ the subalgebra of upper triangular matrices (see the example of Sect 2.4), where $\mathfrak{g}^*$ and $\mathfrak{g}$ are identified by the trace functional $tr(xy)$, and let us choose $f(L) = \frac{1}{2}trL^2$.
Then (3.25) becomes

$$\dot{L} = [L, L_+] \ , \tag{3.26}$$

where $L = L_+ + L_-$ and $L_\pm \in \mathfrak{g}_\pm$.

The various results of this section are of importance in the theory of integrable systems because they furnish conserved quantities in involution. If h is a Casimir function of $(\mathfrak{g}, [\ ,\ ])$, the Hamiltonian vector field X_h^R is tangent to the coadjoint orbits of $(\mathfrak{g}, [\ ,\ ]_R)$. In restriction to an orbit, which is a symplectic manifold, we obtain a Hamiltonian system, for which the restrictions of the Casimir functions of $(\mathfrak{g}^*, \{\ ,\ \})$ are conserved quantities in involution.

For example, by restricting the Hamiltonian system of the preceding example to the adjoint orbit of the matrix

$$\begin{pmatrix} 0 & 1 & 0 & \cdots & 0 \\ 1 & 0 & 0 & \cdots & 0 \\ \vdots & \vdots & \vdots & & \vdots \\ 0 & 0 & \cdots & 0 & 1 \\ 0 & 0 & \cdots & 1 & 0 \end{pmatrix}$$

in $\mathfrak{sl}(n, \mathbb{R})$, we obtain the Toda system, for which $\frac{1}{k}tr(L^k)$ are conserved quantities in involution.

Let us show that Semenov-Tian-Shansky's theorem implies the Adler-Kostant-Symes theorem.
Let us denote the orthogonal of $\mathfrak{g}_+$ (resp., $\mathfrak{g}_-$) in $\mathfrak{g}^*$ by $\mathfrak{g}_+^\perp$ (resp., $\mathfrak{g}_-^\perp$). Then $\mathfrak{g}_+^*$ can be identified with $\mathfrak{g}_-^\perp$, while $\mathfrak{g}_-^*$ can be identified with $\mathfrak{g}_+^\perp$, and $\mathfrak{g}^*$ splits as

$$\mathfrak{g}^* = \mathfrak{g}_-^\perp \oplus \mathfrak{g}_+^\perp \approx \mathfrak{g}_+^* \oplus \mathfrak{g}_-^* \ .$$

If f is a function on $\mathfrak{g}^*$, then

$$d_\xi(f|_{\mathfrak{g}_+^*}) = (d_\xi f)_+ . \tag{3.27}$$

It follows that, for functions f_1 and f_2 on $\mathfrak{g}^*$,

$$\{f_1|_{\mathfrak{g}_+^*}, f_2|_{\mathfrak{g}_+^*}\}_{\mathfrak{g}_+^*} = \{f_1, f_2\}_R|_{\mathfrak{g}_+^*} , \tag{3.28}$$

where $\{ , \}_{\mathfrak{g}_+^*}$ denotes the Lie-Poisson bracket on the dual of the Lie algebra $\mathfrak{g}_+$. In fact, by formula (2.19), both sides evaluated at $\xi \in \mathfrak{g}_+^*$ are equal to $\langle \xi, [(d_\xi f_1)_+, (d_\xi f_2)_+]\rangle$. By the preceding theorem, if f_1 and f_2 are Casimir functions on $\mathfrak{g}^*$, they satisfy $\{f_1, f_2\}_R = 0$, and therefore, by (3.28),

$$\{f_1|_{\mathfrak{g}_+^*}, f_2|_{\mathfrak{g}_+^*}\}_{\mathfrak{g}_+^*} = 0 .$$

Thus we obtain the Adler-Kostant-Symes theorem, namely,

Theorem. Let $\mathfrak{g} = \mathfrak{g}_+ \oplus \mathfrak{g}_-$, where $\mathfrak{g}_+$ and $\mathfrak{g}_-$ are Lie subalgebras of $\mathfrak{g}$, and let f_1 and f_2 be Casimir functions on $\mathfrak{g}^*$. Then the restrictions of f_1 and f_2 to $\mathfrak{g}_-^\perp \approx \mathfrak{g}_+^*$ commute in the Lie-Poisson bracket of $\mathfrak{g}_+^*$.

3.7   Solution by Factorization

We shall show that, when the R-matrix is defined by the decomposition of a Lie algebra into a sum of complementary Lie subalgebras, the problem of integrating dynamical system (3.23) can be reduced to a factorization problem in the associated Lie group. Thus in this case, the Lax equation (3.24) can be solved "by factorization". Actually this scheme is valid in the more general situation where R is a factorizable R-matrix, and this fact explains the terminology.

Let G be a Lie group with Lie algebra $\mathfrak{g} = \mathfrak{g}_+ \oplus \mathfrak{g}_-$, where $\mathfrak{g}_+$ and $\mathfrak{g}_-$ are Lie subalgebras of $\mathfrak{g}$, and let G_+ (resp., G_-) be the connected Lie subgroup of G with Lie algebra $\mathfrak{g}_+$ (resp., $\mathfrak{g}_-$). The solution by factorization of the initial value problem,

$$\begin{cases} \dot\xi = -ad^*_{(d_\xi f)_+}\xi , \\ \xi(0) = \xi_0 , \end{cases}$$

where f is a Casimir function on $\mathfrak{g}^*$, and ξ_0 is an element of $\mathfrak{g}^*$, is the following.

Let $x_0 = d_{\xi_0} f \in \mathfrak{g}$, and assume that $e^{tx_0} \in G$ has been factorized as

$$e^{tx_0} = g_+(t)^{-1} g_-(t) , \tag{3.29}$$

where $g_\pm(t) \in G_\pm, g_\pm(0) = e$, which is possible for $|t|$ small enough. We shall prove that

$$\xi(t) = Ad^*_{g_+(t)}\xi_0 \tag{3.30}$$

solves the preceding initial-value problem.

In fact, because f is Ad^*-invariant, for $g \in G, \eta \in \mathfrak{g}^*$,

$$(d_{Ad^*_g \xi_0} f)(\eta) = \frac{d}{ds} f(Ad^*_g \xi_0 + s\eta)|_{s=0}$$

$$= \frac{d}{ds} f(\xi_0 + s Ad^*_{g^{-1}} \eta)|_{s=0} = (d_{\xi_0} f)(Ad^*_{g^{-1}} \eta),$$

so that

$$d_{Ad^*_g \xi_0} f = Ad_g (d_{\xi_0} f) . \tag{3.31}$$

Now, from $g_+(t)e^{t x_0} = g_-(t)$, we obtain, by differentiating,

$$\dot{g}_+(t) g_+(t)^{-1} + Ad_{g_+(t)} x_0 = \dot{g}_-(t) g_-(t)^{-1} .$$

Using the definition of x_0, and formulæ (3.30) (3.31), we obtain

$$d_{\xi(t)} f = -\dot{g}_+(t) g_+(t)^{-1} + \dot{g}_-(t) g_-(t)^{-1} ,$$

so that

$$(d_{\xi(t)} f)_+ = -\dot{g}_+(t) g_+(t)^{-1} ,$$

while it follows from (3.30) that

$$\dot{\xi}(t) = ad^*_{\dot{g}_+(t) g_+(t)^{-1}} \xi(t).$$

Therefore $\xi(t)$ given by (3.30) solves the given initial-value problem.

3.8   Bibliographical Note

For Poisson manifolds and coadjoint orbits, see Cartier [21], Vaisman [30] or Marsden and Ratiu [12]. For Sect. 3.3, see Faddeev and Takhtajan [23], Babelon and Viallet [20]. Double Lie algebras were introduced by Semenov-Tian-Shansky [18,28], where the theorem of Sect. 3.6 is proved, and the factorization method is explained. For the Toda system, see Kostant [9], Reyman and Semenov-Tian-Shansky [27], Semenov-Tian-Shansky [28], Babelon and Viallet [20]. For the Adler-Kostant-Symes theorem, see Kostant [9], Guillemin and Sternberg [7]. The books by Faddeev and Takhtajan [23] and by Perelomov [25] contain surveys of the Lie-algebraic approach to integrable equations.

4   Poisson Lie Groups

When a Lie group is also a Poisson manifold, it is natural to require that the Poisson structure and the multiplication defining the group structure be compatible in some sense. This idea, when made precise, leads to the notion of a Poisson Lie group. In fact, the introduction of Poisson Lie groups was motivated by the properties of the monodromy matrices of difference equations describing lattice integrable systems. We shall give the definition of Poisson Lie groups and show that the infinitesimal object corresponding to a Poisson Lie group is a Lie bialgebra. The emphasis will be on Poisson Lie groups defined by r-matrices, and we shall study examples on matrix groups.

It has gradually emerged from the physics literature that the group of dressing transformations should be considered as the action of a Poisson Lie group on a Poisson manifold. This action is a Poisson action in a sense that we shall define in Sect. 4.7, and which generalizes the Hamiltonian actions. For such actions one can define a momentum mapping (also known as a non-Abelian Hamiltonian). If G and G^* are dual Poisson Lie groups, the dressing action of G^* on G, and that of G^* on G will be defined and characterized in various ways in Sect. 4.9.

4.1   Multiplicative Tensor Fields on Lie Groups

First recall that in a Lie group G, the left- and right-translations by an element $g \in G$, denoted by λ_g and ρ_g respectively, are defined by

$$\lambda_g(h) = gh \ , \ \rho_g(h) = hg \ ,$$

for $h \in G$. Taking the tangent linear map to λ_g (resp., ρ_g) at point $h \in G$, we obtain a linear map from $T_h G$, the tangent space to G at h, into the tangent space to G at $\lambda_g(h) = gh$ (resp., at $\rho_g(h) = hg$). For any positive integer k, using the k-th tensor power of the tangent map to λ_g (resp., ρ_g), we can map k-tensors at h to k-tensors at gh (resp., hg). If Q_h is a tensor at h, we simply denote its image under this map by $g.Q_h$ (resp., $Q_h.g$). Here and below, we write tensor for smooth contravariant tensor field.

Definition. A tensor Q on a Lie group G is called multiplicative if

$$Q_{gh} = g.Q_h + Q_g.h \ , \tag{4.1}$$

for all $g, h \in G$.

Note that this relation implies that

$$Q_e = 0 \ , \tag{4.2}$$

where e is the unit element of the group G. In fact, setting $h = g = e$ in (4.1), we obtain $Q_e = 2Q_e$.

Let us denote the Lie algebra of G, which is the tangent space $T_e G$ to G at e, by $\mathfrak{g}$. We can associate to any k-tensor Q on the Lie group G a mapping $\rho(Q)$ from G to the k-th tensor power of $\mathfrak{g}$ defined by

$$\rho(Q)(g) = Q_g.g^{-1} \ , \tag{4.3}$$

for $g \in G$. When Q is multiplicative, the mapping $\rho(Q)$ has the following property. For all $g, h \in G$,

$$\rho(Q)(gh) = Ad_g(\rho(Q)(h)) + \rho(Q)(g) \ . \tag{4.4}$$

In fact, by (4.3) and (4.1),

$$\begin{aligned} \rho(Q)(gh) &= Q_{gh}.(gh)^{-1} = (g.Q_h + Q_g.h).h^{-1}.g^{-1} \\ &= g.\rho(Q)(h).g^{-1} + \rho(Q)(g) \ . \end{aligned}$$

Taking into account the definition of the adjoint action of the group on tensor powers of $\mathfrak{g}$, $Ad_g a = g.a.g^{-1}$, for $a \in \otimes^k \mathfrak{g}$, we obtain (4.4). Observe that if $a = a_1 \otimes \cdots \otimes a_k$, then $g.a = g.a_1 \otimes \cdots \otimes g.a_k$.

Definition. *A mapping U from G to a representation space $(V, \mathcal{R})$ of G satisfying*

$$U(gh) = \mathcal{R}(g)(U(h)) + U(g) \tag{4.5}$$

is called a 1-cocycle of G with values in V, with respect to the representation $\mathcal{R}$.

Thus we have proved

Proposition. *If Q is a multiplicative k-tensor on a Lie group G, then $\rho(Q)$ is a 1-cocycle of G with values in the k-th tensor power of $\mathfrak{g}$, with respect to the adjoint action of G.*

As the names suggest, there is a relationship between Lie-group cocycles and Lie-algebra cocycles. The following is a well-known result.

Proposition. *If $U : G \to V$ is a 1-cocycle of G with values in V, with respect to the representation $\mathcal{R}$, then $T_e U : \mathfrak{g} \to V$ defined by*

$$T_e U(x) = \frac{d}{dt} U(\exp tx)|_{t=0} ,$$

for $x \in \mathfrak{g}$, is a 1-cocycle of $\mathfrak{g}$ with values in V, with respect to the representation $d_e \mathcal{R}$ defined by

$$d_e \mathcal{R}(x) = \frac{d}{dt} \mathcal{R}(\exp tx)|_{t=0} .$$

Conversely, if u is a 1-cocyle of $\mathfrak{g}$ with values in V, with respect to a representation σ of $\mathfrak{g}$, and if G is connected and simply connected, there exists a unique 1-cocyle U of G with values in V, with respect to the representation $\mathcal{R}$ such that $d_e \mathcal{R} = \sigma$, satisfying $T_e U = u$.

To any multiplicative k-tensor Q, we associate the 1-cocycle on $\mathfrak{g}$ with respect to the adjoint action of $\mathfrak{g}$ on the k-th tensor power of $\mathfrak{g}$,

$$DQ = T_e(\rho(Q)) . \tag{4.6}$$

This linear map from $\mathfrak{g}$ to $\overset{k}{\otimes} \mathfrak{g}$ coincides with the linearization of Q at e, namely it satisfies, for $x \in \mathfrak{g}$,

$$(DQ)(x) = (\mathcal{L}_X Q)(e) ,$$

where X is any vector field on G defined in a neighborhood of e such that $X_e = x$, and $\mathcal{L}$ is the Lie derivation.

Example. Let $q \in \overset{k}{\otimes} \mathfrak{g}$. Set $Q = q^\lambda - q^\rho$, where

$$q^\lambda(g) = g.q , \quad q^\rho(g) = q.g , \tag{4.7}$$

for $g \in G$. Then Q is a multiplicative k-tensor. Such a multiplicative tensor is called *exact* because its associated 1-cocycle $\rho(Q) : G \to \overset{k}{\otimes} \mathfrak{g}$ is exact, namely

$$\rho(Q)(g) = Ad_g q - q . \tag{4.8}$$

(If an element $q \in \overset{k}{\otimes} \mathfrak{g}$ is considered as a 0-cochain on G, its group coboundary is $g \mapsto Ad_g q - q$.) In this case, $(DQ)(x) = ad_x q$, so that $DQ = \delta q$, the Lie-algebra coboundary of the 0-cochain q, with respect to the adjoint representation.

4.2   Poisson Lie Groups and Lie Bialgebras

Definition. *A Lie group G, with Poisson bivector P, is called a* Poisson Lie group *if P is multiplicative.*

By (4.2), $P_e = 0$, so (G, P) is not a symplectic manifold.

Examples. Obviously, G with the trivial Poisson structure ($P = 0$) is a Poisson Lie group.

The dual $\mathfrak{g}^*$ of a Lie algebra, considered as an Abelian group, with the linear Poisson structure is a Poisson Lie group. In fact, in this case (4.1) reads

$$P_{\xi+\eta} = P_\xi + P_\eta ,$$

for $\xi, \eta \in \mathfrak{g}^*$, and this relation holds since $\xi \mapsto P_\xi$ is linear.

Remark. The condition that the Poisson bivector P be multiplicative is equivalent to the following condition:

$$\{\varphi \circ \lambda_g, \psi \circ \lambda_g\}(h) + \{\varphi \circ \rho_h, \psi \circ \rho_h\}(g) = \{\varphi, \psi\}(gh) ,$$

for all functions φ, ψ on G, and for all g, h in G, which means that the multiplication map from $G \times G$ to G maps Poisson brackets on $G \times G$ to Poisson brackets on G. In other words, the multiplication is a Poisson map from $G \times G$ to G, where $G \times G$ is endowed with the product Poisson structure. (We recall the definition of the product of Poisson manifolds and that of Poisson maps below, in Sect. 4.7.)

Let $r \in \bigwedge^2 \mathfrak{g}$. Then the bivector defined by

$$P = r^\lambda - r^\rho \tag{4.9}$$

is multiplicative. Below we shall derive conditions on r for P to be a Poisson bivector.

When P is a multiplicative bivector, let us set

$$\gamma = DP = T_e(\rho(P)) . \tag{4.10}$$

By the two propositions of Sect. 4.1, we know that γ is a 1-cocycle of $\mathfrak{g}$ with values in $\bigwedge^2 \mathfrak{g}$, with respect to the adjoint representation.

Proofs of the following two propositions will be given in Appendix 1.

Proposition. *If P is a multiplicative Poisson bivector on G, let γ be defined by (4.10). Then ${}^t\gamma : \bigwedge^2 \mathfrak{g}^* \to \mathfrak{g}^*$ is a Lie bracket on $\mathfrak{g}^*$.*

Then $(\mathfrak{g}, \gamma)$ is a Lie bialgebra, which is called the *tangent Lie bialgebra* of (G, P).

Proposition. *Conversely, if $(\mathfrak{g}, \gamma)$ is a Lie bialgebra, there exists a unique (up to isomorphism) connected and simply connected Poisson Lie group (G, P) with tangent Lie bialgebra $(\mathfrak{g}, \gamma)$.*

Here we shall be concerned only with the case where P is defined by $r \in \bigwedge^2 \mathfrak{g}$ by means of (4.9).

Proposition. *The multiplicative bivector P defined by (4.9) is a Poisson bivector if and only if $[\![r, r]\!]$ is Ad-invariant.*

Proof. Recall that $[\![r, r]\!]$ was defined in (2.4). The proof rests on the formulæ $[\![r^\lambda, r^\lambda]\!] = [\![r, r]\!]^\lambda$, $[\![r^\rho, r^\rho]\!] = -[\![r, r]\!]^\rho$, and $[\![r^\lambda, r^\rho]\!] = 0$, whence $\rho([P, P])(g) = \mathrm{Ad}_g([\![r, r]\!]) - [\![r, r]\!]$.

If G is connected, the Ad-invariance under the action of G is equivalent to the ad-invariance under the action of its Lie algebra $\mathfrak{g}$. (See Sect. 1.2 and 3.5 for the definitions of ad- and Ad-actions.) Thus, in this case, $P = r^\lambda - r^\rho$ is a Poisson bivector if and only if r is a solution of the generalized Yang-Baxter equation. In particular, if r is a triangular r-matrix, then P is a Poisson bivector. (See Remark 1, below.) More generally, if r is a quasi-triangular r-matrix with ad-invariant symmetric part s and skew-symmetric part $a \in \bigwedge^2 \mathfrak{g}$, such that $\langle a, a \rangle + \langle s, s \rangle = 0$, then

$$P = r^\lambda - r^\rho = a^\lambda - a^\rho ,$$

and P is a Poisson bivector since

$$-\frac{1}{2}[\![a, a]\!] = \langle a, a \rangle = -\langle s, s \rangle ,$$

which is an ad-invariant element of $\bigwedge^3 \mathfrak{g}$. Then (G, P) is called a *quasi-triangular Poisson Lie group*. If moreover $\underline{s}$ is invertible, then the Poisson Lie group (G, P) is called *factorizable*.

Thus any quasi-triangular r-matrix, in particular a triangular or a factorizable r-matrix, gives rise to a Poisson-Lie structure on G. The Poisson bracket of functions on G thus defined is called the *Sklyanin bracket* or the *quadratic bracket*. The reason for this latter name will appear in the next section.

Remark 1. It is clear from the proof of the preceding proposition that, when r is a triangular r-matrix, both r^λ and r^ρ are also Poisson structures. Moreover, they are *compatible, i.e.,* any linear combination of r^λ and r^ρ is a Poisson structure. However, $r^\lambda - r^\rho$ is the only Poisson-Lie structure in this family.

Remark 2. When r is quasi-triangular, $\langle a, a \rangle = -\langle s, s \rangle$ and therefore both $a^\lambda - a^\rho$ and $a^\lambda + a^\rho$ are Poisson structures. While $a^\lambda - a^\rho$ is a Poisson-Lie structure, $a^\lambda + a^\rho$ is not. In fact, its rank at the unit, e, is the rank of $\underline{a}$, and therefore not 0, unless we are in the trivial case, $a = 0$. If $\underline{a}$ is invertible, $a^\lambda + a^\rho$ is symplectic in a neighborhood of e.

4.3 The Second Russian Formula (Quadratic Brackets)

Let us assume that G is a Lie group of $p \times p$ matrices. Then $\mathfrak{g}$ is a Lie algebra of $p \times p$ matrices and r is a $p^2 \times p^2$ matrix. If L is a point in G, the entries of the left (resp., right) translate of r by L are those of the product of matrices $(L \otimes L)r$ (resp., $r(L \otimes L)$). As in Sect. 3.3, we shall consider each entry a_i^j of L as the restriction to G of the corresponding linear function on the space of $p \times p$ matrices, and we shall denote the table of their pairwise Poisson brackets, in the Poisson-Lie structure on G defined by r, by $\{L \overset{\otimes}{,} L\}$. The differential of such a linear function is constant and coincides with the linear function itself. Therefore the $\overset{j\ \ell}{\underset{i\ k}{}}$ coefficient of $\{L \overset{\otimes}{,} L\}$ is the $\overset{j\ \ell}{\underset{i\ k}{}}$ coefficient of $(L \otimes L)r - r(L \otimes L)$. Whence, for $L \in G$,

$$\{L \overset{\otimes}{,} L\} = [L \otimes L, r] . \tag{4.11}$$

Formula (4.11) is what I call the *second Russian formula*. (It is close to, but different from (3.11) in Sect. 3.3.) It is clear from formula (4.11) that the Poisson brackets of any two entries of matrix L are quadratic functions of the entries of L, which justifies the name "quadratic bracket". (Note that in formula (3.11) in Sect. 3.3, the Poisson bracket is linear.) Formula (4.11) is the basis of the r-matrix method for classical lattice integrable systems.

4.4 Examples

Quasi-Triangular Structure of $SL(2, \mathbb{R})$. Let $G = SL(2, \mathbb{R})$ be the group of real 2×2 matrices with determinant 1. We consider $r = \frac{1}{8}(H \otimes H + 4X \otimes Y)$, the factorizable r-matrix of Example 2 in Sect. 2.3, an element of $\mathfrak{sl}(2, \mathbb{R}) \otimes \mathfrak{sl}(2, \mathbb{R})$, with skew-symmetric part $r_0 = \frac{1}{4}(X \otimes Y - Y \otimes X)$.

Let $L = \begin{pmatrix} a & b \\ c & d \end{pmatrix}$ be a generic element of G. Then $ad - bc = 1$. By (4.11), we obtain

$$\begin{pmatrix} \{a,a\} & \{a,b\} & \{b,a\} & \{b,b\} \\ \{a,c\} & \{a,d\} & \{b,c\} & \{b,d\} \\ \{c,a\} & \{c,b\} & \{d,a\} & \{d,b\} \\ \{c,c\} & \{c,d\} & \{d,c\} & \{d,d\} \end{pmatrix} = [L \otimes L, r_0] ,$$

where

$$L \otimes L = \begin{pmatrix} a^2 & ab & ba & b^2 \\ ac & ad & bc & bd \\ ca & cb & da & db \\ c^2 & cd & dc & d^2 \end{pmatrix} \quad \text{and} \quad r_0 = \frac{1}{4} \begin{pmatrix} 0 & 0 & 0 & 0 \\ 0 & 0 & -1 & 0 \\ 0 & 1 & 0 & 0 \\ 0 & 0 & 0 & 0 \end{pmatrix}.$$

Therefore we find the quadratic Poisson brackets,

$$\{a,b\} = \frac{1}{4}ab \ , \ \{a,c\} = \frac{1}{4}ac \ , \ \{a,d\} = \frac{1}{2}bc \ ,$$

$$\{b,c\} = 0 \ , \quad \{b,d\} = \frac{1}{4}bd \ , \ \{c,d\} = \frac{1}{4}cd \ .$$

Using the Leibniz rule for Poisson brackets, we find

$$\{a, ad - bc\} = \{b, ad - bc\} = \{c, ad - bc\} = \{d, ad - bc\} = 0.$$

Thus $ad - bc$ is a Casimir function for this Poisson structure, which is indeed defined on $SL(2, \mathbb{R})$.

Triangular Structure of $SL(2, \mathbb{R})$. Another Poisson Lie strucutre on $SL(2, \mathbb{R})$ is defined by the triangular r-matrix, considered in Example 3 of Sect. 2.3,

$$r = X \otimes H - H \otimes X = \begin{pmatrix} 0 & -1 & 1 & 0 \\ 0 & 0 & 0 & -1 \\ 0 & 0 & 0 & 1 \\ 0 & 0 & 0 & 0 \end{pmatrix}.$$

Taking into account the constraint $ad - bc = 1$, we find

$$\{a,b\} = 1 - a^2 \ , \ \{a,c\} = c^2 \ , \ \{a,d\} = c(-a+d)$$

$$\{b,c\} = c(a+d) \ , \ \{b,d\} = d^2 - 1 \ , \ \{c,d\} = -c^2 \ .$$

Here also we can show that $ad - bc$ is a Casimir function.

Quasi-Triangular Structure of $SU(2)$. Let $G = SU(2)$, and $\mathfrak{g} = \mathfrak{su}(2)$, with basis

$$X = \frac{1}{2} \begin{pmatrix} 0 & 1 \\ -1 & 0 \end{pmatrix} \ , \ Y = \frac{1}{2} \begin{pmatrix} 0 & i \\ i & 0 \end{pmatrix} \ , \ Z = \frac{1}{2} \begin{pmatrix} i & 0 \\ 0 & -i \end{pmatrix} \ ,$$

and commutation relations

$$[X, Y] = Z \ , \ [Y, Z] = X \ , \ [Z, X] = Y.$$

Then $r = Y \otimes X - X \otimes Y$ is a skew-symmetric r-matrix. In fact

$$\langle r, r \rangle = [r_{12}, r_{13}] + [r_{12}, r_{23}] + [r_{13}, r_{23}]$$

$$= [X \otimes Y \otimes 1 - Y \otimes X \otimes 1, X \otimes 1 \otimes Y - Y \otimes 1 \otimes X]$$

$$+ [X \otimes Y \otimes 1 - Y \otimes X \otimes 1, 1 \otimes X \otimes Y - 1 \otimes Y \otimes X]$$

$$+ [X \otimes 1 \otimes Y - Y \otimes 1 \otimes X, 1 \otimes X \otimes Y - 1 \otimes Y \otimes X]$$

$$= -Z \otimes Y \otimes X + Z \otimes X \otimes Y - X \otimes Z \otimes Y + Y \otimes Z \otimes X + X \otimes Y \otimes Z - Y \otimes X \otimes Z .$$

Now,

$$[X \otimes 1 \otimes 1, \langle r, r \rangle] = Y \otimes Y \otimes X - Y \otimes X \otimes Y + Z \otimes Z \otimes X - Z \otimes X \otimes Z$$

$$[1 \otimes X \otimes 1, \langle r, r \rangle] = -Z \otimes Z \otimes X + X \otimes X \otimes Y - X \otimes Y \otimes X + X \otimes Z \otimes Z$$

$$[1 \otimes 1 \otimes X, \langle r, r \rangle] = Z \otimes X \otimes Z - X \otimes Z \otimes Z - X \otimes Y \otimes Y + Y \otimes X \otimes Y .$$

Thus $ad_X^{(3)} \langle r, r \rangle = 0$, and, similarly, $ad_Y^{(3)} \langle r, r \rangle = 0$, $ad_Z^{(3)} \langle r, r \rangle = 0$. Thus r satisfies the generalized Yang-Baxter equation. (This r-matrix is not triangular.)

The Poisson brackets on $SU(2)$ defined by r, setting $L = \begin{pmatrix} a & b \\ c & d \end{pmatrix} \in SU(2)$, are

$$\{a, b\} = iab , \ \{a, c\} = iac , \ \{a, d\} = 2ibc ,$$

$$\{b, c\} = 0 , \ \{b, d\} = ibd , \ \{c, d\} = icd .$$

The Lie bracket defined by r on $(\mathfrak{su}(2))^*$, the dual of the Lie algebra $\mathfrak{su}(2)$, is given by

$$[Z^*, X^*] = X^* , \ [Z^*, Y^*] = Y^* , \ [X^*, Y^*] = 0 .$$

These examples can be generalized to yield Poisson-Lie structures on each simple Lie group (Drinfeld [16]) and on each compact Lie group (Lu and Weinstein [45]). Very interesting examples arise on loop groups (Drinfeld [16], Reyman and Semenov-Tian-Shansky [27]).

4.5   The Dual of a Poisson Lie Group

Since every Lie bialgebra has a dual Lie bialgebra and a double, we shall study the corresponding constructions at the Lie group level.

If (G, P) is a Poisson Lie group, we consider its Lie bialgebra $\mathfrak{g}$ whose 1-cocycle is $\gamma = DP$ (Sect. 4.2). We denote the dual Lie bialgebra by $(\mathfrak{g}^*, \gamma)$. By the last proposition of Sect. 4.2, we know that there exists a unique connected and simply connected Poisson Lie group with Lie bialgebra $(\mathfrak{g}^*, \gamma)$. We denote it by G^* and we call it the *dual* of (G, P). More generally, any Poisson-Lie group with Lie bialgebra $(\mathfrak{g}^*, \gamma)$ is called *a dual* of (G, P).

If G itself is connected and simply connected, then the dual of G^* is G (since the dual of $\mathfrak{g}^*$ is $\mathfrak{g}$, because $\mathfrak{g}$ is finite-dimensional).

Example 1. If P is the trivial Poisson structure on G ($P = 0$), then the Lie-algebra structure of $\mathfrak{g}^*$ is Abelian and the dual group of G is the Abelian group $\mathfrak{g}^*$ with its linear Poisson structure.

Example 2. Let $G = SL(2, \mathbb{R})$ with the Poisson structure defined by the quasi-triangular r-matrix, with skew-symmetric part $\frac{1}{4}(X \otimes Y - Y \otimes X)$. We have seen in Sect. 1.7 that the dual $\mathfrak{g}^*$ of $\mathfrak{sl}(2, \mathbb{R})$ can be identified with

$$\{(x, y) \in \mathfrak{b}_- \oplus \mathfrak{b}_+ \mid \mathfrak{h}-\text{components of } x \text{ and } y \text{ are opposite}\}.$$

This result extends to $\mathfrak{sl}(n, \mathbb{R})$ equipped with the standard r-matrix. Thus, if $G = SL(n, \mathbb{R})$, then

$$G^* = \{(L_-, L_+) \in B_- \times B_+ \mid \text{diagonal elements of } L_- \text{ and } L_+ \text{ are inverse}\}.$$

Here B_+ (resp., B_-) is the connected component of the group of upper (resp., lower) triangular matrices with determinant 1.

For example, for $n = 2$,

$$B_+ = \left\{ \begin{pmatrix} a & b \\ 0 & \frac{1}{a} \end{pmatrix} \mid a > 0, b \in \mathbb{R} \right\}$$

and

$$B_- = \left\{ \begin{pmatrix} a & 0 \\ c & \frac{1}{a} \end{pmatrix} \mid a > 0, c \in \mathbb{R} \right\},$$

so that

$$G^* = \left\{ \left(\begin{pmatrix} a & b \\ 0 & \frac{1}{a} \end{pmatrix}, \begin{pmatrix} \frac{1}{a} & 0 \\ c & a \end{pmatrix} \right) \mid a > 0, b, c \in \mathbb{R} \right\}.$$

This 3-dimensional Lie group can be identified with

$$SB(2, \mathbb{C}) = \left\{ \begin{pmatrix} \alpha & \beta + i\gamma \\ 0 & \alpha^{-1} \end{pmatrix} \mid \alpha > 0, \beta, \gamma \in \mathbb{R} \right\}.$$

Example 3. The dual of the Lie bialgebra $\mathfrak{su}(2)$ (see Sect. 1.7) can be integrated to a Poisson Lie group. The real Lie group $SB(2, \mathbb{C})$ defined above has the Lie algebra with basis $H = \begin{pmatrix} 1 & 0 \\ 0 & -1 \end{pmatrix}$, $X = \begin{pmatrix} 0 & 1 \\ 0 & 0 \end{pmatrix}$, $X' = iX$, and commutation relations $[H, X] = 2X, [H, X'] = 2X', [X, X'] = 0$, and thus is isomorphic to $\mathfrak{su}(2)^*$. Thus the dual of the Poisson Lie group $SU(2)$ is also the group $SB(2, \mathbb{C})$.

However, $(\mathfrak{sl}(2, \mathbb{R}))^*$ and $(\mathfrak{su}(2))^*$ are not isomorphic as Lie bialgebras, so $(SL(2, \mathbb{R}))^*$ and $(SU(2))^*$, which are both isomorphic as Lie groups to $SB(2, \mathbb{C})$, do not have the same Poisson structure.

More generally (see Sect. 1.7), the dual of the compact form $\mathfrak{k}$ of a complex simple Lie algebra $\mathfrak{g}$ is a Lie algebra $\mathfrak{b}$ such that

$$\mathfrak{g}^{\mathbb{R}} = \mathfrak{k} \oplus \mathfrak{b}.$$

If K is a compact Lie group, with Lie algebra $\mathfrak{k}$, then K is a Poisson Lie group whose dual, K^*, is the connected and simply connected Lie group with Lie algebra $\mathfrak{b}$. Thus, in the Iwasawa decomposition, $G = KAN$, both K and AN are Poisson Lie groups in duality. (See Lu and Weinstein [45].)

4.6   The Double of a Poisson Lie Group

When (G, P) is a Poisson Lie group, its tangent Lie bialgebra $(\mathfrak{g}, \gamma)$ has a double $\mathfrak{d}$, which is a factorizable Lie bialgebra. (See Sect. 2.5.)

The connected and simply connected Lie group $\mathcal{D}$ with Lie algebra $\mathfrak{d}$ is called the *double* of (G, P). Since $\mathfrak{d}$ is a factorizable Lie bialgebra with r-matrix $r_{\mathfrak{d}}$, $\mathcal{D}$ is a factorizable Poisson Lie group, with Poisson structure $P_{\mathcal{D}} = r_{\mathfrak{d}}^{\lambda} - r_{\mathfrak{d}}^{\rho}$, where λ and ρ refer to left and right translations in the Lie group $\mathcal{D}$. More precisely, the double of (G, P) is the Poisson Lie group $(\mathcal{D}, P_{\mathcal{D}})$. Since $\mathfrak{g}$ and $\mathfrak{g}^*$ are Lie subalgebras of $\mathfrak{d}$, G and G^* are Lie subgroups of $\mathcal{D}$.

More generally, any Poisson Lie group with Lie algebra $\mathfrak{d}$ is called *a double* of (G, P).

Example. When G is a Lie group with trivial Poisson structure, we know that its dual is the Abelian Lie group $\mathfrak{g}^*$. A Lie group whose Lie algebra is the double of $\mathfrak{g}$, *i.e.*, the semi-direct product of $\mathfrak{g}$ and $\mathfrak{g}^*$ with the coadjoint representation, is the Lie group T^*G, the cotangent bundle of G.

By Remark 2 of Sect. 4.2, $r_{\mathfrak{d}}^{\lambda} + r_{\mathfrak{d}}^{\rho}$ is also a Poisson structure on $\mathcal{D}$ (but not a Poisson-Lie structure). The Lie group $\mathcal{D}$ equipped with the Poisson structure $r_{\mathfrak{d}}^{\lambda} + r_{\mathfrak{d}}^{\rho}$ is called the *Heisenberg double* of (G, P). This Poisson structure is actually symplectic in a neighborhood of the unit.

If the Poisson Lie group (G, P) itself is defined by a factorizable r-matrix $r \in \mathfrak{g} \otimes \mathfrak{g}$, then we can describe its double in a simple way.

In fact, integrating the Lie-algebra morphisms described in Sects. 2.4 and 2.5, we obtain morphisms of Lie groups. In particular, let G_R be the connected and simply connected Lie group with Lie algebra $\mathfrak{g}_R$. Then the Lie-algebra morphisms R_+ and R_- can be integrated to Lie-group morphisms $\mathcal{R}_+$ and $\mathcal{R}_-$ from G_R to G, and the pair $\mathcal{J} = (\mathcal{R}_+, \mathcal{R}_-)$ defines an embedding of G_R into the direct product $G \times G$. Locally, near the unit, the double of G can be identified with the product of the manifolds G and G_R, which are Lie subgroups of $G \times G$.

4.7   Poisson Actions

Whereas a Hamiltonian action of a Lie group G on a Poisson manifold M is defined as a group action which preserves the Poisson structure, a Poisson action is an action of a Poisson Lie group on a Poisson manifold satisfying a different property expressed in terms of the Poisson bivectors of both the manifold and the group. When the Poisson structure of the group is trivial, *i.e.*, vanishes, we recover the Hamiltonian actions.

Definition. *Let (G, P_G) be a Poisson Lie group and (M, P_M) a Poisson manifold. An action α of G on M is called a* Poisson action *if $\alpha : G \times M \to M$, $(g, m) \mapsto g.m$, is a Poisson map.*

Recall that, setting $\alpha(g, m) = g.m$, α is an action of G on M if, for $g, h \in G, m \in M$,

$$g.(h.m) = (gh).m ,$$

$$e.m = m .$$

Recall also that the Poisson bracket on $G \times M$ is defined by

$$\{\Phi, \Psi\}_{G \times M}(g, m) = \{\Phi(., m), \Psi(., m)\}_G(g) + \{\Phi(g, .), \Psi(g, .)\}_M(m) ,$$

for $\Phi, \Psi \in C^\infty(G \times M)$.

Finally, recall that a Poisson map α from a Poisson manifold (N, P_N) to a Poisson manifold (M, P_M) is a map such that

$$\{\varphi \circ \alpha, \psi \circ \alpha\}_N = \{\varphi, \psi\}_M \circ \alpha, \text{ for } \varphi, \psi \in C^\infty(M) .$$

Thus α is a Poisson action if, for $g, h \in G, m \in M$,

$$\{\varphi, \psi\}_M(g.m) = \{\varphi_m, \psi_m\}_G(g) + \{\varphi^g, \psi^g\}_M(m) , \tag{4.12}$$

where we have set $\varphi_m(g) = \varphi(g.m)$ and $(\varphi^g)(m) = \varphi(g.m)$. As expected, if $\{ , \}_G = 0$, (4.12) reduces to the condition that, for each $g \in G$, the mapping $m \in M \mapsto g.m \in M$ be a Poisson map.

If (G, P) is a Poisson Lie group, the left and right actions of G on itself are Poisson actions. We shall give more examples of Poisson actions in Sect. 4.9.

We shall now give an infinitesimal criterion for Poisson actions. We know that an action α of G on M defines an action α' of $\mathfrak{g}$ on M, $\alpha' : x \in \mathfrak{g} \mapsto x_M$, where x_M is the vector field on M defined by

$$x_M(m) = \frac{d}{dt}(\exp(-tx).m)|_{t=0}.$$

In fact, α' maps $x \in \mathfrak{g}$ to a vector field x_M on M in such a way that

$$[x, y]_M = [x_M, y_M], \text{ for } x, y \in \mathfrak{g} .$$

Moreover we can extend α' to a map from $\bigwedge^2 \mathfrak{g}$ to the bivectors on M by setting

$$(x \wedge y)_M(m) = x_M(m) \wedge y_M(m) ,$$

and, more generally still, we can extend α' to a morphism of associative algebras from $\bigwedge \mathfrak{g}$ to the algebra of fields of multivectors on M.

Let us still denote by $.m$ the exterior powers of the differential at g of the map $\alpha_m : g \mapsto g.m$ from G to M, and by $g.$ those of the map $\alpha^g : m \mapsto g.m$ from

M to M. With these notations, which we shall use in the proof of the following proposition, we can write

$$\alpha'(w)(m) = w_M(m) = (-1)^{|w|}w.m,$$

for any $w \in \bigwedge \mathfrak{g}$, where $|w|$ is the degree of w.

Proposition. *Let (G, P_G) be a connected Poisson Lie group, with associated 1-cocyle of $\mathfrak{g}$,*

$$\gamma = DP_G = T_e(\rho(P_G)) : \mathfrak{g} \to \wedge^2 \mathfrak{g},$$

and let (M, P_M) be a Poisson manifold. The action $\alpha : G \times M \to M$ is a Poisson action if and only if

$$\mathcal{L}_{x_M}(P_M) = -(\gamma(x))_M , \tag{4.13}$$

for all $x \in \mathfrak{g}$, where $\mathcal{L}$ denotes the Lie derivation.

Proof. In fact, condition (4.12) is equivalent to

$$(P_M)_{g.m} = (P_G)_g.m + g.(P_M)_m , \tag{4.14}$$

where we have used the notations introduced above. We shall now show that (4.13) is the infinitesimal form of (4.14). From (4.14), we find

$$\mathcal{L}_{x_M}(P_M)(m) = \frac{d}{dt}(\exp tx.(P_M)_{\exp(-tx).m})|_{t=0}$$

$$= \frac{d}{dt}\left(\exp tx.((P_G)_{\exp(-tx)}.m + \exp(-tx).(P_M)_m)\right)|_{t=0}$$

$$= \frac{d}{dt}(\exp tx.((P_G)_{\exp(-tx)}.m))|_{t=0}.$$

Because α is an action, $\alpha^g \circ \alpha_m = \alpha_m \circ \lambda_g$, where λ_g is the left tanslation by g in G. Thus,

$$\exp tx.((P_G)_{\exp(-tx)}.m) = (\exp tx.(P_G)_{\exp(-tx)}).m .$$

On the other hand, by definition,

$$\gamma(x) = \mathcal{L}_{x^\rho}(P_G)(e) = \frac{d}{dt}(\exp(-tx).(P_G)_{\exp tx})|_{t=0} ,$$

since the flow of a right-invariant vector field acts by left translations. Therefore, we obtain

$$\mathcal{L}_{x_M}(P_M)(m) = -(\gamma(x)).m = -(\gamma(x))_M(m) .$$

This computation also shows that, conversely, (4.13) implies (4.14) when G is connected.

Definition. *A Lie-algebra action $x \mapsto x_M$ is called an* infinitesimal Poisson action *of the Lie bialgebra $(\mathfrak{g}, \gamma)$ on (M, P_M) if it satisfies (4.13).*

4.8   Momentum Mapping

Generalizing the momentum mapping for Hamiltonian actions, we adopt the following definition. Here G^* is again the Poisson Lie group dual to G.

Definition. *A map* $J : M \to G^*$ *is said to be a* momentum mapping *for the Poisson action* $\alpha : G \times M \to M$ *if, for all* $x \in \mathfrak{g}$,

$$x_M = \underline{P}_M(J^*(x^\lambda)) , \qquad (4.15)$$

where x^λ *is the left-invariant differential 1-form on* G^* *defined by the element* $x \in \mathfrak{g} = (T_e G^*)^*$, *and* $J^*(x^\lambda)$ *is the inverse image of* x^λ *under* J.

If, in particular, G is a Lie group with trivial Poisson structure, then $G^* = \mathfrak{g}^*$, the differential 1-form x^λ is the constant 1-form x on $\mathfrak{g}^*$, and

$$J^*(x^\lambda) = d(J(x)), \text{ where } J(x)(m) = \langle J(m), x \rangle .$$

Thus, in this case, we recover the usual definition of a momentum mapping for a Hamiltonian action, $J : M \to \mathfrak{g}^*$, that is

$$x_M = \underline{P}_M(d(J(x))) ,$$

i.e., x_M is the Hamiltonian vector field with Hamiltonian $J(x) \in C^\infty(M)$. Whence the name "non-Abelian Hamiltonian" also given to the momentum mapping in the case of a Poisson action.

4.9   Dressing Transformations

We consider a Poisson Lie group (G, P_G), its dual (G^*, P_{G^*}) and its double $\mathcal{D}$. Their respective Lie algebras are $\mathfrak{g}$, $\mathfrak{g}^*$ and $\mathfrak{d}$.

For each $x \in \mathfrak{g}$, we consider the vector field on G^*,

$$\ell(x) = \underline{P}_{G^*}(x^\lambda) , \qquad (4.16)$$

where x^λ is the left-invariant 1-form on G^* defined by $x \in \mathfrak{g} = (T_e G^*)^*$. Then

Theorem. *(i) The map* $x \mapsto \ell(x) = \underline{P}_{G^*}(x^\lambda)$ *is an action of* $\mathfrak{g}$ *on* G^*, *whose linearization at* e *is the coadjoint action of* $\mathfrak{g}$ *on* $\mathfrak{g}^*$.

(ii) The action $x \mapsto \ell(x)$ *is an infinitesimal Poisson action of the Lie bialgebra* $\mathfrak{g}$ *on the Poisson Lie group* G^*.

A concise proof of this theorem will be given in Appendix 2.

This action is called the *left infinitesimal dressing action* of $\mathfrak{g}$ on G^*. In particular, when G is a trivial Poisson Lie group, its dual group G^* is the Abelian group $\mathfrak{g}^*$, and the left infinitesimal dressing action of $\mathfrak{g}$ on $\mathfrak{g}^*$ is given by the linear vector fields $\ell(x) : \xi \in \mathfrak{g}^* \mapsto -ad_x^* \xi \in \mathfrak{g}^*$, for each $x \in \mathfrak{g}$. We can prove directly that $x \mapsto \ell(x)$ is a Lie-algebra morphism from $\mathfrak{g}$ to the Lie algebra of linear vector fields on $\mathfrak{g}^*$. In fact, applying $\ell([x,y])$ to the linear function $z \in (\mathfrak{g}^*)^*$, we find $\ell([x,y]) \cdot z = \ell(x) \cdot \ell(y) \cdot z - \ell(y) \cdot \ell(x) \cdot z$.

Similarly, the *right infinitesimal dressing action* of $\mathfrak{g}$ on G^* is defined by

$$x \mapsto r(x) = -\underline{P}_{G^*}(x^\rho),$$

where x^ρ is the right-invariant 1-form on G^* defined by $x \in \mathfrak{g}$, and its linearization is the opposite of the coadjoint action of $\mathfrak{g}$ on $\mathfrak{g}^*$.

The dressing vector fields $\ell(x) = \underline{P}_{G^*}(x^\lambda)$ have the following property, called *twisted multiplicativity*,

$$\ell_{uv}(x) = u.\ell_v(x) + \ell_u(Ad^*_{v^{-1}}x).v \ , \tag{4.17}$$

for $u, v \in G^*$, and an analogous property holds for the right dressing vector fields,

$$r_{uv}(x) = r_u(x).v + u.r_v(Ad^*_u x) \ . \tag{4.18}$$

Dually, we can define the *left* and *right infinitesimal dressing actions* of $\mathfrak{g}^*$ on G by

$$\xi \mapsto \underline{P}_G(\xi^\lambda) \text{ and } \xi \mapsto -\underline{P}_G(\xi^\rho).$$

Integrating these infinitesimal Poisson actions, when these vector fields are complete, we obtain Poisson actions of G on G^*, and of G^* on G, called the *left* and *right dressing actions*.

In the rest of this section, we derive the main properties of the dressing transformations.

Proposition. *The symplectic leaves of G (resp., G^*) are the connected components of the orbits of the right or left dressing action of G^* (resp., G).*

Proof. This is clear from the definitions, since either the vector fields $\underline{P}_{G^*}(x^\rho)$ or the vector fields $\underline{P}_{G^*}(x^\lambda)$ span the tangent space to the symplectic leaves of G^*, and similarly for G.

Proposition. *The momentum mapping for the left (resp., right) dressing action of G on G^* is the opposite of the identity map (resp., is the identity map) from G^* to G^*.*

The proof follows from the definitions, and there is a similar statement for the actions of G^* on G.

If the dual G^* is identified with a subset of the quotient of $\mathcal{D}$ under the right (resp., left) action of G, the left (resp., right) dressing action of G on G^* is identified with left- (resp., right-) multiplication by elements of G, and similarly for the actions of G on G^*.

There is an alternate way of defining the dressing actions, which shows their relationship to the factorization problem encountered in Sect. 3.7.

Let g be in G and u in G^*. We consider their product ug in $\mathcal{D}$. Because $\mathfrak{d} = \mathfrak{g} \oplus \mathfrak{g}^*$, elements in $\mathcal{D}$ sufficiently near the unit can be decomposed in a unique way as a product of an element in G and an element in G^* (in this order). Applying this fact to $ug \in \mathcal{D}$, we see that there exist elements ${}^u g \in G$ and $u^g \in G^*$ such that

$$ug = {}^u g \, u^g . \tag{4.19}$$

We thus define locally a left action of G^* on G and a right action of G on G^*. In other words, the action of $u \in G^*$ on $g \in G$ (resp., the action of $g \in G$ on $u \in G^*$) is given by

$$(u,g) \mapsto (ug)_G \quad (\text{resp.}, (u,g) \mapsto (ug)_{G^*}) ,$$

where $(ug)_G$ (resp., $(ug)_{G^*}$) denotes the G-factor (resp., G^*-factor) of $ug \in \mathcal{D}$ as $g'u'$, with $g' \in G, u' \in G^*$.

In the same way, the product $gu \in \mathcal{D}$ can be uniquely decomposed (if it is sufficiently near the unit) into ${}^g u \, g^u$, where ${}^g u \in G^*$ and $g^u \in G$. So, by definition,

$$gu = {}^g u \, g^u. \tag{4.20}$$

In this way, we obtain (locally) a left action of G on G^* and a right action of G^* on G.

Let us show that the left action of G on G^* is indeed a group action. By definition, for $g, h \in G, u \in G^*$,

$$(gh)u = {}^{gh}u \, (gh)^u$$

and

$$g(hu) = g \, ({}^h u \, h^u) = (g \, {}^h u) \, h^u = {}^g({}^h u) \, g^{({}^h u)} \, h^u.$$

These two equations imply

$$^{gh}u = {}^g({}^h u) , \tag{4.21}$$

which is what was to be proved. They also imply the relation

$$(gh)^u = g^{({}^h u)} \, h^u, \tag{4.22}$$

which expresses the twisted multiplicativity property of the dressing transformations of the left action of G on G^*. Analogous results hold for the three other actions which we have defined.

Proposition. *The left and right actions of G on G^* and of G^* on G defined by* (4.19) *and* (4.20) *coincide with the dressing actions.*

Proof. To prove that these actions coincide with the dressing actions defined above, it is enough to show that the associated infinitesimal actions coincide.

Recall that the Lie bracket on the double $\mathfrak{d} = \mathfrak{g} \oplus \mathfrak{g}^*$ satisfies

$$[x, \xi]_{\mathfrak{d}} = -ad_\xi^* x + ad_x^* \xi,$$

for $x \in \mathfrak{g}, \xi \in \mathfrak{g}^*$ (see Sect. 1.6). In the linearization of the left (resp., right) action of G on G^*, the image of $(x, \xi) \in \mathfrak{g} \oplus \mathfrak{g}^*$ is the projection onto $\mathfrak{g}^*$ of $[x, \xi]_\mathfrak{d}$ (resp., $[\xi, x]_\mathfrak{d}$), *i.e.*, the linearized action is the coadjoint action (resp., the opposite of the coadjoint action) of $\mathfrak{g}$ on $\mathfrak{g}^*$.

Similarily the linearized action of the left (resp., right) action of G^* on G is the coadjoint (resp., the opposite of the coadjoint action) of $\mathfrak{g}^*$ on $\mathfrak{g}$.

From relation (4.22) we deduce that the vector fields of the left infinitesimal action of $\mathfrak{g}$ on G^* satisfy the twisted multiplicativity property (4.17). This fact and the fact that the linearized action is the coadjoint action of $\mathfrak{g}$ on $\mathfrak{g}^*$ permit identifying this infinitesimal action with the left infinitesimal dressing action of $\mathfrak{g}$ on G^* (see Lu and Weinstein [45]).

The proofs for the right action of G on G^*, and for the left and right actions of G^* on G are similar.

To conclude, we wish to relate the dressing transformations defined above with the formulæ expressing the dressing of G-valued fields satisfying a zero-curvature equation. (See Faddeev and Takhtajan [29], Babelon and Bernard [34].) The field equation expresses a compatibility condition for a linear system, the Lax representation, equivalent to a nonlinear soliton equation. This nonlinear equation admits a Hamiltonian formulation such that the Poisson brackets of the $\mathfrak{g}$-valued Lax matrix are expressed in terms of a factorizable r-matrix. The dressing transformations act on the G-valued fields and preserve the solutions of the field equation. This action is in fact a Poisson action of the dual group, G^*.

If the Poisson Lie structure of G is defined by a factorizable r-matrix, the double, $\mathcal{D}$, is isomorphic to $G \times G$, with, as subgroups, the diagonal subgroup $\{(g, g)|g \in G\} \approx G$, and $\{(g_+, g_-)|g_\pm = \mathcal{R}_\pm h, h \in G\} \approx G^*$, with Lie algebras $\mathfrak{g}$ and $\{(\underline{r}_+ x, \underline{r}_- x)|x \in \mathfrak{g}\} \approx \mathfrak{g}^*$, respectively (see Sect. 2.5 and 4.6).

In this case, the factorization problems consist in finding group elements $g' \in G$ and $g'_\pm = \mathcal{R}_\pm h', h' \in G$, satisfying

$$(g_+, g_-)(g, g) = (g', g')(g'_+, g'_-) , \tag{4.23}$$

or $g' \in G$ and $g'_\pm = \mathcal{R}_\pm h', h' \in G$, satisfying

$$(g, g)(g_+, g_-) = (g'_+, g'_-)(g', g') . \tag{4.24}$$

Let us write the left action of G^* on G in this case. From $ug = {}^u g\ u^g$, we obtain from (4.23),

$$g_+ g = g' g'_+, \quad g_- g = g' g'_- .$$

Eliminating g', we find that $g'_+{}^{-1} g'_-$ is obtained from $g_+^{-1} g_-$ by conjugation by g^{-1}, and that g'_+, g'_- solve the factorization equation

$$g'_+{}^{-1} g'_- = g^{-1}(g_+{}^{-1} g_-)g . \tag{4.25}$$

It follows that the action of the element $(g_+, g_-) \in G^*$ on $g \in G$ is given by

$$g' = g_+ g g'_+{}^{-1} = g_- g g'_-{}^{-1} , \tag{4.26}$$

where the group elements g'_+ and g'_- solve the factorization equation (4.25).

Similarly for the right action of G^* on G, we obtain the factorization equation

$$g'_+ g'^{-1}_- = g(g_+ g_-^{-1})g^{-1} \,, \qquad (4.27)$$

and the action of $(g_+, g_-) \in G^*$ on $g \in G$ is given by

$$g' = g'^{-1}_+ g g_+ = g'^{-1}_- g g_- \,, \qquad (4.28)$$

where the group elements g'_+ and g'_- solve the factorization equation (4.27).

Thus we recover the formula of the dressing transformation in Faddeev and Takhtajan [29] and in Babelon and Bernard [34]. (In the convention of [29], the g'_- considered here is replaced by its inverse, while in [34], the factorization equation is $g'^{-1}_- g'_+ = g(g_-^{-1} g_+)g^{-1}$.)

4.10   Bibliographical Note

Multiplicative fields of tensors, were introduced by Lu and Weinstein [45]. See Kosmann-Schwarzbach [39], Vaisman [30], Dazord and Sondaz [37]. The results of Sect. 4.2 are due to Drinfeld [15], and are further developed in Kosmann-Schwarzbach [38] [39], Verdier [31], and Lu and Weinstein [45]. For formula (4.11), see e.g., Takhtajan [29]. For the examples of Sect. 4.5, see [45] and Majid [46][24]. Poisson actions were introduced by Semenov-Tian-Shansky [19], who showed that they were needed to explain the properties of the dressing transformations in field theory. Their infinitesimal characterization is due to Lu and Weinstein [45]. The generalization of the momentum mapping to the case of Poisson actions is due to Lu [44], while Babelon and Bernard [34], who call it the "non-Abelian Hamiltonian", have shown that in the case of the dressing transformations of G-valued fields the momentum mapping is given by the monodromy matrix of the associated linear equation. For the properties of the dressing transformations, see Semenov-Tian-Shansky [19], Lu and Weinstein [45], Vaisman [30], Alekseev and Malkin [32] (also, Kosmann-Schwarzbach and Magri [40] for the infinitesimal dressing transformations). A comprehensive survey of results, including examples and further topics, such as affine Poisson Lie groups, is given by Reyman [26].

Appendix 1
The 'Big Bracket' and Its Applications

Let F be a finite-dimensional (complex or real) vector space, and let F^* be its dual vector space. We consider the exterior algebra of the direct sum of F and F^*, $\bigwedge(F \oplus F^*) = \overset{\infty}{\underset{n=-2}{\oplus}} \left(\underset{p+q=n}{\oplus} (\bigwedge^{q+1} F^* \otimes \bigwedge^{p+1} F) \right).$

We say that an element of $\bigwedge(F \oplus F^*)$ is of bidegree (p, q) and of degree $n = p + q$ if it belongs to $\bigwedge^{q+1} F^* \oplus \bigwedge^{p+1} F$. Thus elements of the base field are of bidegree $(-1, -1)$, elements of F (resp., F^*) are of bidegree $(0, -1)$ (resp., $(-1, 0)$), and a linear map $\mu : \bigwedge^2 F \to F$ (resp., $\gamma : F \to \bigwedge^2 F$) can be considered to be an element of $\bigwedge^2 F^* \otimes F$ (resp., $F^* \otimes \bigwedge^2 F$) which is of bidegree $(0,1)$ (resp., $(1,0)$).

Proposition. *On the graded vector space $\bigwedge(F \oplus F^*)$ there exists a unique graded Lie bracket, called the* big bracket, *such that*

- *if $x, y \in F$, $[x, y] = 0$,*
- *if $\xi, \eta \in F^*$, $[\xi, \eta] = 0$,*
- *if $x \in F, \xi \in F^*$, $[x, \xi] = \langle \xi, x \rangle$.*
- *if $u, v, w \in \bigwedge(F \oplus F^*)$ are of degrees $|u|, |v|$ and $|w|$, then*

$$[u, v \wedge w] = [u, v] \wedge w + (-1)^{|u||v|} v \wedge [u, w] .$$

This last formula is called the *graded Leibniz rule*. The following proposition lists important properties of the big bracket.

Proposition. *Let $[\,,\,]$ denote the big bracket. Then*

i) $\mu : \bigwedge^2 F \to F$ is a Lie bracket if and only if $[\mu, \mu] = 0$.
ii) ${}^t\gamma : \bigwedge^2 F^* \to F^*$ is a Lie bracket if and only if $[\gamma, \gamma] = 0$.
iii) Let $\mathfrak{g} = (F, \mu)$ be a Lie algebra. Then γ is a 1-cocycle of $\mathfrak{g}$ with values in $\bigwedge^2 \mathfrak{g}$, where $\mathfrak{g}$ acts on $\bigwedge^2 \mathfrak{g}$ by the adjoint action, if and only if $[\mu, \gamma] = 0$.

The Dual and the Double of a Lie Bialgebra. By the graded commutativity of the big bracket,

$$[\gamma, \mu] = [\mu, \gamma].$$

This equality proves the proposition of Sect. 1.5 without any computation.

To prove the theorem of Sect. 1.6, we write, by the bilinearity and graded skew-symmetry of the big bracket,

$$[\mu + \gamma, \mu + \gamma] = [\mu, \mu] + 2[\mu, \gamma] + [\gamma, \gamma].$$

Using the bigrading of $\bigwedge(F \oplus F^*)$, we see that conditions

$$[\mu + \gamma, \mu + \gamma] = 0$$

and

$$[\mu, \mu] = 0, \ [\mu, \gamma] = 0, \ [\gamma, \gamma] = 0$$

are equivalent. The first is equivalent to the fact that $\mu + \gamma$ defines a Lie-algebra structure on $F \oplus F^*$ which leaves the canonical scalar product invariant and is such that F and F^* are Lie subalgebras, and the second is equivalent to the defining relations for the Lie bialgebra $(\mathfrak{g}, \gamma)$, where $\mathfrak{g} = (F, \mu)$. Therefore in the finite-dimensional case, Lie bialgebras are in 1-1 correspondence with Manin triples.   Q.E.D.

The following lemma is basic.

Lemma. *Let $\mathfrak{g} = (F, \mu)$ be a Lie algebra. Then*
a) $d_\mu : a \mapsto [\mu, a]$ is a derivation of degree 1 and of square 0 of the graded Lie algebra $\bigwedge(F \oplus F^*)$,
b) if $a \in \bigwedge F$, then $d_\mu a = -\delta a$, where δ is the Lie algebra cohomology operator,
c) for $a, b \in \bigwedge F$, let us set

$$[\![a, b]\!] = [[a, \mu], b].$$

Then $[\![\ , \ ]\!]$ is a graded Lie bracket of degree 1 on $\bigwedge F$ extending the Lie bracket of $\mathfrak{g}$. If $a = b$ and $a \in \bigwedge^2 F$, this bracket coincides with the quantity introduced in (2.4).

The bracket $[\![\ , \ ]\!]$ is called the algebraic Schouten bracket of the exterior algebra of $\mathfrak{g}$.

Coboundary Lie Bialgebras. We can now prove the second proposition in Sect. 2.1. Let $\gamma = \delta a = -d_\mu a = -[\mu, a]$, where $a \in \bigwedge^2 \mathfrak{g}$. By the graded Jacobi identity,

$$[\gamma, \gamma] = [d_\mu a, d_\mu a] = [[\mu, a], [\mu, a]] = [\mu, [a, [\mu, a]]] - [[\mu, [\mu, a]], a].$$

The second term vanishes because $[\mu, \mu] = 0$, and therefore $[\mu, [\mu, a]] = 0$. Now, by part c) of the lemma, we obtain

$$[\gamma, \gamma] = d_\mu [\![a, a]\!] ,$$

and therefore, by *(ii)* above, $\gamma = \delta a$ is a Lie algebra structure on $\mathfrak{g}^*$ if and only if $[\![a, a]\!]$ is *ad*-invariant.

The Tangent Lie Bialgebra of a Poisson Lie Group and the Integration Theorem. We now prove the propositions of Sect. 4.2. Let (G, P) be a Poisson Lie group, and let $\gamma = DP = T_e(\rho(P))$. We can show that, because P is multiplicative,

$$[DP, DP] = -D[\![P, P]\!].$$

To prove this equality we decompose P as a sum of decomposable bivectors and we use the biderivation property of the Schouten bracket.

It follows from this relation that, if $[\![P, P]\!] = 0$, then $[\gamma, \gamma] = 0$. Conversely, let $(\mathfrak{g}, \gamma)$ be a Lie bialgebra. Then γ is a Lie algebra 1-cocycle and it can be integrated into a Lie group 1-cocycle, Γ, on the connected and simply connected Lie group G with Lie algebra $\mathfrak{g}$. For $g \in G$, let $P_g = \Gamma(g).g$. We thus define a multiplicative bivector, P, on G . Moreover, P is a Poisson bivector. In fact,

$[\![P, P]\!]$ is multiplicative (as the Schouten bracket of a multiplicative bivector) and, by the above relation, $[\gamma, \gamma] = 0$ implies that $D[\![P, P]\!] = 0$. This is enough (see Lu and Weinstein [45]) to prove that $[\![P, P]\!] = 0$.

Manin Pairs. If $\mathfrak{p}$ is a finite-dimensional Lie algebra with an invariant, nondegenerate scalar product, and if $\mathfrak{a}$ is an isotropic Lie subalgebra of $\mathfrak{p}$ of maximal dimension, then $(\mathfrak{p}, \mathfrak{a})$ is called a *Manin pair*. (The dimension of $\mathfrak{p}$ is necessarily even, and $\dim \mathfrak{a} = \frac{1}{2} \dim \mathfrak{p}$). If $\mathfrak{b}$ is only an isotropic subspace (not necessarily a Lie subalgebra) complementary to $\mathfrak{a}$, the corresponding structure on $\mathfrak{a}$ is called a *Lie quasi-bialgebra* or a *Jacobian quasi-bialgebra*. Lie quasi-bialgebras are generalizations of Lie bialgebras, which were defined by Drinfeld as the classical limit of quasi-Hopf algebras. The double of a Lie quasi-bialgebra is again a Lie quasi-bialgebra.

Twilled Lie Algebras. A *twilled Lie algebra* (also called a *double Lie algebra*, but this is a definition different from that in Sect. 3.6, or a *matched pair of Lie algebras*) is just a Lie algebra that splits as the direct sum of two Lie subalgebras. In a twilled Lie algebra, each summand acts on the other by 'twisted derivations'. The double of a Lie bialgebra is a twilled Lie algebra in which the two summands are in duality, and the actions by twisted derivations are the coadjoint actions. There is a corresponding notion of a twilled Lie group (or *double group* or *matched pair of Lie groups*), in which each factor acts on the other. These actions have the property of twisted multiplicativity as in (4.22), and the vector fields of the associated infinitesimal action have a property of twisted multiplicativity as in (4.17). The double of a Poisson Lie group, G, is a twilled Lie group, with factors G and G^*, and the dressing actions described in Sect. 4.9 are the action of one factor on the other.

Bibliographical Note. For the definition of the 'big bracket', see Kostant and Sternberg [10]. For its use in the theory of Lie bialgebras, see Lecomte et Roger [42], and see Kosmann-Schwarzbach [39] on which this Appendix is based. See [39] and Bangoura and Kosmann-Schwarzbach [36] for applications of the big bracket to the case of Lie quasi-bialgebras. For twilled Lie algebras and Lie groups, see Kosmann-Schwarzbach and Magri [40], Majid [24,46] and Lu and Weinstein [45].

Appendix 2
The Poisson Calculus and Its Applications

We recall some basic facts from Poisson calculus, and we prove that the dressing vector fields define infinitesimal Poisson actions of $\mathfrak{g}$ on G^* and of $\mathfrak{g}^*$ on G. The notations are those of Sects. 3 and 4. If (M, P) is a Poisson manifold, we denote the space of smooth functions on M by $C^\infty(M)$. We further denote by $\{\ ,\ \}$ the Poisson bracket defined by P, and we set

$$(\underline{P}\xi)(\eta) = P(\xi, \eta).$$

For any Poisson manifold (M, P) there is a Lie bracket $[\ ,\ ]_P$ defined on the vector space of differential 1-forms,

$$[\![\xi, \eta]\!]_P = \mathcal{L}_{\underline{P}\xi}\eta - \mathcal{L}_{\underline{P}\eta}\xi - d(P(\xi, \eta)). \tag{A.1}$$

This bracket is $\mathbb{R}$-linear, and it satisfies

$$[\![\xi, f\eta]\!]_P = f[\![\xi, \eta]\!]_P + (\mathcal{L}_{\underline{P}\xi}f)\eta, \tag{A.2}$$

for $f \in C^\infty(M)$. In fact, this Lie bracket is characterized by (A.2) together with the property,

$$[\![df, dg]\!]_P = d(\{f, g\}), \tag{A.3}$$

for $f, g \in C^\infty(M)$.

The linear mapping $\underline{P}$ from differential 1-forms to vector fields is a Lie-algebra morphism, $i.e.$, it satisfies

$$[\underline{P}\xi, \underline{P}\eta] = \underline{P}([\![\xi, \eta]\!]_P). \tag{A.4}$$

Mapping $\underline{P}$ can be extended to a $C^\infty(M)$-linear mapping $\bigwedge\underline{P}$ from differential forms of all degrees to fields of multivectors (skew-symmetric contravariant tensor fields), setting

$$(\wedge\underline{P})(\xi_1 \wedge \ldots \wedge \xi_q) = \underline{P}\xi_1 \wedge \ldots \wedge \underline{P}\xi_q. \tag{A.5}$$

The Schouten bracket $[\![\ ,\ ]\!]$ is a graded Lie bracket on the vector space of fields of multivectors, with its grading shifted by 1, extending the Lie bracket of vector fields and satisfying a graded version of the Leibniz rule, (A.8) below. More precisely, for Q, Q', Q'' multivectors of degrees q, q', q'', respectively,

$$[\![Q, Q']\!] = -(-1)^{(q-1)(q'-1)}[\![Q', Q]\!], \tag{A.6}$$

$$[\![Q, [\![Q', Q'']\!]]\!] = [\![[\![Q, Q']\!], Q'']\!] + (-1)^{(q-1)(q'-1)}[\![Q', [\![Q, Q'']\!]]\!], \tag{A.7}$$

$$[\![Q, Q' \wedge Q'']\!] = [\![Q, Q']\!] \wedge Q'' + (-1)^{(q-1)q'}Q' \wedge [\![Q, Q'']\!]. \tag{A.8}$$

The following properties are satisfied:

$$[\![X, f]\!] = X \cdot f = \langle df, X \rangle, \tag{A.9}$$

for any vector field X, and

$$[\![P, f]\!] = -\underline{P}(df). \tag{A.10}$$

When P is a bivector, the bracket $[\![P, P]\!]$ coincides with the quantity introduced in (3.4).

Moreover, if P is a Poisson bivector, the mapping $Q \mapsto [\![P, Q]\!]$ is a derivation of degree 1 and of square 0 of the associative, graded commutative algebra of multivectors, which we denote by d_P and which we call the *Lichnerowicz-Poisson differential.*

Proposition. *The linear map $\bigwedge(-\underline{P})$ intertwines the Lichnerowicz-Poisson differential and the de Rham differential of forms, d.*

Proof. We have to show that, for any q-form ξ,

$$d_P((\wedge^q \underline{P})(\xi)) = -(\wedge^{q+1}\underline{P})(d\xi). \qquad (A.11)$$

For $q = 0$, this is just (A.10). When $q = 1$ and $\xi = df$, where $f \in C^\infty(M)$, then both sides vanish, since the Hamiltonian vector field $\underline{P}(df)$ leaves P invariant.

If $f \in C^\infty(M)$, then

$$d_P((\wedge^q \underline{P})(f\xi)) + (\wedge^{q+1}\underline{P})d(f\xi)$$

$$= [\![P, f]\!] \wedge (\wedge^q \underline{P})(\xi) + f[\![P, (\wedge^q \underline{P})(\xi)]\!] + (\wedge^{q+1}\underline{P})(df \wedge \xi) + f(\wedge^{q+1}\underline{P})(d\xi)$$

$$= f\left([\![P, (\wedge^q \underline{P})(\xi)]\!] + (\wedge^{q+1}\underline{P})(d\xi)\right).$$

Therefore, (A.11) holds for all 1-forms. Since d (resp., d_P) is a derivation of the associative, graded commutative algebra of differential forms (resp., multivectors), formula (A.11) holds for each integer $q \geq 0$.

We also recall the following fact from Lie-group theory.

Lemma. *Let G be a Lie group with Lie algebra $\mathfrak{g}$. Then, for $\xi \in \mathfrak{g}^*$, $d(\xi^\lambda) = (\delta\xi)^\lambda$ and $d(\xi^\rho) = -(\delta\xi)^\rho$, where $\delta\xi$ is the Lie algebra coboundary of the 1-cochain ξ on $\mathfrak{g}$ with values in $\mathbb{R}$.*

Explicitly,

$$\delta\xi(x_1, x_2) = -\langle \xi, [x_1, x_2]\rangle$$

so that

$$\delta\xi = -\,{}^t\mu(\xi)\,,$$

where $\mu : \bigwedge^2 \mathfrak{g} \to \mathfrak{g}$ is the Lie bracket on $\mathfrak{g}$.

If μ is considered as an element in $\bigwedge^2 \mathfrak{g}^* \otimes \mathfrak{g}$, then the linear mapping defined in terms of the big bracket (see Appendix 1), $d_\mu : \alpha \mapsto [\mu, \alpha]$ is a derivation of degree 1 and square 0 of the exterior algebra $\bigwedge(F \oplus F^*)$ and

$$d_\mu\xi = [\mu, \xi] = {}^t\mu(\xi) = -\delta\xi.$$

On a Poisson Lie group, the left-invariant 1-forms and the right-invariant 1-forms are Lie subalgebras of the space of differential 1-forms equipped with Lie bracket (A.1). More precisely

Proposition. *Let (G, P) be a Poisson Lie group with tangent Lie bialgebra $(\mathfrak{g}, \gamma)$ and set $[\xi, \eta]_{\mathfrak{g}^*} = {}^t\gamma(\xi \otimes \eta)$. Then, for all $\xi, \eta \in \mathfrak{g}^*$,*

$$[\![\xi^\lambda, \eta^\lambda]\!]_P = ([\xi, \eta]_{\mathfrak{g}^*})^\lambda \text{ and } [\![\xi^\rho, \eta^\rho]\!]_P = ([\xi, \eta]_{\mathfrak{g}^*})^\rho. \tag{A.12}$$

In other words, mappings $\xi \mapsto \xi^\lambda$ and $\xi \mapsto \xi^\rho$ are Lie-algebra morphisms from $(\mathfrak{g}^*, \gamma)$ to the Lie algebra of differential 1-forms on (G, P).

Properties of the Dressing Actions. We are now able to give the proof of the theorem of Section 4.9.

1. We prove that $\ell : x \mapsto \underline{P}_{G^*}(x^\lambda)$ is a Lie-algebra morphism from $\mathfrak{g}$ to the vector fields on G^* with the usual Lie bracket of vector fields. In fact, by (A.12) applied to (G^*, P_{G*}) and (A.4),

$$\ell_{[x,y]} = \underline{P}_{G^*}([x, y]^\lambda) = \underline{P}_{G^*}([\![x^\lambda, y^\lambda]\!]_{P_{G*}}) = [\underline{P}_{G^*}(x^\lambda), \underline{P}_{G^*}(y^\lambda)] = [\ell_x, \ell_y].$$

Similarly, $r_{[x,y]} = -[r_x, r_y]$.

2. We show that the linearization of the dressing action of G on G^* is the coadjoint action of $\mathfrak{g}$ on $\mathfrak{g}^*$.

By definition, the linearization at a fixed point m_0 of the action α of Lie group G on a manifold M is the map $x \in \mathfrak{g} \mapsto \dot\alpha(x) \in \mathrm{End}(T_{m_0}M)$ which is the differential of the linearized action of G on $T_{m_0}M$. Therefore,

$$\dot\alpha(x)(v) = (\mathcal{L}_{x_M}V)(m_0),$$

where V is a vector field on M with value v at m_0. (Thus the endomorphism $\dot\alpha(x)$ of $T_{m_0}M$ associated with $x \in \mathfrak{g}$ is the linearization of the vector field $-x_M$, and the assignment $x \in \mathfrak{g} \mapsto \dot\alpha(x) \in \mathrm{End}(T_{m_0}M)$ is a morphism of Lie algebras, being the composition of two antimorphisms.)

Applying this fact to $M = G^*$, with α the dressing action of G on G^*, and $x_{G^*} = \ell_x = \underline{P}_{G^*}(x^\lambda)$, we find, for $\xi \in T_eG^* = \mathfrak{g}^*$,

$$x \cdot \xi = \mathcal{L}_{\ell_x}(\Xi)(e),$$

where Ξ is a vector field on G^* with value ξ at $e \in G^*$. We choose $\Xi = \xi^\rho$, and for $y \in \mathfrak{g}$, we compute,

$$\langle \mathcal{L}_{\ell_x}\xi^\rho, y^\lambda \rangle(e) = -\langle \mathcal{L}_{\xi^\rho}(\underline{P}_{G^*}(x^\lambda)), y^\lambda \rangle(e)$$

$$= -\mathcal{L}_{\xi^\rho}\langle \underline{P}_{G^*}(x^\lambda), y^\lambda \rangle(e) + \langle \underline{P}_{G^*}(x^\lambda), \mathcal{L}_{\xi^\rho}(y^\lambda) \rangle(e) = -\mathcal{L}_{\xi^\rho}(P_{G^*}(x^\lambda, y^\lambda))(e)$$

$$= -\langle \xi, \mu(x, y) \rangle = \langle ad_x^*\xi, y \rangle.$$

Therefore, $x \cdot \xi = ad_x^*\xi$.

A similar proof shows that the linearization of the right dressing action of G on G^* is $(x, \xi) \in \mathfrak{g} \times \mathfrak{g}^* \mapsto -ad_x^*\xi \in \mathfrak{g}^*$, and the proofs in the dual situation are identical.

3. To prove that $x \mapsto \ell_x = \underline{P}_{G^*}(x^\lambda)$ is an infinitesimal Poisson action we use (4.13). Thus we have to show that

$$\mathcal{L}_{\ell_x}(P_{G^*}) = -(\gamma(x))_{G^*},$$

where γ is the linearization of P_{G^*} at e defining the Lie bracket of $\mathfrak{g}^*$, and

$$(\gamma(x))_{G^*} = \wedge^2 \underline{P}_{G^*}((\gamma(x))^\lambda).$$

Now by relation (A.11) and the Lemma,

$$\mathcal{L}_{\ell_x}(P_{G^*}) = [\![\underline{P}_{G^*}x^\lambda, P_{G^*}]\!] = -d_{P_{G^*}}(\underline{P}_{G^*}x^\lambda) = \wedge^2 \underline{P}_{G^*}d(x^\lambda) = -\wedge^2 \underline{P}_{G^*}(\gamma(x)^\lambda).$$

Q.E.D.

The proofs for the right dressing action and in the dual case are similar. Thus, using the basic general properties of the Poisson calculus, we have obtained in the above formula a one-line proof of the Poisson property of the dressing actions.

Bibliographical Note. For the Poisson calculus see Vaisman [30], or the earlier articles and book, Weinstein [50], K.H. Bhaskara and K. Viswanath, *Calculus on Poisson Manifolds*, Bull. London Math. Soc., **20**, 68-72 (1988) and *Poisson Algebras and Poisson Manifolds*, Pitman Research Notes in Math., Longman 1988, and Kosmann-Schwarzbach and Magri [41]. (In [50], mapping π is the opposite of $\underline{P}$ defined here, while the bracket $\{\ ,\ \}$ coincides with $[\![\ ,\ ]\!]_P$. In [41], mapping $\underline{P}$ and the bracket of 1-forms are opposite to the ones defined in this Appendix). See also Lu and Weinstein [45].

Acknowledgments

I gratefully acknowledge the support of CIMPA and of CEFIPRA-Indo-French Center for the Promotion of Advanced Research which permitted my stay at Pondicherry University in 1996 when I delivered the six lectures from which this text has been drawn.

Selected Bibliography

Background on Manifolds, Lie Algebras and Lie Groups, and Hamiltonian Systems

1. R. Abraham, J.E. Marsden and T. Ratiu, *Manifolds, Tensor Analysis and Applications*, Springer-Verlag 1988
2. V.I. Arnold, *Mathematical Methods of Classical Mechanics*, 2nd ed., Springer-Verlag 1989 (in Russian, Nauka, Moscow 1974; French translation, Editions Mir, Moscou 1976)
3. O. Barut and R. Rączka, *Theory of Group Representations and Applications*, 2nd ed., World Scientific 1986

4. J.G.F. Belinfante and B. Kolman, *A Survey of Lie Groups and Lie Algebras with Applications and Computational Methods*, 3rd ed., SIAM Philadelphia 1992

5. Y. Choquet-Bruhat, C. DeWitt-Morette, M. Dillard-Bleick, *Analysis, Manifolds and Physics*, Part I (1982), Part II (1989), North-Holland

6. W.D. Curtis and F.R. Miller, *Differential Manifolds and Theoretical Physics*, Academic Press 1985

7. V. Guillemin and S. Sternberg, *Symplectic Techniques in Physics*, Cambridge University Press 1984

8. A.A. Kirillov, *Elements of the Theory of Representations*, Springer-Verlag 1975 (in Russian, Nauka, Moscow 1971; French translation, Editions Mir, Moscou 1974)

9. B. Kostant, *The solution to the generalized Toda lattice and representation theory*, Adv. Math. **34**, 195–338 (1979)

10. B. Kostant and S. Sternberg, *Symplectic reduction, BRS cohomology and infinite-dimensional Clifford algebras*, Ann. Phys. (N.Y.) **176**, 49–113 (1987)

11. J.-L. Koszul, *Homologie et cohomologie des algèbres de Lie*, Bull. Soc. Math. France **78**, 1–63 (1950)

12. J.E. Marsden and T.S. Ratiu, *Introduction to Mechanics and Symmetry*, Springer-Verlag 1994; second edition 1999

13. M. Postnikov, *Lectures in Geometry, Semester V, Lie Goups and Lie Algebras*, Mir, Moscow 1986 (in Russian, Nauka, Moscow 1982; French translation, Editions Mir, Moscou 1985)

14. D.H. Sattinger and O.L. Sattinger, *Lie Groups and Algebras with Applications to Physics, Geometry and Mechanics*, Springer-Verlag 1986

A. Fundamental Articles on Poisson Lie Groups

15. V.G. Drinfeld, *Hamiltonian Lie groups, Lie bialgebras and the geometric meaning of the classical Yang-Baxter equation*, Sov. Math. Dokl. **27**, n°1, 68–71 (1983)

16. V.G. Drinfeld, *Quantum groups*, in Proc. Intern. Cong. Math. Berkeley 1986, vol. 1, Amer. Math. Soc. (1987), pp. 798–820

17. I.M. Gelfand and I.Y. Dorfman, *Hamiltonian operators and the classical Yang-Baxter equation*, Funct. Anal. Appl. **16**, n°4, 241–248 (1982)

18. M.A. Semenov-Tian-Shansky, *What is a classical r-matrix?*, Funct. Anal. Appl. **17**, n°4, 259–272 (1983)

19. M.A. Semenov-Tian-Shansky, *Dressing transformations and Poisson group actions*, Publ. RIMS (Kyoto) **21**, 1237–1260 (1985)

B. Books and Lectures on Poisson Manifolds, Lie Bialgebras, r-Matrices, and Poisson Lie Groups

20. O. Babelon and C.-M. Viallet, *Integrable Models, Yang-Baxter equation and quantum groups*, SISSA Lecture Notes, 54 EP (1989)

21. P. Cartier, *Some fundamental techniques in the theory of integrable systems*, in Lectures on Integrable Systems, In Memory of J.-L. Verdier, Proc. of the CIMPA School on Integrable Systems, Nice (France) 1991, O. Babelon, P. Cartier, Y. Kosmann-Schwarzbach, eds., World Scientific 1994, pp. 1–41. (Introduction to symplectic and Poisson geometry)

22. V. Chari and A. Pressley, *A Guide to Quantum Groups*, Cambridge University Press 1994. (Chaps. 1, 2, 3 and Appendix on simple Lie algebras)

23. L. Faddeev and L. Takhtajan, *Hamiltonian Methods in the Theory of Solitons*, Springer-Verlag 1987

24. S. Majid, *Foundations of Quantum Group Theory*, Cambridge University Press 1995. (Chap. 8)

25. A. Perelomov, *Integrable Systems of Classical Mechanics and Lie Algebras*, Birkhäuser 1990

26. A.G. Reyman, *Poisson structures related to quantum groups*, in Quantum Groups and their Applications in Physics, Intern. School "Enrico Fermi" (Varenna 1994), L. Castellani and J. Wess, eds., IOS, Amsterdam 1996, pp. 407–443

27. A. Reyman and M.A. Semenov-Tian-Shansky, *Group-theoretical methods in the theory of finite-dimensional integrable systems*, in Dynamical Systems VII, V.I. Arnold and S.P. Novikov, eds., Springer-Verlag 1994 (Encycl. of Mathematical Sciences, vol. 16), pp. 116–225

28. M.A. Semenov-Tian-Shansky, *Lectures on R-matrices, Poisson-Lie groups and integrable systems*, in Lectures on Integrable Systems, In Memory of J.-L. Verdier, Proc. of the CIMPA School on Integrable Systems, Nice (France) 1991, O. Babelon, P. Cartier, Y. Kosmann-Schwarzbach, eds., World Scientific 1994, pp. 269–317. (Lectures on group-theoretical methods for integrable systems, Lie algebras, *r*-matrices, Poisson Lie groups)

29. L.A. Takhtajan, *Elementary course on quantum groups*, in Lectures on Integrable Systems, In Memory of J.-L. Verdier, Proc. of the CIMPA School on Integrable Systems, Nice (France) 1991, O. Babelon, P. Cartier, Y. Kosmann-Schwarzbach, eds., World Scientific 1994, pp. 319–347 (Introduction to Poisson Lie groups and quantum groups)

30. I. Vaisman, *Lectures on the Geometry of Poisson Manifolds*, Birkhäuser 1994

31. J.-L. Verdier, *Groupes quantiques, d'après V.G. Drinfel'd*, Séminaire Bourbaki, exposé 685, Astérisque **152–153**, Soc. Math. Fr. 1987, pp. 305–319

C. Further Developments on Lie Bialgebras, *r*-Matrices and Poisson Lie Groups

32. A.Y. Alekseev and A.Z. Malkin, *Symplectic structures associated to Poisson-Lie groups*, Comm. Math. Phys. **162**, 147–173 (1994)

33. R. Aminou, Y. Kosmann-Schwarzbach, *Bigèbres de Lie, doubles et carrés*, Ann. Inst. Henri Poincaré, Phys. Théor., **49**A, n°4, 461–478 (1988)

34. O. Babelon and D. Bernard, *Dressing symmetries*, Comm. Math. Phys. **149**, 279–306 (1992)

35. O. Babelon and C.-M. Viallet, *Hamiltonian structures and Lax equations*, Phys. Lett B **237**, 411–416 (1990)

36. M. Bangoura and Y. Kosmann-Schwarzbach, *The double of a Jacobian quasi-bialgebra*, Lett. Math. Phys. **28**, 13–29 (1993)

37. P. Dazord and D. Sondaz, *Groupes de Poisson affines*, in Symplectic Geometry, Groupoids and Integrable Systems, P. Dazord and A. Weinstein, eds., Springer-Verlag 1991, pp. 99–128

38. Y. Kosmann-Schwarzbach, *Poisson-Drinfeld groups*, in Topics in Soliton Theory and Exactly Solvable Nonlinear Equations, M. Ablowitz, B. Fuchssteiner and M. Kruskal, eds., World Scientific 1987, pp. 191–215

39. Y. Kosmann-Schwarzbach, *Jacobian quasi-bialgebras and quasi-Poisson Lie groups*, Contemporary Mathematics **132**, 459–489 (1992)

40. Y. Kosmann-Schwarzbach and F. Magri, *Poisson-Lie groups and complete inte-grability, I. Drinfeld bigebras, dual extensions and their canonical representations*, Ann. Inst. Henri Poincaré, Phys. Théor., **49**A, n°4, 433–460 (1988)

41. Y. Kosmann-Schwarzbach and F. Magri, *Poisson-Nijenhuis structures*, Ann. Inst. Henri Poincaré, Phys. Théor., **53**A, n°1, 35–81 (1990)

42. P. Lecomte and C. Roger, *Modules et cohomologie des bigèbres de Lie*, Comptes rendus Acad. Sci. Paris **310**, série I, 405–410 and **311**, série I, 893–894 (1990)

43. L.C. Li and S. Parmentier, *Nonlinear Poisson structures and r-matrices*, Comm. Math. Phys. **125**, 545–563 (1989)

44. J.H. Lu, *Momentum mappings and reduction of Poisson actions*, in Symplectic Geometry, Groupoids and Integrable Systems, P. Dazord and A.Weinstein, eds., Springer-Verlag 1991, pp. 209–226

45. J.H. Lu and A. Weinstein, *Poisson Lie groups, dressing transformations and Bruhat decompositions*, J. Diff. Geom. **31**, 501–526 (1990)

46. S. Majid, *Matched pairs of Lie groups associated to solutions of the Yang-Baxter equations*, Pacific J. Math. **141**, 311–319 (1990)

47. N.Y. Reshetikhin and M.A. Semenov-Tian-Shansky, *Quantum R-matrices and fac-torization problems*, J. Geom. Phys. **5**, 533–550 (1988)

48. C. Roger, *Algèbres de Lie graduées et quantification*, in Symplectic Geometry and Mathematical Physics, P. Donato et al., eds., Birkhäuser 1991, pp. 374–421

49. Y. Soibelman, *On some problems in the theory of quantum groups*, Advances in Soviet Mathematics **9**, 3–55 (1992)

50. A. Weinstein, *Some remarks on dressing transformations*, J. Fac. Sci. Univ. Tokyo IA, **35**, n°1, 163–167 (1988)

D. Bibliographical Note Added in the Second Edition

The theory of Lie bialgebras, r-matrices and Poisson Lie groups has continued to expand in the interval between the first and second editions of this book. Especially noteworthy is the expansion of the theory into the study of Lie bialgebroids and Poisson groupoids. More than fifty papers have appeared on Lie bialgebras and bialgebroids, while some thirty other papers deal primarily with Poisson Lie groups and Poisson groupoids. We mention only four useful references, where many more references can be found.

– Chapter I of the book *Algebras of functions on quantum groups. Part I*, by L.I. Korogodski and Y.S. Soibelman (Mathematical Surveys and Monographs, 56, American Mathematical Society, Providence, RI, 1998) covers much of the material presented here.

– The article *Loop groups, R-matrices and separation of variables* by J. Harnad, in Integrable Systems: from classical to quantum, CRM Proc. Lecture Notes, 26, Amer. Math. Soc., Providence, RI, 2000 p. 21–54, reviews the basics of the theory of classical R-matrices and presents important applications to the study of integrable systems.

– In the same volume, p. 165–188, *Characteristic systems on Poisson Lie groups and their quantization*, by N. Reshetikhin, includes a review of impor-tant results in the theory of Lie bialgebras, of Poisson Lie groups and their symplectic leaves, and presents applications to the theory of some dynamical systems generalizing the Toda system.

- The paper *Symplectic leaves of complex reductive Poisson-Lie groups*, by M. Yakimov, Duke Math. J. **112**, 453–509 (2002), is an in-depth study of the geometry of some Poisson Lie groups using Lie theory .

The first and third references will also be useful for leads to the vast literature concering the quantization of Lie bialgebras.

Analytic and Asymptotic Methods for Nonlinear Singularity Analysis: A Review and Extensions of Tests for the Painlevé Property

M.D. Kruskal[1], N. Joshi[2], and R. Halburd[3]

[1] Department of Mathematics, Rutgers University, New Brunswick NJ 08903, USA, kruskal@math.rutgers.edu

[2] Department of Pure Mathematics, University of Adelaide, Adelaide SA 5005, Australia; now at School of Mathematics and Statistics F07, University of Sydney, NSW 2006, Australia, nalini@maths.usyd.edu.au

[3] Department of Applied Mathematics, University of Colorado, Boulder CO 80309-529, USA; now at Department of Mathematical Sciences, Loughborough University, Loughborough Leicestershire LE11 3TU, UK, r.g.halburd@lboro.ac.uk

Abstract. The integrability (solvability via an associated single-valued linear problem) of a differential equation is closely related to the singularity structure of its solutions. In particular, there is strong evidence that all integrable equations have the Painlevé property, that is, all solutions are single-valued around all movable singularities. In this expository article, we review methods for analysing such singularity structure. In particular, we describe well known techniques of nonlinear regular-singular-type analysis, i.e., the Painlevé tests for ordinary and partial differential equations. Then we discuss methods of obtaining sufficiency conditions for the Painlevé property. Recently, extensions of *irregular* singularity analysis to nonlinear equations have been achieved. Also, new asymptotic limits of differential equations preserving the Painlevé property have been found. We discuss these also.

1 Introduction

A differential equation is said to be integrable if it is solvable (for a sufficiently large class of initial data) via an associated (single-valued) linear problem. A famous example is the Korteweg-de Vries equation (KdV),

$$u_t + 6uu_x + u_{xxx} = 0, \tag{1.1}$$

where the subscripts denote partial differentiation.

The KdV equation was discovered to be integrable by Gardner, Greene, Kruskal, and Miura [21]. (Its method of solution is called the *inverse scattering transform* (IST) method; see the paper by Mark Ablowitz in the present collection.) Since this discovery, a large collection of nonlinear equations (see [1]) has been identified to be integrable. These range over many dimensions and include not

M.D. Kruskal, N. Joshi, and R. Halburd, Analytic and Asymptotic Methods for Nonlinear Singularity Analysis, Lect. Notes Phys. **638**, 175–208 (2004)
http://www.springerlink.com/

just partial differential equations (PDEs) but also differential-difference equations, integro-differential equations, and ordinary differential equations (ODEs).

Six classical nonlinear second-order ODEs, called the Painlevé equations, are prototypical examples of integrable ODEs. They possess a characteristic singularity structure, i.e., all movable singularities of all solutions are poles. Movable here means that the singularity's position varies as a function of initial values. A differential equation is said to have the Painlevé property if all solutions are single-valued around all movable singularities. (See comments below and in Sect. 2 on variations of this definition.) Thefore, the Painlevé equations possess the Painlevé property. Painlevé [55], Gambier [20], and R. Fuchs [19] identified these equations (under some mild conditions) as the only ones (of second order and first degree) with the Painlevé property whose general solutions are new transcendental functions.

Integrable equations are rare. Perturbation of such equations generally destroys their integrability. On the other hand, any constructive method of identifying the integrability of a given system contains severe shortcomings. The problem is that if a suitable associated linear problem cannot be found it is unclear whether the fault lies with the lack of integrability of the nonlinear system or with the lack of ingenuity of the investigator. So the identification of integrability has come to rely on other evidence, such as numerical studies and the singularity structure of the system.

There is strong evidence [60, 61] that the integrability of a nonlinear sytem is intimately related to the singularity structure admitted by the system in its solutions. Dense multi-valuedness (branching) around movable singularities of solutions is an indicator of nonintegrability [62]. The Painlevé property excludes such branching and has been proposed as a pointer to integrability.

The complex singularity structure of solutions was first used by Kowalevskaya [39, 42] to identify an integrable case of the equations of motion for a rotating top. Eighty eight years later, this connection was reobserved in the context of integrable PDEs by Ablowitz and Segur [5], and Ablowitz, Ramani, and Segur [3, 4]. Their observations led to the following conjecture.

The ARS Conjecture: *Any ODE which arises as a reduction of an integrable PDE possesses the Painlevé property, possibly after a transformation of variables.*

For example, the sine-Gordon equation,

$$u_{xt} = \sin u, \tag{1.2}$$

which is well known to be integrable [1, 2], admits the simple scaling symmetry

$$x \to \lambda x, \quad t \to \lambda^{-1} t. \tag{1.3}$$

To find a reduction with respect to the symmetry (1.3), restrict the search to the subspace of solutions that are invariant under (1.3) by introducing new variables z, w such that

$$u(x,t) = w(z), \quad z = xt.$$

This gives

$$zw'' + w' = \sin w,\tag{1.4}$$

where the prime denotes differentiation with respect to z. To investigate the Painlevé property, this equation must first be transformed to one that is rational (or possibly algebraic) in w. (Otherwise, the nonlinear analogue of Fröbenius analysis used to investigate the Painlevé property cannot find a leading-order term to get started. See Sect. 2.) Introduce the new dependent variable $y :=$ $\exp(iw)$. Then (1.4) becomes

$$z(yy'' - y'^2) + yy' = \frac{1}{2}y(y^2 - 1).$$

This equation (a special case of the third Painlevé equation) can be shown to have the Painlevé property (see Sect. 2).

There is now an overwhelming body of evidence for the ARS conjecture. A version that is directly applicable to PDEs, rather than their reductions, was given by Weiss, Tabor, and Carnevale (WTC) [59]. The ARS conjecture and its variant by WTC are now taken to be almost self-evident because they have been formally verified for every known analytic soliton equation [45, 51, 58] (where analytic means that the equation is, or may be written to be, locally analytic in the dependent variable and its derivatives). Previously unknown integrable versions of the soliton equations [15, 28, 30] have been identified by the use of the conjecture. The conjecture has also been extrapolated to identify integrable ODEs [9, 57].

Rigorous results supporting the conjecture exist for ODEs with special symmetries or symplectic structure [60, 61]. Necessary conditions for possessing the Painlevé property [17, 29] have also been derived for general semilinear analytic second-order PDEs. No new integrable PDEs were found – suggesting that in this class at least, there is a one-to-one correspondence (modulo allowable transformations) between integrable equations and those possessing the Painlevé property. Moreover, proofs of weakened versions of the conjecture exist [5, 49]. These results point strongly to the truth of the ARS conjecture. Nevertheless, the conjecture has not yet been proved.

The main aim of this paper is to describe methods for investigating the singularity structure of solutions of ODEs and PDEs. These may be divided into two classes, those that parallel methods for analysing *regular* singular points and those that parallel techniques for *irregular* singular points of linear ODEs.

The first class of methods has been widely used formally. The most popular procedure is to expand every solution of the differential equation of interest in an infinite series near a movable singularity of the equation [46], i.e., the solution $u(z)$ is expanded as

$$u(z) = \sum_{n=0}^{\infty} a_n(z - z_0)^{n+\rho},\tag{1.5}$$

where z_0 is the arbitrary location of a singularity and ρ is the leading power that needs to be found. Such an expansion is often called a Painlevé expansion. The equation is assumed to have the Painlevé property if the series is self-consistent, single-valued, and contains a sufficient number of degrees of freedom to describe all possible solutions or the general solution. These demands yield necessary conditions for the Painlevé property to hold. The series and the expansion techniques are analogues of the usual Fröbenius (or Fuchsian) expansion procedure for linear ODEs. This procedure was extended to PDEs by [59].

These techniques are, in general, not sufficient to prove that a differential equation has the Painlevé property. For example, even if the only possible formal solutions are Laurent series, the poles indicated by these series may accumulate elsewhere to give rise to a worse (branched) singularity.

Painlevé gave sufficient conditions to show that his eponymous equations have the Painlevé property. However, his proof is not widely understood. We describe briefly here an alternative, direct, method of proof due to Joshi and Kruskal [34]. To gain sufficiency, we showed that the solutions of the Painlevé equations possess convergent Laurent expansions around every movable singularity, and moreover (in any given bounded region) the radius of convergence of each series is uniformly bounded below. In other words, the poles of any solution cannot coalesce to form a more complicated singularity elsewhere (in the finite plane).

For nonlinear PDEs, the question of how to get sufficient conditions for the Painlevé property is still open. Nevertheless, partial results are now known. These make the WTC analogue of the Fröbenius method rigorous and go some way toward proving that a given PDE has the Painlevé property. We describe the results due to Joshi and Petersen [35, 36] and Joshi and Srinivasan [38] in Sect. 2. An alternative approach to the convergence of the Painlevé expansions for PDEs has also been developed by Kichenassamy and Littman [40, 41].

Another difficulty with the analysis of singularity structure is that the Painlevé expansions can miss some solutions. This may happen, for example, when the number of degrees of freedom in the series is less than the order of the differential equation. Perturbations of such series often reveal that the missing degrees of freedom lie in terms that occur (paradoxically) before the leading term. For reasons explained in Sect. 2, such terms are called *negative resonances*. In other cases, perturbations reveal no additional degrees of freedom at all. We call the latter series *defective*.

How can we deduce the singular behaviour of solutions that are missed by the Painlevé expansions? We provide an answer based on irregular-singular analysis for linear ODEs [8] and illustrate it through two important examples. The first example is the Chazy equation [1], a third-order ODE, whose general solutions have movable natural barriers. The Painlevé expansion of the solution of the Chazy equation contains only two arbitrary constants. The second example is a fourth-order ODE first studied by Bureau. This example has two families of Painlevé expansions, one of which has negative resonances and one that is defective. In Sect. 3, we show how the Painlevé expansions can be extended through exponential (or WKB-type) perturbations.

Conte, Fordy, and Pickering [16] have followed an alternative approach. Their perturbations of Painlevé expansions involve Laurent series with no leading term (i.e., an infinite number of negative powers). As pointed out by one of us, this is well defined only in an annulus where the expansion variable is lower-bounded away from the singularity. Conte *et al.* demand that each term of such a perturbation must be single-valued. Therefore, their procedure requires a possibly infinite number of conditions to be checked for the Painlevé property. Our approach overcomes this problem.

For linear differential equations, in general, the analysis near an irregular singularity yields asymptotic results, i.e., asymptotic behaviours along with their domains of asymptotic validity near the singularity. The latter is crucial in this description. For example, it is well known that the Airy function $Ai(x)$ which solves the ODE

$$y'' = xy,$$

has the asymptotic behaviour

$$Ai(x) \approx \frac{1}{2\sqrt{\pi}x^{1/4}} \exp\left(-2x^{3/2}/3\right) \quad \text{as } |x| \to \infty, \quad |\arg x| < \pi$$

near the irregular singular point at infinity. (See [8,52].) Note that the asymptotic behaviour of $Ai(x)$ is apparently multivalued but the function itself is single-valued everywhere. (In fact, $Ai(x)$ is entire, i.e., it is analytic throughout the whole complex x-plane.)

The resolution of this apparent paradox lies in the angular width of the sector of validity of the above behaviour, which is strictly less than 2π. To describe $Ai(x)$ in the whole plane, we need its asymptotic behaviour in regions that include the line $|\arg(x)| = \pi$. (Such behaviours are well known. See e.g. [8].) These, together with the behaviour given above, show that the analytic continuation of $Ai(x)$ along a large closed curve around infinity is single-valued. Therefore, the global asymptotic description is not actually multivalued.

On the other hand, suppose an asymptotic behaviour is multivalued and its sector of validity extends further than 2π. Then there are (at least) two asymptotic descriptions of a solution at the same place (near an irregular singularity). This violates the uniqueness of the asymptotic description of a solution, unless the solution is itself multivalued. Therefore, such an asymptotic behaviour is an indication that the solution cannot satisfy the requirements of the Painlevé property. The Bureau equation we study in Sect. 3 provides an example of this case.

Such results form an important extension of the usual tests for the Painlevé property. However, there is no denying that many fundamental questions remain open in this area, even at a formal level. For example, the Painlevé property is easily destroyed by straightforward transformations of the dependent variable(s). (E.g., a solution $u(z)$ of an ODE with movable simple poles is transformed to a function $w(z)$ with movable branch points under $u \mapsto w^2$.) An extension of

the Painlevé property called the poly-Painlevé property has been proposed by
Kruskal [45,58] to overcome these difficulties. It allows solutions to be branched
around movable singularities so long as a solution is not densely valued at a
point. However, such developments lie outside the scope of this paper and we
refer the reader to [45] for further details and references.

Other major problems remain. One is to extend the classification work of
Painlevé and his colleagues to other classes of differential equations. Cosgrove
has accomplished the most comprehensive extensions in recent times [17]. The
universal method of classification, called the α-method, is based on asymptotic
ideas (see Sect. 2). Asymptotic limits of differential equations can illuminate
such studies.

The Painlevé equations are well known to have as asymptotic limits other
equations with the Painlevé property. These limits are called *coalescence* limits
because singularities of an equation merge under the limit. In Sect. 4, we describe
the well known coalescence limits of the Painlevé equations and show that these
limits also occur for integrable PDEs.

Throughout this article, solutions of differential equations are assumed to be
complex-valued functions of complex variables.

2   Nonlinear-Regular-Singular Analysis

In this section, we survey the main techniques used to study the Painlevé property. These range from the α-method to the widely used formal test known as
the Painlevé test.

Consider the second-order linear ODE

$$u''(z) + p(z)u'(z) + q(z)u(z) = 0,$$

where primes denote differentiation with respect to z. Fuchs' theorem [8] states
that u can only be singular (nonanalytic) at points where p and q are singular.
Such singularities are called *fixed* because their positions are determined *a priori*
(before solving the equation) and their locations remain unchanged throughout
the space of all possible solutions.

However, fixed singularities are not the only possibilities for nonlinear equations. Consider the Riccati equation,

$$u''(z) + u^2(z) = 0.$$

It has the general solution

$$u(z) = \frac{1}{z - z_0},$$

where z_0 is an arbitrary constant. If, for example, the initial condition is $u(0) = 1$,
then $z_0 = -1$. If the initial condition is changed to $u(0) = 2$, then z_0 moves to
$z_0 = -1/2$. In other words, the location of the singularity at z_0 *moves* with
initial conditions. Such singularities are called *movable*.

Table 1. Examples of possible singular behaviour

	Equation	General Solution	Singularity Type
1.	$y' + y^2 = 0$	$y = (z - z_0)^{-1}$	simple pole
2.	$2yy' = 1$	$y = \sqrt{z - z_0}$	branch point
3.	$y'' + y'^2 = 0$	$y = \ln(z - z_0) + k$	logarithmic branch point
4.	$yy'' + y'^2(y/y' - 1) = 0$	$y = k \exp\left([z - z_0]^{-1}\right)$	isolated essential singularity
5.	$(1 + y^2)y'' + (1 - 2y)y'^2 = 0$	$y = \tan\left(\ln(k[z - z_0])\right)$	nonisolated essential singularity
6.	$\left(y'' + y^3 y'\right)^2 = y^2 y'^2 \left(4y' + y^4\right)$	$y = k\tan\left[k^3(x - x_0)\right]$ or $y = \left((4/3)/(x - x_0)\right)^{1/3}$	pole branch point

Nonlinear equations exhibit a vast range of types of movable singularities. Some examples are given in the Table 1 (where k and z_0 are arbitrary constant parameters).

A normalized ODE, i.e., one that is solved for the highest derivative, such as

$$y^{(n)} = F\left(y^{(n-1)}, \dots, y', y, z\right), \tag{2.6}$$

gives rise to possible singularities in its solutions wherever F becomes singular. (Where F is analytic, and regular initial data are given, standard theorems show that an analytic solution must exist.) Note that these singularities may include the points at infinity in y (or its derivatives) and z, which we denote by $y = \infty$ (or $y' = \infty$, etc), $z = \infty$. The singularities of F, therefore, denote possible singularities in the solution(s). They may be divided into two classes: those given by values of z alone, and those involving values of y or its derivatives. The former are determined *a priori* for all solutions. Therefore, they can only give rise to *fixed* singularities. To find movable singularities, we therefore need to investigate the singular values of F that involve y (or its derivatives). Similar statements can be made in the case of PDEs.

For example, the Riccati equation,

$$y' = -y^2/z =: F(y, z),$$

has a right side F with two singularities given by $z = 0$ and $y = \infty$. The general solution is

$$y(z) = \frac{1}{\log(z/z_0)}.$$

It is clear that $z = 0$ is a fixed singularity (it stays the same for all initial conditions) whereas z_0 denotes a movable singularity where y becomes unbounded.

Singularities of nonlinear ODEs need not only occur at points where y is unbounded. Example 2 of Table 1 indicates possible movable singularities at points where $y = 0$. The solution shows that these are actually movable branch points.

These considerations show that singular values of the normalized differential equation lie at the base of the solutions' singularity structure. Techniques for investigating singularity structure usually focus on these singular values.

In the first three subsections below, we describe common definitions of the Painlevé property, and the two major techniques known as the α-method and the Painlevé test for deriving necessary conditions for the Painlevé property. In the subsequent three subsections below, we discuss the need for sufficiency conditions, the direct method of proving the Painlevé property, and convergence-type results for PDEs.

2.1   The Painlevé Property

The actual definition of the Painlevé property has been subject to some variation. There are three definitions in the literature.

Definition 21. *An ODE is said to possess*

1. *the* specialized Painlevé property *if all movable singularities of all solutions are poles.*
2. *the* Painlevé property *if all solutions are single-valued around all movable singularities.*
3. *the* generalized Painlevé property *if the general solution is single-valued around all movable singularities.*

(The qualifiers "specialized" and "generalized" are not usually used in the literature.) The first property defined above clearly implies the others. This property was also the first one investigated (by ARS) in recent times. It is the property possessed by the six Painlevé equations.

The second, more general, definition above is the one used by Painlevé in his work on the classification of ODEs. It allows, for example, movable unbranched essential singularities in any solution. Of the examples in Table 1, equations 1 and 4 have the Painlevé property; equation 1 also has the specialized Painlevé property. The remaining equations have neither property.

The third property is the most recently proposed variation, although there is evidence that Chazy assumed it in investigating ODEs of higher (≥ 1) degree or order (≥ 2). The sixth example given in Table 1 satisfies neither of the first two properties above because the special solution $\left(4/3/(x-x_0)\right)^{1/3}$ has movable branch points around which the solution is multivalued. However, it does satisfy the generalized Painlevé property because the general solution $k\tan\left[k^3(x-x_0)\right]$ is meromorphic.

Most of the known techniques for investigating the Painlevé property have their origin in the classical work of Painlevé and his colleagues. They classified ODEs of the form

$$u'' = F(z; u, u'),\tag{2.7}$$

where F is rational in u and u' and analytic in z, according to whether or not they possess the Painlevé property.

They discovered that every equation possessing the Painlevé property could either be solved in terms of known functions (trigonometric functions, elliptic functions, solutions of linear ODEs, etc.) or transformed into one of the six equations now called the Painlevé equations (P_I–P_{VI}). They have standard forms that are listed below. (They are representatives of equivalence classes under Möbius transformations.) Their general solutions are higher transcendental functions.

The Painlevé Equations

$$u'' = 6u^2 + z,$$

$$u'' = 2u^3 + zu + \alpha,$$

$$u'' = \frac{1}{u}u'^2 - \frac{1}{z}u' + \frac{1}{z}(\alpha u^2 + \beta) + \gamma u^3 + \frac{\delta}{u},$$

$$u'' = \frac{1}{2u}u'^2 + \frac{3}{2}u^3 + 4zu^2 + 2(z^2 - \alpha)u + \frac{\beta}{u},$$

$$u'' = \left\{\frac{1}{2u} + \frac{1}{u-1}\right\}u'^2 - \frac{1}{z}u' + \frac{(u-1)^2}{z^2}\left(\alpha + \frac{\beta}{u}\right) + \frac{\gamma u}{z} + \frac{\delta u(u+1)}{u-1},$$

$$u'' = \frac{1}{2}\left\{\frac{1}{u} + \frac{1}{u-1} + \frac{1}{u-z}\right\}u'^2 - \left\{\frac{1}{z} + \frac{1}{u-1} + \frac{1}{u-z}\right\}u'$$
$$+ \frac{u(u-1)(u-z)}{z^2(z-1)^2}\left\{\alpha + \frac{\beta z}{u^2} + \frac{\gamma(z-1)}{(u-1)^2} + \frac{\delta z(z-1)}{(u-z)^2}\right\},$$

Two main procedures were used in this work. The first is known as the α-method and the second is now called Painlevé analysis. Painlevé described the α-method in the following way.

Considérons une équation différentielle dont le coefficient différentiel est une fonction (holomorphe pour $\alpha = 0$) d'un paramètre α. Si l'équation a ses points critiques fixes pour α quelconque (mais $\neq 0$), il en est de même, a fortiori pour $\alpha = 0$, et le développement de l'intégrale $y(x)$, suivant les puissances de α, a comme coëfficients des fonctions de x à points critiques fixes.

(This extract is taken from footnote 3 on p.11 of [55]. In Painlevé's terminology, a critical point of a solution is a point around which it is multivalued.) In other words, suppose a parameter α can be introduced into an ODE in such a way that it is analytic for $\alpha = 0$. Then if the ODE has the Painlevé property for $\alpha \neq 0$, it must also have this property for $\alpha = 0$. We illustrate this method for the classification problem for first-order ODEs below.

The second procedure, called Painlevé analysis, is a method of examining the solution through formal expansions in neighbourhoods of singularities of the ODE. In particular, the procedure focusses on formal series expansions of the solution(s) in neighbourhoods of generic (arbitrary) points (not equal to fixed singularities). The series expansion is based on Fröbenius analysis and usually takes the form given by (1.5).

As mentioned in the Introduction, this procedure was extended to PDEs by WTC (Weiss, Tabor, and Carnevale) [59]. For PDEs, the above definitions of the Painlevé property continue to hold under the interpretation that a *movable singularity* means a *noncharacteristic analytic movable singularity manifold*.

A noncharacteristic manifold for a given PDE is a surface on which we can freely specify Cauchy data. The linear wave equation,

$$u_{tt} - u_{xx} = 0, \tag{2.8}$$

has the general solution

$$u(x,t) = f(t - x) + g(t + x),$$

where f and g are arbitrary. By a suitable choice of f and g we can construct a solution u with any type of singularity on the curves $t - x = k_1$, $t + x = k_2$, for arbitrary constants k_1, and k_2. These lines are characteristic manifolds for (2.8). This example illustrates why the Painlevé property says nothing about the singular behaviour of solutions on characteristic singularity manifolds.

The WTC procedure is to expand the solutions $u(x,t)$ of a PDE as

$$u(x,t) = \sum_{n=0}^{\infty} u_n(x,t)\Phi^{n+\rho}, \tag{2.9}$$

near a noncharacteristic analytic movable singularity manifold given by $\Phi = 0$. (This extends in an obvious way to functions of more than two variables.) The actual expansion can be simplified by using information specific to the PDE about its characteristic directions. For example, for the KdV equation, noncharacteristic means that $\Phi_x \neq 0$. Hence by using the implicit function theorem near the singularity manifold, we can write

$$\Phi(x, t) = x - \xi(t),$$

where $\xi(t)$ is an arbitrary function. This is explored further in Sect. 2.3.2 below.

In some cases, the form of the series equation (1.5) (or (2.9)) needs modification. A simple example of this is the ODE

$$u''' = 2(u')^3 + 1. \tag{2.10}$$

Here $v = u'$ is a Jacobian elliptic function with simple poles of residue ± 1. (See [7].) Hence a series expansion of $u(z)$ around such a singularity z_0, say, must start with $\pm \log(z - z_0)$. The remainder of the series is a power series expansion in powers of $z - z_0$. In such cases, the Painlevé property holds for the new variable v.

2.2   The α-Method

In this section, we illustrate the α-method by using it to find all ODEs of the form

$$u' = \frac{P(z, u)}{Q(z, u)} \tag{2.11}$$

possessing the Painlevé property, where P and Q are analytic in z and polynomial in u (with no common factors). The first step of the α-method is to introduce a small parameter α through a transformation of variables in such a way that the resulting ODE is analytic in α. However, the transformation must be suitably chosen so that the limit $\alpha \to 0$ allows us to focus on a movable singularity. This is crucial for deducing necessary conditions for the Painlevé property.

We accomplish this by using dominant balances of the ODE near such a singularity. (See [8,43] for a definition and discussion of the method of dominant balances.)

If Q has a zero of multiplicity m at $u = a(z)$ then, after performing the transformation $u(z) \mapsto u(z) + a(z)$, (2.11) has the form

$$u^m u' = f(z, u), \tag{2.12}$$

where f is analytic in u at $u = 0$ and $f(z, 0) \not\equiv 0$ (since P and Q have no common factors). Choose z_0 so that $\kappa := f(z_0, 0) \neq 0$ and define the transformation

$$u(z) = \alpha U(Z), \quad z = z_0 + \alpha^n Z,$$

where n is yet to be determined and α is a small (but nonzero) parameter. Note that this is designed to focus on solutions that become close to the singular value $u = 0$ of the equation somewhere in the z-plane.

Equation (2.12) then becomes

$$\alpha^{m+1-n}U^m\frac{dU}{dZ} = f(z_0 + \alpha^n Z, \alpha U) = \kappa + O(\alpha).$$

This equation has a dominant balance when $n = m+1$. In this case the limit as $\alpha \to 0$ gives

$$U^m\frac{dU}{dZ} = \kappa$$

which has the exact solution

$$U(Z) = \{(m+1)\kappa Z + C\}^{\frac{1}{m+1}},$$

where C is a constant of integration. This solution has a movable branch point at $Z = -C/(\kappa(m+1))$ for all $m > 0$. Therefore, (2.11) cannot possess the Painlevé property unless $m = 0$, i.e., Q must be independent of u. That is, to possess the Painlevé property, (2.11) must necessarily be of the form

$$u' = a_0(z) + a_1(z)u + a_2(z)u^2 + \cdots + a_N(z)u^N, \tag{2.13}$$

for some nonnegative integer N.

The standard theorems of existence and uniqueness fail for this equation wherever u becomes unbounded. To investigate what happens in this case, we transform to $v := 1/u$. (Note that the Painlevé property is invariant under such a transformation.) Equation (2.13) then becomes

$$v' + a_0(z)v^2 + a_1(z)v + a_2(z) + a_3(z)v^{-1} + \cdots + a_N(z)v^{2-N} = 0.$$

But this equation is of the form (2.11), so it can only possess the Painlevé property if $N = 2$.

In summary, for (2.11) to possess the Painlevé property, it must necessarily be a Riccati equation,

$$u' = a_0(z) + a_1(z)u + a_2(z)u^2. \tag{2.14}$$

To show that this is also sufficient, consider the transformation

$$u = -\frac{1}{a_2(z)}\frac{w'}{w}$$

which linearizes (2.14)

$$a_2 w'' - (a_2' + a_1 a_2)w' + a_0 a_2^2 w = 0.$$

By Fuchs theorem [8], the singularities of any solution w can only occur at the singularities of $(a_2' + a_1 a_2)/a_2$ or $a_0 a_2$. These are fixed singularities. Hence the only movable singularities of u occur at the zeroes of w. Since w is analytic at its zeroes, it follows that u is meromorphic around such points. That is, (2.14) has the Painlevé property.

2.3   The Painlevé Test

Here we illustrate the widely used formal tests for the Painlevé property for ODEs and PDEs by using examples.

ODEs. Consider a class of ODEs given by

$$u'' = 6u^n + f(z), \tag{2.15}$$

where f is (locally) analytic and $n \geq 1$ is an integer (the cases $n = 0$ or 1 correspond to linear equations).

Standard theorems that yield analytic solutions fail for this equation wherever the right side becomes singular, i.e. where either $f(z)$ or u becomes unbounded. We concentrate on the second possibility to find movable singularities. This means that the hypothesized expansion equation (1.5) must start with a term that blows up at z_0. To find this term, substitute

$$u(z) \sim c_0(z - z_0)^p, \qquad z \to z_0,$$

where $\Re(p) < 0$, $c_0 \neq 0$, into (2.15). This gives the dominant equation

$$c_0 p(p-1)(z - z_0)^{p-2} + O\left((z - z_0)^{p-1}\right) = 6c_0^2(z - z_0)^{np} + O\left((z - z_0)^{np+1}\right). \tag{2.16}$$

The largest terms here must balance each other (otherwise there is no such solution). Since $c_0 \neq 0$ and $p \neq 0$ or 1, we get

$$p - 2 = np, \quad \Rightarrow \quad p = \frac{-2}{n-1}.$$

If p is not an integer, then u is branched at z_0. Hence, the only $n > 1$ for which (2.15) can possess the Painlevé property are $n = 2$ and $n = 3$.

We will only consider the case $n = 2$ here for conciseness. The case $n = 3$ is similar. (The reader is urged to retrace the following steps for the case $n = 3$.)

If $n = 2$ then $p = -2$ (which is consistent with our assumption that $\Re(p) < 0$). Then (2.16) becomes

$$6c_0(z - z_0)^{-4} = 6c_0^2 + O\left((z - z_0)^{-3}\right),$$

which gives $c_0 = 1$. Hence the hypothesized series expansion for u has the form

$$u(z) = \sum_{n=0}^{\infty} c_n(z - z_0)^{n-2}. \tag{2.17}$$

The function $f(z)$ can also be expanded in a power series in $z - z_0$ because, by assumption, it is analytic. Doing so and substituting expansion (2.17) into

(2.15) gives

$$\sum_{n=0}^{\infty} (n-2)(n-3)c_n(z-z_0)^{n-4}$$

$$= 6 \sum_{i,j=0}^{\infty} c_i c_j (z-z_0)^{i+j-4} + \sum_{m=0}^{\infty} \frac{1}{m!} f^{(m)}(z_0)(z-z_0)^m$$

$$= 6(z-z_0)^{-4} + 12c_1(z-z_0)^{-3}$$
$$+6(c_1^2 + 2c_2)(z-z_0)^{-2} + 12(c_3 + c_1 c_2)(z-z_0)^{-1}$$
$$+ \sum_{n=4}^{\infty} \left\{ 6 \sum_{m=0}^{n} c_m c_{n-m} + \frac{1}{(n-4)!} f^{(n-4)}(z_0) \right\} (z-z_0)^{n-4}.$$

Equating coefficients of like powers of $(z-z_0)$ we get $c_1 = 0$, $c_2 = 0$, $c_3 = 0$, and

$$(n-2)(n-3)c_n = 6 \sum_{m=0}^{n} c_m c_{n-m} + \frac{1}{(n-4)!} f^{(n-4)}(z_0), \quad (n \geq 4).$$

Note that c_n appears on both sides of this equation. Solving for c_n, we find

$$(n+1)(n-6)c_n = 6 \sum_{m=1}^{n-1} c_m c_{n-m} + \frac{1}{(n-4)!} f^{(n-4)}(z_0). \qquad (2.18)$$

For each $n \neq 6$, this relation defines c_n in terms of $\{c_m\}_{0 \leq m < n}$. However, for $n = 6$, the coefficient of c_n vanishes and (2.18) fails to define c_6. If the right side also vanishes, c_6 is arbitrary. However, if the right side does not vanish there is a contradiction which implies that the series (2.17) cannot be a formal solution of (2.15).

In that second case, the expansion can be modified to yield a formal solution by inserting appropriate logarithmic terms starting at the index $n = 6$. (This is also the case for Fröbenius expansions when the indicial exponents differ by an integer – see [8]. See [41] also for a rigorous study of equations admitting such algebraico-logarithmic expansions in several variables.) In such a case, logarithmic terms appear infinitely often in the expansion and cannot be transformed away (as in (2.10) above). They therefore indicate multivaluedness around movable singularities.

That is, (2.15) fails the Painlevé test unless the right side of (2.18) vanishes at $n = 6$. This condition reduces to

$$f''(z_0) = 0.$$

However, since z_0 is arbitrary, this implies that $f'' \equiv 0$. That is, $f(z) = az+b$ for some constants a, b. If $a = 0$, this equation can be solved in terms of (Weierstrass) elliptic functions. Otherwise, translating z and rescaling u and z gives

$$u'' = 6u^2 + z, \qquad (2.19)$$

which is the first Painlevé equation.

The index of the free coefficient, c_6, in the above expansion is called a *resonance*. The expansion contains two arbitrary constants, c_6 and z_0, which indicates that it captures the generic singular behaviour of a solution (because the equation is second-order).

There is a standard method for finding the location of resonances which avoids calculation of all previous coefficients. We illustrate this method here for P_I. After determining the leading order behaviour, substitute the perturbation

$$u \sim (z - z_0)^{-2} + \cdots + \beta(z - z_0)^{r-2} ,$$

where $r > 0$, into (2.19). Here β plays the role of the arbitrary coefficient. To find a resonance r, we collect terms in the equation that are linear in β and demand that the coefficient of β vanishes. This is equivalent to demanding that β be free. The resulting equation,

$$(r + 1)(r - 6) = 0,$$

is called the resonance equation and is precisely the coefficient of c_r on the left side of (2.18).

The positive root $r = 6$ is precisely the resonance we found earlier. The negative root $r = -1$, often called the universal resonance, corresponds to the translation freedom in z_0. (Consider $z_0 \mapsto z_0 + \epsilon$. Taking $|\epsilon| < |z - z_0|$, and expanding in ϵ shows that $r = -1$ does correspond to an arbitrary perturbation.)

Note, however, that $r = -1$ is not always a resonance. For example, consider an expansion that starts with a nonzero constant term such as

$$1 + a_1(z - z_0) + \dots .$$

Perturbation of z_0 does not add a term corresponding to a simple pole to this expansion.

If any resonance is not an integer, then the equation fails the Painlevé test. The role played by other negative integer resonances is not fully understood. We explore this issue further through irregular singular point theory in Sect. 3.

For each resonance, the *resonance condition* needs to be verified, i.e., that the equation at that index is consistent. These give rise to necessary conditions for the Painlevé property. If all nonnegative resonance conditions are satisfied and all formal solutions around all generic arbitrary points z_0 are meromorphic, the equation is said to pass the Painlevé test.

This procedure needs to be carried out for evey possible singularity of the normalized equation. For example, the sixth Painlevé equation, P_{VI}, has four singular values in u, i.e., $u = 0, 1, z$ and ∞. The expansion procedure outlined above needs to be carried out around arbitrary points where u approaches each such singular value. (Table 2 in Sect. 4 lists all singular values of the Painlevé equations.)

PDEs. In this subsection, we illustrate the WTC series expansion technique with an example. Consider the variable coefficient KdV equation,

$$u_t + f(t)uu_x + g(t)u_{xxx} = 0. \tag{2.20}$$

Let $\phi(x, t)$ be an arbitrary holomorphic function such that $S := \{(x, t) : \phi(x, t)\} = 0$ is noncharacteristic. The fact that S is noncharacteristic for (2.20) means that

$$\phi_x \neq 0 \tag{2.21}$$

on S. By the implicit function theorem, we have $\phi(x, t) = x - \xi(t)$ locally, for some arbitrary function $\xi(t)$.

We begin by substituting an expansion of the form

$$u(x, t) = \sum_{n=0}^{\infty} u_n(t)\phi^{n+\alpha} \tag{2.22}$$

into (2.20). The leading order terms give $\alpha = -2$. Equating coefficients of powers of ϕ gives

$$n = 0 : u_0 = -12g/f, \tag{2.23}$$
$$n = 1 : u_1 = 0, \tag{2.24}$$
$$n = 2 : u_2 = \xi'/f \tag{2.25}$$
$$n = 3 : u_3 = u_0'/(fu_0), \tag{2.26}$$
$$n \geq 4 : (n+1)(n-4)(n-6)gu_n \tag{2.27}$$
$$= -f \sum_{k=0}^{n-4}(k+1)u_{n-k-3}u_{k+3} \tag{2.28}$$
$$+(n-4)\xi'u_{n-2} - u_{n-3}' \tag{2.29}$$

Arbitrary coefficients can enter at $n = 4, 6$ if the recursion relation is consistent. Consistency at $n = 6$ is equivalent to

$$\left(\frac{u_0'}{fu_0}\right)^2 + \frac{1}{f}\left(\frac{u_0'}{fu_0}\right)_t = 0.$$

This implies that

$$g(t) = f(t)\left\{a_0 \int^t f(s)ds + b_0\right\},$$

where a_0 and b_0 are arbitrary constants. In this case, (2.20) can be transformed exactly into the KdV equation (see Grimshaw [22]).

In particular, for the usual form of the KdV ($f(t) = 6$, $g(t) = 1$) we have the formal series expansion

$$u(x,t) = \frac{-2}{\phi^2} + \frac{\xi'(t)}{6} + u_4(t)\phi^2$$
$$-\xi''(t)\phi^3 + u_6(t)\phi^4 + O(\phi^5). \qquad (2.30)$$

Questions of convergence and well-posedness, i.e. continuity of the solution as the arbitrary functions ($\xi(t)$, $u_4(t)$, $u_6(t)$) vary, are discussed in Sect. 2.6 below.

2.4 Necessary versus Sufficient Conditions for the Painlevé Property

The methods we described above can only yield necessary conditions for the Painlevé property. Here we illustrate this point with an example (due to Painlevé) which does not possess the Painlevé property, but for which the Painlevé test indicates only meromorphic solutions.

Consider the ODE

$$(1 + u^2)u'' + (1 - 2u)u'^2 = 0 \qquad (2.31)$$

(see Ince [27]). The singularities of this equation are $u = \pm i$, $u = \infty$ and $u' = \infty$. Series expansions can be developed for solutions exhibiting each of the above singular behaviours and the equation passes the Painlevé test. This equation, however, has the general solution

$$u(z) = \tan\{\log[k(z - z_0)]\},$$

where k and z_0 are constants. For $k \neq 0$, u has poles at

$$z = z_0 + k^{-1}\exp\{-(n + 1/2)\pi\}$$

for every integer n. These poles accumulate at the movable point z_0, giving rise to a movable branched nonisolated essential singularity there. This example clearly illustrates the fact that passing the Painlevé test is not a guarantee that the equation actually possesses the Painlevé property.

This danger arises also for PDEs. The PDE

$$w_t = (1 + w^2)w_{xx} + (1 - 2w)w_x,$$

under the assumption

$$w(x,t) =: u(x) ,$$

reduces to the ODE above.

To be certain that a given differential equation possesses the Painlevé property, we must either solve it explicitly or implicitly (possibly through transformations into other equations known to have the property), or develop tests for sufficiency. Most results in the literature rely on the former approach. In the next section, we develop a method for testing sufficiency.

2.5   A Direct Proof of the Painlevé Property for ODEs

In this section we outline a direct proof given in (Joshi and Kruskal [31, 34]) that the Painlevé equations indeed possess the Painlevé property. The proof is based on the well known Picard iteration method (used to prove the standard theorems of existence and analyticity of solutions near regular points) modified to apply near singular points of the Painlevé equations. A recommended simple example for understanding the method of proof is

$$u'' = 6u^2 + 1$$

which is solved by Weierstrass elliptic functions.

Consider the initial value problem for each of the six Painlevé equations with regular data for u and u' given at an ordinary point $z_1 \in \mathbb{C}$. (The point z_1 cannot equal 0 for P_{III}, P_V or P_{VI}, and cannot equal 1 for P_{VI} – see Table 2. Also, $u(z_1)$ cannot equal the values given in the third column of Table 2.)

Standard theorems yield a (unique) solution U in any region in which the Lipschitz condition holds. However, they fail where the right side becomes unbounded, i.e. at its singular values. (See e.g. [14].) Since our purpose is to study the behaviour of the solution near its movable singularities, and these lie in the finite plane, we confine our attention to an arbitrarily large but bounded disk $|z| < B$ (where B is real and say > 1). For P_{III}, P_V, and P_{VI} this must be punctured at the finite fixed singularities. Henceforth we concentrate on P_I for simplicity.

The ball $|z| < B$ contains two types of regions. Around each movable singularity, we select a neighbourhood where the largest terms in the equation are sufficiently dominant over the other terms. We refer to these as *special regions*. Outside these special regions, the terms remain bounded. Therefore, the ball resembles a piece of Swiss cheese, the holes (which may not be circular in general but in this case turn out to be nearly circular) being the special regions where movable singularities reside and the solid cheese being free of any singularity.

Starting at z_1 in (the cheese-like region of) the ball, we continue the solution U along a ray until we encounter a point z_2 on the edge of a special region. Inside the region, we convert P_I to an integral equation by operating successively on the equation as though only the dominant terms were present.

Table 2. Fixed and movable singularities of the Painlevé equations

Equation	Fixed Singularities (z-value)	Movable Singularities (u-value)
P_I	$\{\infty\}$	$\{\infty\}$
P_{II}	$\{\infty\}$	$\{\infty\}$
P_{III}	$\{0, \infty\}$	$\{0, \infty\}$
P_{IV}	$\{\infty\}$	$\{0, \infty\}$
P_V	$\{0, \infty\}$	$\{0, 1, \infty\}$
P_{VI}	$\{0, 1, \infty\}$	$\{0, 1, z, \infty\}$

The dominant terms of

$$u'' = 6u^2 + z, \tag{2.32}$$

are u'' and $6u^2$. Integrating these dominant terms after multiplying by their integrating factor u', we get

$$\frac{u'^2}{2} = 2u^3 + zu - \int_{z_2}^{z} u \, dz + \bar{k}, \tag{2.33}$$

with

$$\bar{k} := \int_{z_1}^{z_2} u \, dz + k,$$

where the constant k (kept fixed below) is determined explicitly by the initial conditions.

Since u is large, u' does not vanish (according to (2.33)) and, therefore, there is a path of steepest ascent starting at z_2. We will use this idea to find a first point in the special region where u becomes infinite.

Let d be an upper bound on the length of the path of integration from z_2 to z, and assume that $A > 0$ is given such that

$$A^2 > 4B, A^2 > 4\pi B, A^2 > 4d, A^3 > 4|k|.$$

Assume that $|u| \geq A$ at z_2. Then (2.33) gives $u'(z_2) \neq 0$. Taking the path of integration to be the path of steepest ascent, we can show that

$$|u'| > \sqrt{2}|u|^{3/2},$$

and that the distance to a point where $|u|$ becomes infinite is

$$d < \sqrt{2}A^{-1/2}.$$

(See page 193 of [34] for details.) So there is a first singularity encountered on this path which we will call z_0.

Now integrating the dominant terms once more (by dividing by $2u^3$, taking the square root and integrating from z_0) we get

$$u = \left(\int_{z_0}^{z} \left\{ 1 + \frac{1}{2u^3} \left(\bar{k} + zu - \int_{z_2}^{z} u \, dz \right) \right\}^{1/2} dz \right)^{-2}. \tag{2.34}$$

Substituting a function of the form

$$u(x,t) = \frac{1}{(z - z_0)^2} + f(z)$$

where $f(z)$ is analytic at z_0 into the right side of (2.34) returns a function of the same form. Notice that, therefore, no logarithmic terms can arise.

It is worth noting that the iteration of the integral equation (2.34) gives the same expansion that we would have obtained by the Painlevé test. In particular, it generates the appropriate formal solution without any assumptions of its form, and it points out precisely where logarithmic terms may arise without additional investigations. (For example, try iteration of the integral equation with the term z on the right side of (2.32) replaced by z^2, i.e. with $zu - \int_{z_2}^{z} u\,dz$ replaced by $z^2 u - 2 \int_{z_2}^{z} zu\,dz$ in (2.34).)

The remainder of the proof is a demonstration that the integral equation (2.34) has a unique solution meromorphic in a disk centred at z_0, that its radius is lower-bounded by a number that is independent of z_0, and that the solution is the same as the continued solution U. The uniformity of the lower bound is crucial for the proof. Uniformity excludes the possibility that the movable poles may accumulate to form movable essential singularities as in example (2.31).

Since the analytic continuation of U is accomplished along the union of segments of rays and circular arcs (skirting around the boundaries of successive special regions encountered on such rays) and these together with the special regions cover the whole ball $|z| < B$, we get a proof that the first Painlevé transcendent is meromorphic throughout the ball.

2.6   Rigorous Results for PDEs

Sufficient results for the Painlevé property of PDEs have been harder to achieve than ODEs. This is surprising because such results are lacking even for the most well known integrable PDEs. In this section, we describe some partial results towards this direction for the KdV equation.

Definition 22. *The* WTC *data for the KdV equation is the set* $\{\xi(t),\, u_4(t),\, u_6(t)\}$ *of arbitrary functions describing this Painlevé expansion.*

The following theorem proves that the series (2.30) converges for analytic WTC data.

Theorem 1. *(Joshi and Petersen [35, 36]) Given an analytic manifold* $S := \{(x,t) : x = \xi(t)\}$*, with* $\xi(0) = 0$*, and two arbitrary analytic functions*

$$\lim_{x \to \xi(t)} \left(\frac{\partial}{\partial x}\right)^4 [w(x,t)(x - \xi(t))^2], \qquad \lim_{x \to \xi(t)} \left(\frac{\partial}{\partial x}\right)^6 [w(x,t)(x - \xi(t))^2]$$

there exists in a neighbourhood of $(0,0)$ *a meromorphic solution of the KdV equation (1.1) of the form*

$$w(x,t) = \frac{-12}{\left(x - \xi(t)\right)^2} + h(x,t) \,,$$

where h *is holomorphic.*

The next theorem provides us with a useful lower bound on the radius of convergence of this series.

Theorem 2. *(Joshi and Srinivasan [38]) Given WTC data $\xi(t)$, $u_4(t)$, $u_6(t)$ analytic in the ball $B_{2\rho+\epsilon}(0) = \{t \in \mathbf{C} : |t| < 2\rho + \epsilon\}$, let*

$$M = \sup_{|t|=2\rho} \{1, |\xi(t)|, |u_4(t)|, |u_6(t)|\}.$$

The radius of convergence $R_\rho = R$ of the power series (1.5) satisfies

$$R \geq \frac{\min\{1, \rho\}}{10M}.$$

This lower bound is used in [38] to prove the well-posedness of the WTC Cauchy problem. That is, the locally meromorphic function described by the convergent series (2.30) has continuous dependence on the WTC data in the sup norm.

To date there is no proof that the Korteweg-de Vries equation possesses the Painlevé property. The main problem lies in a lack of methods for obtaining the global analytic description of a locally defined solution in the space of several complex variables. However, some partial results have been obtained.

The usual initial value problem for the KdV equation is given on the characteristic manifold $t = 0$. Well known symmetry reductions of the KdV equation, e.g., to a Painlevé equation, suggest that a generic solution must possess an infinite number of poles. WTC-type analysis shows that these can occur on non-characteristic manifolds which intersect $t = 0$ transversely. These results suggest that only very special solutions can be entire, i.e., have no singularities, on $t = 0$.

Joshi and Petersen [37] showed that if the initial value

$$u(x,0) = u_0(x) := \sum_{n=0}^{\infty} a_n x^n,$$

is entire in x and, moreover, the coefficients a_n are real and nonnegative then there exists no solution that is holomorphic in any neighbourhood of the origin in $\mathbf{C}^2$ unless

$$u_0(x) = a_0 + a_1 x.$$

This result can be extended to the case of more general a_n under a condition on the growth of the function $u_0(x)$ as $x \to \infty$.

3   Nonlinear-Irregular-Singular Point Analysis

The Painlevé expansions cannot describe all possible singular behaviours of solutions of differential equations. In this section, we describe some extensions based on irregular-singular-point theory for linear equations.

The Painlevé expansions at their simplest are Laurent series with a leading term, and may, therefore, fail to describe solutions that possess movable isolated essential singularities. Consider the ODE

$$3u'u''' = 5(u'')^2 - (u')^2 \frac{u''}{u} - \frac{(u')^4}{u^2},$$

which has the general solution

$$u(z) = \alpha \exp\left\{\beta(z - z_0)^{-1/2}\right\}.$$

Clearly u has a branched movable essential singularity. As suggested in [44], the Painlevé test can be extended to capture this behaviour by considering solutions that become exponentially large near z_0. To do this we expand

$$u(z) = a_{-1}(z)e^{S(z)} + a_0(z) + a_1(z)e^{-S(z)} + a_2(z)e^{-2S(z)} + a_3(z)e^{-3S(z)} + \ldots,$$

where S and the a_n are generalized power series and S grows faster than any logarithm as z approaches z_0.

In other words, generalized expansions (those involving logarithms, powers, exponentials and their compositions) are necessary if we are to describe all possible singularities. These are asymptotic expansions which may fail to converge. We show in this section that they can nevertheless yield analytic information about solutions. We illustrate this with two main examples. The first is the Chazy equation and the second a fourth-order equation studied by Bureau.

3.1   The Chazy Equation

In this subsection, we examine the Chazy equation,

$$y''' = 2yy'' - 3(y')^2. \tag{3.35}$$

This equation is exactly solvable through the transformation [12, 13]

$$z(t) := \frac{u_2(t)}{u_1(t)}, \quad y(z(t)) = \frac{6}{u_1(t)}\frac{du_1(t)}{dt} = 6\frac{d}{dt}\log u_1(t),$$

where u_1 and u_2 are two independent solutions of the hypergeometric equation,

$$t(t - 1)\frac{d^2u}{dt^2} + \left(\frac{1}{2} - \frac{7}{6}t\right)\frac{du}{dt} - \frac{u}{144} = 0.$$

Following the work of Halphen [26], Chazy noted that the function $z(t)$ maps the upper half t-plane punctured at 0, 1, and ∞ to the interior of a circular triangle with angles $\pi/2$, $\pi/3$, and 0 in the z-plane (see, for example, Nehari [50]). The analytic continuation of the solutions u_1 and u_2 through one of the intervals $(0, 1)$, $(1, \infty)$, or $(-\infty, 0)$ corresponds to an inversion of the image triangle across one of its sides into a complementary triangle. Continuing this process indefinitely leads to a tessellation of either the interior or the exterior of a circle on the Riemann sphere. This circle is a *natural barrier* in the sense that the solution can be analytically extended up to but not through it.

We will see below that any solution of (3.35) is single-valued everywhere it is defined. The general solution, however, possesses a movable natural barrier, i.e., a closed curve on the Riemann sphere whose location depends on initial conditions and through which the solution cannot be analytically continued.

Leading order analysis of (3.35) shows that, near a pole,

$$y \sim -6(z - z_0)^{-1}, \quad \text{or} \quad y \sim A(z - z_0)^{-2},$$

where A is an arbitrary (but nonzero) constant. On calculating successive terms in this generalized series expansion we only find the exact solution,

$$y(z) = \frac{A}{(z - z_0)^2} - \frac{6}{z - z_0}. \tag{3.36}$$

This solution has only two arbitrary constants, A and z_0, and clearly cannot describe all possible solutions of (3.35) which is third-order. That is, solutions of the form (3.36) fail to capture the generic behaviour of the full space of solutions.

The absent degree of freedom may lie in a perturbation of this solution. Applying the usual procedure for locating resonances, i.e., substituting the expression

$$y(z) \sim \frac{-6}{(z - z_0)^2} + \cdots + \beta(z - z_0)^{r-2}$$

into (3.35) and demanding that β be free, we find that

$$(r + 1)(r + 2)(r + 3) = 0,$$

i.e. we must have $r = -1$, $r = -2$ or $r = -3$. The case $r = -1$ corresponds to the fact that A is arbitrary in (3.36). The case $r = -2$ corresponds to the freedom in z_0. The case $r = -3$, however, indicates something more.

Since the usual Fröbenius-type series fails to describe the general solution near a singular point, we turn to be a nonlinear analogue of irregular singular point theory. Arguing from analogy with the linear theory (see, for example, Bender and Orszag [8]) we seek a solution of the form

$$y(z) = \frac{A}{(z - z_0)^2} - \frac{6}{(z - z_0)} + \exp S(z), \tag{3.37}$$

where $\exp S(z)$ is regarded as small in a region near z_0 (generically, z_0 will be on the boundary of this region).

Since the Chazy equation is autonomous we can, without loss of generality, take $z_0 = 0$. For simplicity we also take $A = 1/2$. Substituting the expansion (3.37) into (3.35) gives

$$S''' + 3S'S'' + S'^3$$
$$= \left(\frac{1}{z^2} - \frac{12}{z} \right) (S'' + S'^2) + 6 \left(\frac{1}{z^3} - \frac{6}{z^2} \right) S'$$
$$+ 6 \left(\frac{1}{z^4} - \frac{4}{z^3} \right) + 2 \left(S'' + S'^2 \right) e^S - 3S'^2 e^S. \tag{3.38}$$

To ensure that $\exp(S)$ is exponential rather than algebraic, we must assume that $S'' \ll S'^2$, $S''' \ll S'^3$. Using these assumptions along with $\exp S \ll 1$, $S' \gg 1$, and $z \ll 1$, (3.38) gives

$$S' \sim \frac{1}{z^2} - \frac{2}{z}.$$

Integration yields

$$y(z) \sim \frac{1}{2z^2} - \frac{6}{z} + \frac{k}{z^2}e^{-1/z},$$

where k is an arbitrary constant; k represents the third degree of freedom that was missing from the Laurent series (3.36). Extending to higher orders in $\exp S(z)$, we obtain the double series

$$y(z) = \frac{1}{2z^2} - \frac{6}{z}$$
$$+ \frac{k}{z^2}e^{-1/z}\left(1 + O(z)\right) + \frac{k^2}{8z^2}e^{-2/z}\left(1 + O(z)\right) \qquad (3.39)$$
$$+ O\left(\frac{e^{-3/z}}{z^2}\right).$$

It can be shown that this series is convergent in a half-plane, given here by $\Re(1/z) > 0$.

This asymptotic series is valid wherever

$$|k\exp(-1/z)| \ll 1. \qquad (3.40)$$

Suppose that k is small. Put

$$z = -\xi + \eta,$$

where $\xi > 0$, to see whether the half-plane of validity can be extended. Then the condition (3.40) becomes

$$\frac{\xi}{\xi^2 + \eta^2} \ll \log\left(\frac{1}{|k|}\right).$$

By completing squares (after multiplying out the denominator) this can be rewritten as

$$(\xi - \delta)^2 + \eta^2 \gg \delta^2,$$

where

$$\delta := -\frac{1}{2\log|k|} > 0.$$

So, asymptotically, the region of validity of the series (3.39) lies outside a circle of radius δ centered at $-\delta$. This is the circular barrier present in the general solution of the Chazy equation. In summary, the exponential (or WKB-type) approach has led to a three parameter solution. Morover, this description is valid in a region bounded by a circular curve where it diverges.

3.2 The Bureau Equation

Bureau partially extended the classification work of Painlevé and his colleagues to fourth-order equations. However, there were cases whose Painlevé property could not be determined within the class of techniques developed by Painlevé's school. One of these was

$$u'''' = 3u''u - 4u'^2, \tag{3.41}$$

which we will call the Bureau equation. In this subsection, we show that the general solution is multivalued around movable singularities by using exponential or WKB-type expansions based on irregular-singular-point theory.

It has been pointed out by several authors that (3.41) possesses two families of Painlevé expansions,

$$u \approx \frac{a_\nu}{z^\nu} \tag{3.42}$$

(where we have shifted $z - z_0$ to z by using the equation's translation invariance), distinguished by

$$\nu = 2, \quad a_2 = 60$$
$$\nu = 3, \quad a_3 \text{ arbitrary}$$

with resonances given by

$$u \approx \frac{a_\nu}{z^\nu} + \ldots + kz^{r-\nu}$$
$$\nu = 2 \ \Rightarrow \ r = -3, -2, -1, 20$$
$$\nu = 3 \ \Rightarrow \ r = -1, 0.$$

The case $\nu = 2$ has a full set of resonances (even though two are negative resonances other than the universal one). However, the case $\nu = 3$ is defective because its perturbation (in the class of Painlevé expansions) yields no additional degrees of freedom to the two already present in a_3 and z_0. It is, in fact, given by the two-term expansion

$$u = \frac{a}{z^3} + \frac{60}{z^2}.$$

We concentrate on this defective expansion in the remainder of this subsection. Since this expansion allows no perturbation in the class of conventional Painlevé expansions (which are based on regular-singular-point theory), we turn to perturbations of the form based on irregular-singular-point theory. Consider

$$u = \frac{a}{z^3} + \frac{60}{z^2} + \hat{u},$$

where

$$\hat{u} = \exp(S(z)), \ S' \gg \frac{1}{z}, z \ll 1.$$

(The assumption on S' is to assure that $\exp(S)$ is exponential rather than algebraic.) Substituting this into (3.41), we get

$$S'^4 = \frac{3a}{z^3}S'^2 + \left(\frac{3a}{z^3}S'' + \frac{24a}{z^4}S' - 6S'^2S''\right)$$
$$+ \left(\frac{36a}{z^5} + \frac{180}{z^2}S'^2 - 4S'S''' - 3S''^2\right)$$
$$+ \left(\frac{960}{z^3}S' + \frac{180}{z^2}S'' - S''''\right) + \frac{1080}{z^4}$$
$$+ \left(3S'' - S'^2\right)e^S \tag{3.43}$$

The condition that $\exp(S)$ not be algebraic also implies that $S'^2 \gg S''$, $S'^3 \gg S'''$, and $S'^4 \gg S^{(IV)}$. So dividing (3.43) by S'^2, taking the square root of the equation and expanding we get

$$S' = \frac{(3a)^{1/2}}{z^{3/2}} + \frac{31}{4z} + O(z^{-1/2}), \tag{3.44}$$

where we have used recursive substitution of the leading values of S' and S'' to get the term of order $1/z$. That is, we get

$$S = -2\frac{(3a)^{1/2}}{z^{1/2}} + \frac{31}{4}\log z + \text{const} + o(1).$$

Take one such solution, with say $a = 1/3$. Then the perturbed solution has expansion

$$u = \frac{1}{3z^3} + \frac{60}{z^2} + k_\pm z^{31/4}\exp\left(-2/z^{1/2}\right)\left(1 + o(1)\right), \tag{3.45}$$

where $k_\pm$ is an arbitrary constant. Note that there are two exponentials here (due to the two branches of the square root of z) and, therefore, $k_\pm$ represents two degrees of freedom. In the following, we consider only one of these solutions by fixing a branch of the square root in the exponential, say the one that is positive real on the positive real semi-axis in the z-plane. For short, we write $k_+ = k$.

Now consider the domain (or sector) of validity of this solution. Note that the neglected terms in its expansion contain a series of powers of $\exp(S)$ due to the nonlinear terms in (3.43). Therefore, for the expansion to be asymptotically valid, this exponential term must be bounded, i.e.,

$$\left|kz^{31/4}\exp\left(-2/z^{1/2}\right)\right| < 1$$
$$\Rightarrow |k|\exp\left(\Re\left(-2/z^{1/2} + (31/4)\ln(z)\right)\right) < 1. \tag{3.46}$$

We show below that the domain of validity given by this inequality contains a punctured disk (on a Riemann surface) whose angular width is larger than 2π.

Assume there is a branch cut along the negative semi-axis in the z-plane with $\arg(z) \in (-\pi, \pi]$. Consider $z^{1/2}$ in polar coordinates, i.e., $z^{1/2} = re^{i\theta}$, where $-\pi/2 < \theta < \pi/2$. The positive branch will then give real part

$$\Re\left(-2/z^{1/2} + (31/4)\ln(z)\right) = -2\frac{1}{r}\cos(\theta) + \frac{31\ln r}{4}.$$

Let $K := \ln|k|/2$. To satisfy (3.46), we must have

$$-\frac{2}{r}\cos(\theta) + \frac{31\ln r}{4} + 2K + o(1) < 0$$

for $r \ll 1$, i.e.,

$$-\cos(\theta) < -\frac{31r\ln r}{8} - Kr\left(1 + o(1)\right). \tag{3.47}$$

Since r is small (note $\ln r < 0$), this can only be violated near $\theta = \pm\pi/2$. Fix r small. Expand $\theta = \pi/2 + \epsilon$. Then (3.47) gives

$$\epsilon < -\frac{31r\ln r}{8} - Kr\left(1 + o(1)\right) + O(\epsilon^3). \tag{3.48}$$

In particular, ϵ can be negative, so long as $|\epsilon| < 1$. A similar calculation can be made near $-\pi/2$.

These results show that the asymptotic validity of the solution given by (3.45) can be extended to a domain which is a disk of angular width $> 2\pi$. The small angular overlap is given by a sector of angular width 2ϵ where ϵ is bounded by $O(r\ln r)$ according to (3.48).

Let z_s be a point in this overlapping wedge with small modulus. At such a point, we have two asymptotic representations of u, one given by a prior choice of branch of $z_s^{1/2}$ and the other given by analytic continuation across the branch cut. If the true solution is single-valued in this domain, the choice of two asymptotic representations violates uniqueness. Therefore, the true solution must itself be multivalued. In other words, the exponential expansion shows that Bureau's equation cannot have the Painlevé property.

4   Coalescence Limits

In this section we examine asymptotic limits of integrable equations that preserve the Painlevé property. In the case of ODEs, such limits form the basis of Painlevé's α-test. They are useful in the identification of nonintegrable equations and may be useful for indentifying new integrable equations as limits of others.

Painlevé [56] noted that under the transformation

$$z = \epsilon^2 x - 6\epsilon^{-10},$$
$$u = \epsilon y + \epsilon^{-5},$$
$$\alpha = 4\epsilon^{-15},$$

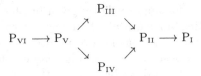

Fig. 1. Asymptotic limits among the Painlevé equations

P_{II} becomes

$$y''(x) = 6y^2 + x + \epsilon^6 \left\{ 2y^3 + xy \right\}. \tag{4.49}$$

In the limit as ϵ vanishes, (4.49) becomes P_I. We write the above limiting process as $P_{II} \longrightarrow P_I$. Painlevé gave a series of such asymptotic limits which are summarized in Fig. 1.

Each of these asymptotic limits coalesces the singular u-values of the Painlevé equation (see Table 2), i.e., they coalesce movable singularities. In [24], Halburd and Joshi proved that in the $P_{II} \longrightarrow P_I$ limit, simple poles of opposite residue coalesce to form the double poles in solutions of P_I. They also proved that all solutions of P_I can be obtained as limits of solutions of (4.49):

Theorem 3. *Choose $x_0, \alpha, \beta \in \mathbf{C}$. Let y_I and y be maximally extended solutions of P_I and (4.49) respectively, both satisfying the initial value problem given by*

$$y(x_0) = \alpha, \qquad y'(x_0) = \beta.$$

Let $\Omega \subset \mathbf{C}$ be the domain of analyticity of y_I. Given any compact $K \subset \Omega$, $\exists r_K > 0$ such that y is analytic in (x, ϵ) for $x \in K$ and $|\epsilon| < r_K$. Moreover, $y \to y_I$ in the sup norm as $\epsilon \to 0$.

The series of asymptotic (or *coalescence*) limits given in Fig. 1 by no means represents a complete list of such limits. The singular u-values of P_{IV} are 0 and ∞, corresponding to the zeros and poles of the solutions, respectively (the solutions of P_{IV} are meromorphic [34, 55]). The standard coalescence limit in which P_{IV} becomes P_{II} merges poles and zeros. However, the general solution of P_{IV} contains simple poles of oppositely signed residues which may be able to merge pairwise. An asymptotic limit coalescing these simple poles to form double poles does exist [32].

To see this, consider a transformation in which regions near infinity (where the poles are close to each other) are mapped to the finite plane in the limit $\epsilon \to 0$. It is necessary to rescale u to counter a cancellation of the oppositely signed poles. This leads to new variables x and $w(x)$ given by

$$u(z) = \epsilon^p w(x),$$
$$z = N + \epsilon^q x,$$

where $p, q > 0$, $\epsilon \ll 1$ and $N \gg 1$ is to be found in terms of ϵ. Then P_{IV} becomes

$$w_{xx} = \frac{w_x^2}{2w} + \frac{3}{2}\epsilon^{2(p+q)}w^3 + 4N\epsilon^{p+2q}w^2 + 8\epsilon^{p+3q}xw^2 + 2(N^2 - \alpha)\epsilon^{2q}w$$

$$+4N\epsilon^{3q}xw + 2\epsilon^{4q}x^2w + \frac{\beta\epsilon^{2(q-p)}}{w}. \tag{4.50}$$

The only maximal dominant balance, i.e., a limiting state of the equation in which a maximal number of largest terms remains [8], occurs when

$$q = p, \quad \text{and} \quad \alpha =: N^2 + a\epsilon^{-2q},$$

where a is a constant. Then, setting the largest terms $N\epsilon^{p+2q}$ and $N\epsilon^{3q}$ to unity gives $N = \epsilon^{-3p}$. Without loss of generality, redefine $\epsilon^p \mapsto \epsilon$. Then we get

$$N = \epsilon^{-3} \quad \text{and} \quad \alpha = \epsilon^{-6} + a\epsilon^{-2}.$$

Equation (4.50) then becomes

$$w_{xx} = \frac{w_x^2}{2w}4w^2 + (2x - a)w + \frac{\beta}{w} + \frac{3}{2}\epsilon^4 w^3 + 8\epsilon^4 xw^2 + 2\epsilon^4 x^2 w,$$

or, in the limit $\epsilon \to 0$,

$$w_{xx} = \frac{w_x^2}{2w}4w^2 + (2x - a)w + \frac{\beta}{w},$$

which is equation (XXXIV) (see p. 340 of Ince [27]) in the Painlevé-Gambier classification of second-order differential equations (after a simple scaling and transformation of variables).

Coalescence limits also exist among PDEs. For example, it is straightforward to derive the transformation [25]

$$\tau = t;$$

$$\xi = x + \frac{3}{2\epsilon^2}t;$$

$$u(x, t) = \epsilon U(\xi, \tau) - \frac{1}{2\epsilon},$$

which maps the modified Korteweg-de Vries equation (mKdV),

$$u_t - 6u^2 u_x + u_{xxx} = 0,$$

to

$$U_\tau - 6\epsilon^2 U^2 U_\xi + 6UU_\xi + U_{\xi\xi\xi} = 0,$$

which becomes the usual KdV equation in the limit $\epsilon \to 0$. An alternative method for obtaining the above asymptotic limit is to use the $P_{II} \to P_I$ limit. The mKdV equation is invariant under the scaling symmetry,

$$u \mapsto \lambda^{-1}u, \quad t \mapsto \lambda^3 t, \quad x \mapsto \lambda x.$$

Define the canonical variables

$$z = \frac{x}{(3t)^{1/3}}, \quad w = \frac{1}{3}\log t, \quad u(x,t) = (3t)^{-1/3}y(z,w).$$

In terms of these variables, the mKdV equation becomes

$$(\underbrace{y_{zz} - 2y^3 - zy - \alpha}_{P_{II}})_z + y_w = 0,$$

where we have included the constant α to emphasize its relation to P_{II}. Now apply the asymptotic transformation used in the $P_{II} \to P_I$ limit, to determine how y and z transform, and transform w in such a way that it remains in the limiting form of the equation as $\epsilon \to 0$. In this way we arrive at an equation equivalent to the KdV equation, which has a reduction to P_I.

In [25] it is shown that the system

$$\begin{aligned}
E_x &= \rho, \\
\tilde{E}_x &= \tilde{\rho}, \\
2N_t &= -(\rho\tilde{E} + \tilde{\rho}E), \\
\rho_t &= NE, \\
\tilde{\rho}_t &= N\tilde{E},
\end{aligned} \tag{4.51}$$

admits a reduction to the full P_{III} (P_{III} with all four constants $\delta \neq 0$, α, β, γ arbitrary). We note that if $\tilde{E} = E^*$ and $\tilde{\rho} = \rho^*$, where a star denotes complex conjugation, we recover the unpumped Maxwell-Bloch equations,

$$E_x = \rho, \quad 2N_t + \rho E^* + \rho^* E = 0, \quad \rho_t = NE.$$

Consider solutions of system (4.51) of the form

$$\begin{aligned}
E &= t^{-1}\varepsilon(z)w, \\
\tilde{E} &= t^{-1}\tilde{\varepsilon}(z)w^{-1}, \\
N &= n(z), \\
\rho &= r(z)w, \\
\tilde{\rho} &= \tilde{r}(z)w^{-1},
\end{aligned}$$

where $z := \sqrt{xt}$ and $w := (x/t)^k$, k constant. Then

$$y(z) := \frac{\varepsilon(z)}{zr(z)}$$

solves P_{III} with constants given by

$$\alpha = 2(r\tilde{\varepsilon} - \tilde{r}\varepsilon + 4k), \quad \beta = 4(1 + 2k), \quad \gamma = 4(n^2 + r\tilde{r}), \quad \delta = -4.$$

Note that by rescaling y we can make δ any nonzero number.

Using the procedure outlined for mKdV $\to$ KdV, it can be shown [25] that the $P_{III} \to P_{II}$ limit induces an asymptotic limit in which the generalized unpumped Maxwell-Bloch system (4.51) becomes the dispersive water-wave equation (DWW)

$$u_{xxxx} + 2u_t u_{xx} + 4u_x u_{xt} + 6u_x^2 u_{xx} + u_{tt} = 0,$$

which is known to admit a reduction to P_{II} (Ludlow and Clarkson [47]). The $P_{II} \to P_I$ limit then gives DWW $\to$ KdV. Ludlow and Clarkson [47] have shown that DWW also admits a symmetry reduction to the full P_{IV}. The $P_{IV} \to P_{II}$ limit then induces an asymptotic limit that maps DWW back to itself in a nontrivial way. The limit $P_{IV} \to P_{34}$, outlined above, induces another limit in which DWW is mapped to the KdV. All six Painlevé equations are known to arise as reductions of the self-dual Yang-Mills (SDYM) equations (Mason and Woodhouse [48]). Asymptotic limits between the Painlevé equations can be used to induce similar limits within the SDYM system (Halburd [23]).

Acknowledgements

The work reported in this paper was partially supported by the Australian Research Council. The authors also gratefully acknowledge with thanks the efforts of the organizing committee, particularly Dr. K.M. Tamizhmani, and CIMPA in arranging the Winter School on Nonlinear Systems at Pondicherry.

References

1. Ablowitz, M.J. and Clarkson, P.A. (1991) "Solitons, Nonlinear Evolution Equations and Inverse Scattering," Lond. Math. Soc. Lecture Notes Series **149**, Cambridge University Press, Cambridge

2. Ablowitz, M.J. and Segur, H. (1981) "Solitons and the Inverse Scattering Transform," SIAM, Philadelphia

3. Ablowitz, M.J., Ramani, A. and Segur, H. (1978) *Nonlinear Evolution Equations and Ordinary Differential Equations of Painlevé Type*, Lett. Nuovo Cim. **23**, 333–338

4. Ablowitz, M.J., Ramani, A. and Segur, H. (1980) *A Connection between Nonlinear Evolution Equations and Ordinary Differential Equations of P-type. I and II*, J. Math. Phys. **21**, 715–721 and 1006–1015

5. Ablowitz, M.J. and Segur, H. (1977) *Exact Linearization of a Painlevé Transcendent*, Phys. Rev. Lett. **38**, 1103–1106

6. Arnold, V.I. (1984) *Mathematical Methods of Classical Mechanics*, Springer-Verlag, New York

7. Abramowitz, M. and Stegun, I., eds. (1972) *Handbook of Mathematical Functions*, Dover, New York

8. Bender, C.M. and Orszag, S.A. (1978) "Advanced Mathematical Methods for Scientists and Engineers," McGraw-Hill, New York

9. Bountis, T., Segur, H., and Vivaldi, F. (1982) *Integrable Hamiltonian systems and the Painlevé property*, Phys. Rev. A **25**, 1257–1264

10. Boutroux, P. (1913) *Recherches sur les transcendents de M. Painlevé et l'étude asymptotique des équations différentielles du second ordre*, Ann. École Norm. Supér. **30**, 255–375; **31**, 99–159

11. Bureau, F.J. (1964) *Differential equations with fixed critical points*, Annali di Matematica pura ed applicata, **LXVI**, 1 – 116

12. Chazy, J. (1910) *Sur les équations différentielles dont l'intégrale générale possède une coupure essentielle mobile*, C.R. Acad. Sc. Paris **150**, 456–458

13. Chazy, J. (1911) *Sur les équations différentielles du troisième et d'ordre supérieur dont l'intégrale générale a ses points critiques fixés*, Acta Math. **34**, 317–385

14. Coddington, E.A. and Levinson, N. (1955) *Theory of Ordinary Differential Equations*, McGraw-Hill, New York

15. Clarkson, P.A. and Cosgrove, C.M. (1987) *Painlevé analysis of the non-linear Schrödinger family of equations*, J. Phys. A: Math. Gen. **20**, 2003–2024

16. Conte, R., Fordy, A., and Pickering, A. (1993) *A perturbative Painlevé approach to nonlinear differential equations*, Physica D, **69** 33–58

17. Cosgrove, C. M., (1993) *Painlevé classification of all semilinear PDEs of the second order, I Hyperbolic equations in two independent variables, II Parabolic and higher dimensional equations*, Stud. Appl. Math. **89** 1–61, 95–151

18. Fordy, A. and Pickering, A. (1991) *Analysing Negative Resonances in the Painlevé Test*, Phys. Lett. A **160**, 347–354

19. Fuchs, R. (1907) Math. Annalen **63** 301–321

20. Gambier B. (1910) *Sur les équations différentielles du second ordre et du premier degré dont l'intégrale générale est à points critiques fixes*, Acta Math. **33**, 1–55

21. Gardner, C.S., Greene, J.M., Kruskal, M.D., and Miura, R.M. (1967) *Method for Solving the Korteweg-de Vries Equation*, Phys. Rev. Lett. **19**, 1095–1097

22. Grimshaw, R. (1979) *Slowly Varying Solitary Waves. I. Korteweg-de Vries Equation*, Proc. R. Soc. Lond. A, **368**, 359–375

23. Halburd, R. (1996) *Integrable Systems; their singularity structure and coalescence limits*, PhD thesis, University of New South Wales

24. Halburd, R. and Joshi, N. (1996) *The Coalescence Limit of the Second Painlevé Equation*, Stud. Appl. Math. **97**, 1–15

25. Halburd, R. and Joshi, N. (1996) *Coalescence Limits of Integrable Partial Differential Equations*, in preparation

26. Halphen, G. (1881) *Sur un système d'équations différentielles*, C.R. Acad. Sc. Paris **92**, 1101–1103 and 1404–1407 [Œuvres (1918), Paris **2**, 475–477 and 478–481]

27. Ince, E.L. (1956) "Ordinary Differential Equations," Dover, New York

28. Hlavatý, L. (1987) *The Painlevé analysis of damped KdV equation*, J. Phys. Soc. Japan **55**, 1405–1406

29. Hlavatý, L. (1991) *Painlevé classification of PDEs*, in Proc. of NATO Advanced Research Workshop, P. Winternitz and D. Levi, eds, Ste Adèle, Quebec (Sept. 2–7, 1990), Plenum Publishing Co

30. Joshi, N. (1987) *Painlevé property of general variable-coefficient versions of the Korteweg-deVries and nonlinear Schrödinger equations*, Phys. Lett. A **125**, 456–460

31. Joshi, N. and Kruskal, M. (1992) *A Direct Proof that Solutions of the First Painlevé Equation have no Movable Singularities except Poles*, in "Nonlinear Evolution Equations and Dynamical Systems" (Baia Verde, 1991), Boiti, M., Martina, L., and Pempinelli, F., eds, World Sci. Publishing, River Edge, NJ, 310–317

32. Joshi, N. and Kruskal, M.D. (1993) *A New Coalescence of Movable Singularities in the Fourth Painlevé Equation*, University of New South Wales Applied Mathematics Report, AM93/11

33. Joshi, N. and Kruskal, M.D. (1993) *A Local Asymptotic Method of Seeing the Natural Barrier of the Solutions of the Chazy Equation*, in "Applications of Analytic and Geometric Methods to Nonlinear Differential Equations "(Exeter, 1992), NATO Adv. Sci. Inst. Ser. C, Math. Phys. Sci. 413, Kluwer Acad. Publ., Dordrecht, 331–339

34. Joshi, N. and Kruskal, M.D. (1994) *A Direct Proof that Solutions of the Six Painlevé Equations have no Movable Singularities Except Poles*, Stud. Appl. Math. **93**, 187–207

35. Joshi, N. and Petersen, J.A. (1996) *Complex blow-up in Burgers' equation: an iterative approach*, Bull. Austral. Math. Soc. **54**, 353–362

36. Joshi, N. and Petersen, J.A. (1994) *A Method for Proving the Convergence of the Painlevé Expansions of Partial Differential Equations*, Nonlinearity **7**, 595–602

37. Joshi, N. and Petersen, J.A. (1996) *Nonexistence Results Consistent with the Global Painlevé Property of the Korteweg-de Vries Equation* (to appear)

38. Joshi, N. and Srinivasan, G.K. (1997) *The Radius of Convergence and the Well-Posedness of the Painlevé Expansions of the Korteweg-de Vries Equation*, Nonlinearity (to appear)

39. Kowalevski, S. (1889a) *Sur le problème de la rotation d'un corps solide autour d'un point fixe*, Acta Math. **12**, 177–232

40. Kichenassamy, S. and Littman, W. (1993) *Blow-up surfaces for nonlinear wave equations I, II*, Comm. PDEs **18**, 431–452 and 1869–1899

41. Kichenassamy, S. and Srinivasan, G. (1995) *The structure of WTC expansions and applications*, J. Phys. A **28**, 1977–20004

42. Kowalevski, S. (1889b) *Sur une propriété d'un système d'équations différentielles qui définit la rotation d'un corps solide autour d'un point fixe*, Acta Math. **14**, 81–93

43. Kruskal, M.D. (1963) *Asymptotology*, in "Mathematical Models in Physical Sciences", Dobrot, S., ed., Prentice-Hall, Englewood Cliffs, New Jersey, 17–48

44. Kruskal, M.D. (1992) *Flexibility in Applying the Painlevé Test*, in "Painlevé Transcendents – their Asymptotics and Physical Applications", Levi, D. and Winternitz, P., eds., NATO ASI Series B: Physics **278**, Plenum Press, New York

45. Kruskal, M.D. and Clarkson, P.A. (1992) *The Painlevé-Kowalevski and Poly-Painlevé Tests for Integrability*, Stud. Appl. Math. **86**, 87-165

46. Kruskal, M.D. and Joshi, N. (1991) *Soliton theory, Painlevé property and integrability*, in *Chaos and Order*, Proceedings of the Miniconference of the Centre for Mathematical Analysis, Australian National University (Feb. 1-3, 1990), Joshi, N. and Dewar, R.L., eds., World Scientific Publishing Co., Singapore, 82–96

47. Ludlow, D.K. and Clarkson, P.A. (1993) *Symmetry Reductions and Exact Solutions for a Generalised Boussinesq Equation* in "Applications of Analytic and Geometric Methods to Nonlinear Differential Equations", P.A. Clarkson, ed., NATO ASI Series C **413**, Kluwer, Dordrecht, 415–430

48. Mason, L.J. and Woodhouse, N.M.J. (1993) *Self-Duality and the Painlevé Transcendents* Nonlinearity **6**, 569–581

49. McLeod, J.B. and Olver, P.J. (1983) *The Connection Between Partial Differential Equations Soluble by Inverse Scattering and Ordinary Differential Equations of Painlevé Type*, SIAM J. Math. Anal. **14** , 488–506

50. Nehari, Z. (1952) "Conformal Mapping," McGraw-Hill, New York
51. Newell, A.C., Tabor, M., and Zheng, Y., (1987) *A unified approach to Painlevé expansions*, Physica D **29**, 1–68
52. Olver, F.W.J. (1992) "Asymptotics and Special Functions," Academic Press, London
53. Painlevé, P. (1888) *Sur les équations différentielles du premier ordre*, C.R. Acad. Sc. Paris **107**, 221–224, 320–323, 724–726
54. Painlevé, P. (1900) *Mémoire sur les équations différentielles dont l'intégrale générale est uniforme*, Bull. Soc. Math. France **28**, 201–261
55. Painlevé, P. (1902) *Sur les équations différentielles du second ordre et d'ordre supérieur dont l'intégrale générale est uniforme*, Acta Math. **25**, 1–85
56. Painlevé, P. (1906) *Sur les équations différentielles du second ordre à points critiques fixes*, C. R. Acad. Sc. Paris **143**, 1111–1117
57. Ramani, A., Dorizzi, B., and Grammaticos, B. (1984) *Integrability and the Painlevé property for low-dimensional systems*, J. Math. Phys. **25**, 878–883
58. Ramani, A., Grammaticos, B., and Bountis, T. (1989) *The Painlevé property and singularity analysis of integrable and non-integrable systems*, Phys. Rep. **180**, 159–245
59. Weiss, J., Tabor, M. and Carnevale, G. (1983) *The Painlevé property for partial differential equations*, J. Math. Phys. **24**, 522–526
60. Yoshida, H. (1983), *Necessary condition for the existence of algebraic first integrals: I, II*, Celes. Mech. **31**, 363–379; 381–399
61. Ziglin, S. (1983), *Branching of solutions and nonexistence of first integrals in Hamiltonian mechanics; I, II*, Func. Anal. Appl. **16**, 181–189; **17**, 6–17
62. Ziglin, S. (1982), *Self-intersection of the complex separatrices and the nonexistence of the integrals in the Hamiltonian systems with one-and-half degrees of freedom*, J. Appl. Math. Mech. **45**, 411–413

Eight Lectures on Integrable Systems

F. Magri[1], P. Casati[1], G. Falqui[2], and M. Pedroni[3]

[1] Dipartimento di Matematica e Applicazioni, Università di Milano-Bicocca, Via degli Arcimboldi 8, 20126 Milano, Italy, `magri@matapp.unimib.it`, `casati@matapp.unimib.it`

[2] SISSA, Via Beirut 2/4, 34014 Trieste, Italy, `falqui@sissa.it`

[3] Dipartimento di Matematica, Università di Genova, Via Dodecaneso 35, 16146 Genova, Italy, `pedroni@dima.unige.it`

These lectures provide an introduction to the theory of integrable systems from the point of view of Poisson manifolds. In classical mechanics, an integrable system is a dynamical system on a symplectic manifold $\mathcal{M}$ which admits a complete set of constants of motion which are in involution. These constants are usually constructed by means of a symmetry group G acting symplectically on the phase–space. The point of view adopted in these lectures is to replace the group G by a "Poisson action" of the algebra of observables on $\mathcal{M}$ defined by a second Poisson bracket on $\mathcal{M}$. The development of this idea leads to the concept of a bihamiltonian manifold, which is a manifold $\mathcal{M}$ equipped with a pair of "compatible" Poisson brackets. The geometry of these manifolds builds upon two main themes: the Marsden–Ratiu (henceforth, MR) reduction process and the concept of generalized Casimir functions. The MR reduction clarifies the complicated geometry of a bihamiltonian manifold and defines suitable reduced phase–spaces, while the generalized Casimir functions constitute the Hamiltonians of the integrable systems. These three concepts (bihamiltonian manifolds, MR reduction, and generalized Casimir functions) are introduced, in this order, in the first three lectures. A simple example (the KdV theory) is used to show how they may be applied in practice.

The next three lectures, which are less pedagogical in nature, aim to provide a glimpse of the general theory of soliton equations. The purpose is to show that the range of applications of the bihamiltonian theory may be pushed far enough to encompass significant classes of examples. The first of these three lectures introduces the class of Poisson manifolds relevant to the theory of soliton equations, and it shows how to perform the MR reduction of these manifolds. The reduced spaces are called Gel'fand–Dickey (henceforth, GD) manifolds: they are the phase–spaces where the soliton equations are defined. The next lecture shows how to construct the generalized Casimir functions of the GD manifolds, and it introduces the GD equations. The sixth lecture, finally, is a quick survey of the relation between the GD theory and the KP theory. The unconventional point of view adopted in this paper argues that the KP equations arise as local conservation laws associated with the GD equations. Since these three lectures aim to stress the ideas of the bihamiltonian approach to soliton equations rather

F. Magri, P. Casati, G. Falqui, and M. Pedroni, Eight Lectures on Integrable Systems, Lect. Notes Phys. **638**, 209–250 (2004)
`http://www.springerlink.com/`
© Springer-Verlag Berlin Heidelberg 2004

than the technical details, they lack proofs. However, references are given which may help fill the gaps.

The final two lectures deal with the applications of the bihamiltonian theory to finite–dimensional integrable systems. The Calogero systems have been chosen as an example. We introduce a special class of bihamiltonian manifolds called Poisson–Nijenhuis manifolds, and we prove a theorem stating the existence of a privileged coordinate system on such a manifold. These coordinates are used to introduce the concept of "extended Lax representation" and to exhibit the bihamiltonian structure of the Calogero systems.

1st Lecture: Bihamiltonian Manifolds

The geometrical setting currently accepted in the theory of integrable systems is known as the theory of "Hamiltonian systems with symmetries" (see, e.g., [1]). The basic objects of this theory are a symplectic manifold $\mathcal{M}$ and a Lie group G acting symplectically on $\mathcal{M}$. We recall their definitions. A symplectic manifold $\mathcal{M}$ is a manifold endowed with a closed, nondegenerate 2–form, ω. This form allows to associate vector fields to functions. The Hamiltonian vector fields are the inverse images of the exact one–forms,

$$\omega(X, \cdot) = -df, \tag{1.1}$$

and will be denoted by X_f. The correspondence between functions and vector fields defined by ω is a morphism of Lie algebras,

$$[X_f, X_g] = X_{\{f,g\}}, \tag{1.2}$$

provided that we define the Poisson bracket on the ring of C^∞ scalar–valued functions $f : \mathcal{M} \to \mathbb{R}$,

$$\{f, g\} := \omega(X_f, X_g). \tag{1.3}$$

This bracket is a bilinear, skew–symmetric composition law on functions. It is nondegenerate,

$$\{\cdot, f\} = 0 \implies f = \text{const.} \ , \tag{1.4}$$

and it satisfies both the Leibniz rule and the Jacobi identity:

$$\{f, gh\} = \{f, g\}h + g\{f, h\} \tag{1.5}$$
$$\{f, \{g, h\}\} + \{g, \{h, f\}\} + \{h, \{f, g\}\} = 0. \tag{1.6}$$

A symplectic action of a Lie group G on $\mathcal{M}$ is a rule which associates a diffeomorphism, $\Phi_a : \mathcal{M} \to \mathcal{M}$, to every element a of the group G, provided that the following two conditions are satisfied:

- the diffeomorphisms Φ_a define a representation of the group G on $\mathcal{M}$,

$$\Phi_a \circ \Phi_b = \Phi_{a \cdot b} \qquad \Phi_e = \text{id}_\mathcal{M}, \tag{1.7}$$

- they preserve the Poisson bracket,

$$\{f \circ \Phi_a, g \circ \Phi_a\} = \{f, g\} \circ \Phi_a. \qquad (1.8)$$

The group action allows us to associate a vector field X_A to every element A of the Lie algebra $\mathfrak{g}$ of the group G. These vector fields, called the *generators* of the symplectic action, give a representation of the Lie algebra $\mathfrak{g}$ on the manifold $\mathcal{M}$,

$$[X_A, X_B] = X_{[A,B]_\mathfrak{g}}, \qquad (1.9)$$

are derivations of the Poisson bracket,

$$\{X_A(f), g\} + \{f, X_A(g)\} - X_A(\{f, g\}) = 0, \qquad (1.10)$$

and, under weak additional conditions on the action Φ_a, they are Hamiltonian. The *momentum mapping* is the map $J : \mathcal{M} \to \mathfrak{g}^*$ which allows us to compute their Hamiltonian functions,

$$h_A(m) = \langle J(m), A \rangle_\mathfrak{g}. \qquad (1.11)$$

It also allows us to define a reduction of the manifold $\mathcal{M}$ (the so–called *Marsden–Weinstein reduction*). This reduction process considers a submanifold $\mathcal{S}$ of $\mathcal{M}$ (a level surface of the momentum mapping corresponding to a regular value of J), a foliation E of $\mathcal{S}$ (the orbits of the subgroup of G which preserves $\mathcal{S}$), and the reduced phase–space, $\mathcal{N} = \mathcal{S}/E$. The reduction theorem asserts that $\mathcal{N}$ remains a symplectic manifold. Furthermore, if $h : \mathcal{M} \to \mathbb{R}$ is a given Hamiltonian function on $\mathcal{M}$ preserved by the group action,

$$h \circ \Phi_a = h, \qquad (1.12)$$

then, according to the reduction theorem, the Hamiltonian vector field X_h is tangent to $\mathcal{S}$ and can be projected to $\mathcal{N}$. The projected vector field is still Hamiltonian and its Hamiltonian is the function h, seen as a function on $\mathcal{N}$. Finally, according to the (Hamiltonian) *Noether theorem*, the momentum mapping $J : \mathcal{M} \to \mathfrak{g}^*$ is constant along the trajectories of the vector field X_h.

The geometric scheme used in these lectures is similar to the previous one, with two important differences. The manifold $\mathcal{M}$ is a Poisson manifold rather than a symplectic one, and the symmetry group G is replaced by a second Poisson bracket on $\mathcal{M}$. Manifolds endowed with a pair of "compatible" Poisson brackets are called *bihamiltonian manifolds*. On a bihamiltonian manifold the second bracket plays the role of the symmetry group G on a symplectic manifold. Indeed, it defines a Poisson reduction (the *Marsden–Ratiu reduction scheme*) and a "momentum map" obeying the Noether theorem. These constructions will be described in the next two lectures. This one is devoted to illustrating the concept of a bihamiltonian manifold.

We recall that a *Poisson manifold* $\mathcal{M}$ is a manifold endowed with a Poisson bracket. A Poisson bracket is a bilinear skew–symmetric composition law on C^∞–functions on $\mathcal{M}$ satisfying the Leibniz rule and the Jacobi identity. It may be degenerate, i.e., there may exist nonconstant functions f which are in involution with all the functions defined on the Poisson manifold,

$$\{\cdot, f\} = 0. \tag{1.13}$$

These functions are called Casimir functions. The *Poisson tensor* is the bivector field P on $\mathcal{M}$ defined by

$$\{f, g\} = \langle df, P dg \rangle. \tag{1.14}$$

It has to be considered as a linear skew–symmetric map $P : T^*\mathcal{M} \to T\mathcal{M}$. In local coordinates, its components $P^{jk}(x^1, \ldots, x^n)$ are the Poisson brackets of the coordinate functions,

$$P^{jk}(x^1, \ldots, x^n) = \{x^j, x^k\}. \tag{1.15}$$

The Hamiltonian vector fields on a Poisson manifold $\mathcal{M}$ are the images under P of the exact one–forms,

$$X_f = P df. \tag{1.16}$$

Locally, this means that

$$X_f^j(x^1, \ldots, x^n) = P^{jk}(x^1, \ldots, x^n) \frac{\partial f}{\partial x^k}. \tag{1.17}$$

This is also a morphism of Lie algebras,

$$[X_f, X_g] = X_{\{f,g\}}. \tag{1.18}$$

The kernel of this morphism is formed by the Casimir functions. The image of this morphism is the distribution $C = \langle X_f \rangle$ spanned by the Hamiltonian vector fields. According to an important theorem in the theory of Poisson manifolds [2], the distribution C is integrable, and its maximal integral leaves are symplectic submanifolds of $\mathcal{M}$ which are called *symplectic leaves*. The symplectic two–form ω on a given symplectic leaf $\mathcal{S}$ is defined by

$$\omega(X, Y) = \langle \alpha, P\beta \rangle, \tag{1.19}$$

where (X, Y) is any pair of vector fields tangent to $\mathcal{S}$ and α and β are 1–forms on $\mathcal{M}$ related to them according to

$$X = P\alpha, \qquad Y = P\beta. \tag{1.20}$$

Exercise 1.1. Show that the components of a Poisson tensor satisfy the cyclic condition,

$$\sum_l \left(P^{jl} \frac{\partial P^{km}}{\partial x^l} + P^{kl} \frac{\partial P^{mj}}{\partial x^l} + P^{ml} \frac{\partial P^{jk}}{\partial x^l} \right) = 0. \tag{1.21}$$

Exercise 1.2. Suppose that $\mathcal{M}$ is an affine space $\mathcal{A}$. Call V the vector space associated to $\mathcal{A}$. Define a bivector field on $\mathcal{A}$ as a mapping $P : \mathcal{A} \times V^* \to V$ which satisfies the skew–symmetry condition,

$$\langle \alpha, P_u\beta \rangle = -\langle \beta, P_u\alpha \rangle,$$

for every pair of covectors (α, β) in V^*. Denote the directional derivative at u of the mapping $P(\cdot, \alpha)$ along the vector v by $P'_u(\alpha; v)$,

$$P'_u(\alpha; v) = \frac{d}{dt} P_{u+tv}\alpha \mid_{t=0} . \tag{1.23}$$

Show that the bivector P is a Poisson bivector if and only if it satisfies the following cyclic condition,

$$\langle \alpha, P'_u(\beta; P_u\gamma) \rangle + \langle \beta, P'_u(\gamma; P_u\alpha) \rangle + \langle \gamma, P'_u(\alpha; P_u\beta) \rangle = 0. \tag{1.24}$$

Let $\mathcal{M}$ be a manifold endowed with two Poisson brackets, $\{\cdot, \cdot\}_0$ and $\{\cdot, \cdot\}_1$. We say that $\mathcal{M}$ is a *bihamiltonian manifold* if the linear combination,

$$\{f, g\}_\lambda := \{f, g\}_1 + \lambda \{f, g\}_0, \tag{1.25}$$

of these brackets satisfies the Jacobi identity for any value of the real parameter λ. This means that the cyclic compatibility condition,

$$\{f, \{g, h\}_0\}_1 + \{h, \{f, g\}_0\}_1 + \{g, \{h, f\}_0\}_1 +$$
$$+ \{f, \{g, h\}_1\}_0 + \{h, \{f, g\}_1\}_0 + \{g, \{h, f\}_1\}_0 = 0, \tag{1.26}$$

holds for any triple of functions (f, g, h) on $\mathcal{M}$. In this case, the bracket $\{\cdot, \cdot\}_\lambda$ is called the *Poisson pencil* defined by $\{\cdot, \cdot\}_0$ and $\{\cdot, \cdot\}_1$ on $\mathcal{M}$. An *exact bihamiltonian manifold* is a Poisson manifold $(\mathcal{M}, \{\cdot, \cdot\}_1)$, endowed with a vector field X such that

$$\{f, g\}_0 := \{X(f), g\}_1 + \{f, X(g)\}_1 - X(\{f, g\}_1) \tag{1.27}$$

is a second Poisson bracket on $\mathcal{M}$. This bracket is automatically compatible with $\{\cdot, \cdot\}_1$, and therefore $\mathcal{M}$ is a bihamiltonian manifold. Exact bihamiltonian manifolds may be compared with exact symplectic manifolds, where the symplectic 2–form ω is exact, $\omega = d\theta$. In the setting of Poisson manifolds, the vector field X plays the same role which the Liouville 1–form θ plays in the setting of symplectic geometry. It will be referred to as the *characteristic vector field* of $\mathcal{M}$.

Exercise 1.3. Denote the Hamiltonian vector fields with respect to P_0 and P_1 by $X_f = P_0 df$ and by $Y_f = P_1 df$, respectively. Show that compatibility condition (1.26) may also be written in the form

$$L_{X_f}(P_1) + L_{Y_f}(P_0) = 0, \tag{1.28}$$

where L_X denotes the Lie derivative along the vector field X.

Exercise 1.4. Show that condition (1.27) entails compatibility condition (1.26). Show that it can be written in the form, $P_0 = L_X P_1$.

The presence of two compatible Poisson tensors allows us to define a sequence of nested distributions, C_k on $\mathcal{M}$, by the following iterative procedure. Let $C_0 = C$ be the characteristic distribution of P_0. If C_k is the kth distribution, let us call L_{k+1} the pre–image of C_k with respect to P_1,

$$L_{k+1} = P_1^{-1}(C_k), \tag{1.29}$$

and let us define C_{k+1} to be the image of L_{k+1} with respect to P_0,

$$C_{k+1} = P_0(L_{k+1}). \tag{1.30}$$

These distributions are integrable as a consequence of the compatibility condition that relates P_0 to P_1. (This statement will not be proved in these lectures.) Their maximal integral leaves are called the *characteristic submanifolds* of $\mathcal{M}$. The integral leaves of C_1 are contained in the integral leaves of C_0, and so on. Hence, they are nested, the one into the other. The functions which are constant along the distributions C_k are called *generalized Casimir functions*.

Example 1.5. A remarkable class of exact bihamiltonian manifolds includes the duals of Lie algebras. We denote by $(X, Y, \dots)$ the elements in $\mathfrak{g}$, and by $(\alpha, \beta, \dots)$ those in $\mathfrak{g}^*$. We recall that a 2–cocycle is a bilinear, skew–symmetric map $\omega : \mathfrak{g} \times \mathfrak{g} \to \mathbb{R}$, satisfying the condition

$$\omega(X, [Y, Z]) + \omega(Y, [Z, X]) + \omega(Z, [X, Y]) = 0. \tag{1.31}$$

A coboundary is a special cocycle defined by

$$\omega_\beta(X, Y) = \langle \beta, [X, Y] \rangle, \tag{1.32}$$

where β is any fixed element in $\mathfrak{g}^*$. Furthermore, we observe that the differential at a point α of a function $f : \mathfrak{g}^* \to \mathbb{R}$ is an element $df(\alpha)$ of $\mathfrak{g}$. Taking these facts for granted, we define a bracket,

$$\{f, g\}_\lambda(\alpha) = \omega(df(\alpha), dg(\alpha)) + \langle \alpha + \lambda\beta, [df(\alpha), dg(\alpha)] \rangle, \tag{1.33}$$

for any value of the real parameter λ. One can show that it satisfies the Jacobi identity for any choice of ω and β. Therefore, $\mathfrak{g}^*$ is a bihamiltonian manifold. More precisely, it is an *exact* bihamiltonian manifold. The characteristic vector field is the constant vector field,

$$\dot{\alpha} = \beta. \tag{1.34}$$

The Poisson tensors P_0 and P_1 are given by

$$X_f(\alpha) = P_0(df(\alpha)) = \text{ad}^*_{df(\alpha)} \beta \qquad (1.35)$$
$$Y_f(\alpha) = P_1(df(\alpha)) = \Omega(df(\alpha)) + \text{ad}^*_{df(\alpha)} \alpha, \qquad (1.36)$$

where $\Omega : \mathfrak{g} \to \mathfrak{g}^*$ is the linear, skew–symmetric map associated with the 2–cocycle ω, and ad^*_X is the generator of the coadjoint action. We shall return to these Poisson tensors in the next two lectures.

Exercise 1.6. Show that (1.33) defines a Poisson tensor by using condition (1.24). Show that $\dot{\alpha} = \beta$ is the characteristic vector field by evaluating condition (1.27) on the linear functions on $\mathfrak{g}^*$.

References

1. R. Abraham and J.E. Marsden, *Foundations of Mechanics*, 2nd edition (Benjamin, Reading, Mass. 1978)
2. P. Libermann and C.M. Marle, *Symplectic Geometry and Analytical Mechanics* (Reidel, Dordrecht 1987)

2nd Lecture: Marsden–Ratiu Reduction

Marsden and Ratiu have devised a reduction scheme for a Poisson manifold by analogy to the symplectic case [2]. The reduction scheme considers a submanifold $\mathcal{S}$ of $\mathcal{M}$, a foliation E of $\mathcal{S}$, and the quotient space $\mathcal{N} = \mathcal{S}/E$. The foliation E is defined by the intersection with $\mathcal{S}$ of a distribution D in $\mathcal{M}$, defined only at the points of $\mathcal{S}$. These elements are left unspecified in the reduction theorem, provided that they satisfy two suitable conditions.

The reduction scheme of Marsden and Ratiu may be easily implemented in the case of a bihamiltonian manifold. In this case the geometry of the manifold naturally determines $\mathcal{S}$ and D. The submanifold $\mathcal{S}$ is a symplectic leaf of the first Poisson tensor P_0, while the distribution D is spanned by the vector fields which are Hamiltonian with respect to P_1 and whose Hamiltonians are the Casimir functions of P_0,

$$D = \langle Y_f : f \text{ is a Casimir of } P_0 \rangle. \qquad (2.1)$$

It can be shown that the assumptions of the reduction theorem are satisfied in this case [1]. Therefore we can state the following

Proposition 2.1. The quotient space $\mathcal{N} = \mathcal{S}/E$ is a bihamiltonian manifold. Let $\pi : \mathcal{S} \to \mathcal{N}$ be the projection onto the quotient and $i : \mathcal{S} \to \mathcal{M}$ the inclusion map. Then, on $\mathcal{N}$ there exists a unique Poisson pencil $\{\cdot, \cdot\}^\lambda_\mathcal{N}$ such that

$$\{f, g\}^\lambda_\mathcal{N} \circ \pi = \{F, G\}^\lambda_\mathcal{M} \circ i \qquad (2.2)$$

for any pair of functions F and G which extend the functions f and g of $\mathcal{N}$ to $\mathcal{M}$, and are constant on D. Technically, this means that the function F satisfies the conditions

$$F \circ i = f \circ \pi \tag{2.3}$$

$$Y_k(F) := \{F, k\}_1 = 0 \tag{2.4}$$

for any function k whose differential, at the points of $\mathcal{S}$, belongs to the kernel of P_0.

The aim of this lecture is to show how this theorem may be used to construct interesting and nontrivial examples of bihamiltonian manifolds. In particular, we shall construct the phase–space of the KdV theory.

The bihamiltonian manifolds relevant for the theory of soliton equations are spaces of C^∞–maps from the circle S^1 into a simple Lie algebra $\mathfrak{g}$. In this lecture we consider $\mathfrak{g} = \mathfrak{sl}(2, \mathbb{C})$, and we denote such a map by

$$S = \begin{pmatrix} p & r \\ q & -p \end{pmatrix}. \tag{2.5}$$

The entries of this matrix are periodic functions of the coordinate x on the circle. This matrix must be considered as a point of our manifold $\mathcal{M}$. We denote a tangent vector to $\mathcal{M}$ at the point S by

$$\dot{S} = \begin{pmatrix} \dot{p} & \dot{r} \\ \dot{q} & -\dot{p} \end{pmatrix}, \tag{2.6}$$

and a covector at the point S by

$$V = \begin{pmatrix} \frac{1}{2}\alpha & \beta \\ \gamma & -\frac{1}{2}\alpha \end{pmatrix}. \tag{2.7}$$

They are arbitrary loops from S^1 into $\mathfrak{g}$. We adopt the convention that the value of the covector V on the tangent vector $\dot{S}$ is given by

$$\langle V, \dot{S} \rangle = \int_{S^1} \mathrm{Tr}(V\dot{S}) \, dx. \tag{2.8}$$

This allows us to identify $\mathfrak{g}^*$ with $\mathfrak{g}$. The space $\mathcal{M}$ is an infinite–dimensional Lie algebra endowed with a canonical cocycle [3],

$$\omega(\dot{S}_1, \dot{S}_2) = \int_{S^1} \mathrm{Tr}\left(\dot{S}_1 \frac{d\dot{S}_2}{dx}\right) dx. \tag{2.9}$$

The linear map $\Omega : \mathfrak{g} \to \mathfrak{g}^*$ associated with this cocycle is

$$\Omega(V) = \frac{dV}{dx}. \tag{2.10}$$

According to the general construction explained at the end of the previous lecture, the space $\mathcal{M}$ is endowed with two Poisson tensors P_0 and P_1 defined by

$$P_0(V) = [A, V] \tag{2.11}$$
$$P_1(V) = V_x + [V, S], \tag{2.12}$$

where V_x denotes the derivative of the loop V with respect to the coordinate x on S^1, and A is the constant matrix,

$$A = \begin{pmatrix} 0 & 0 \\ 1 & 0 \end{pmatrix}. \tag{2.13}$$

In a component–wise form, these tensors are given by:

$$\begin{aligned} \dot{p} &= -\beta \\ \dot{q} &= \alpha \\ \dot{r} &= 0, \end{aligned} \tag{2.14}$$

and by

$$\begin{aligned} \dot{p} &= \tfrac{1}{2}\alpha_x + q\beta - r\gamma \\ \dot{q} &= \gamma_x + 2p\gamma - q\alpha \\ \dot{r} &= \beta_x - 2p\beta + r\alpha \end{aligned} \tag{2.15}$$

respectively.

Exercise 2.1. Use condition (1.24) of the first lecture to show directly that (2.11) and (2.12) define compatible Poisson tensors. Note the importance of the boundary conditions in the infinite–dimensional case. Finally, prove formulas (2.14) and (2.15).

We note that the vector fields defined by the first bivector P_0 are tangent to the affine hyperplanes $r = r_0$, where r_0 is a given periodic function. Therefore the symplectic leaves of P_0 are affine hyperplanes. We set

$$S = \begin{pmatrix} p & 1 \\ q & -p \end{pmatrix}. \tag{2.16}$$

The kernel of P_0 is formed by the covectors satisfying $\alpha = \beta = 0$. Therefore the distribution D is spanned by the vector fields

$$\dot{p} = -\gamma, \qquad \dot{q} = \gamma_x + 2p\gamma, \qquad \dot{r} = 0. \tag{2.17}$$

In this particular example, D is already tangent to S, so the foliation E is defined by the integral leaves of the distribution (2.17). By eliminating γ among these equations, we obtain the constraint

$$(p_x + p^2 + q)^{\cdot} = 0. \tag{2.18}$$

Therefore the leaves of E are the submanifolds of S defined by the equation

$$p_x + p^2 + q = u, \qquad (2.19)$$

where u is a given periodic function. This computation shows that the quotient space $\mathcal{N} = S/E$ is the space of functions on S^1, and that (2.19) yields the canonical projection $\pi : S \to \mathcal{N}$.

Exercise 2.2. This exercise is a digression regarding the foliation E. Consider the set of matrices

$$T = \begin{pmatrix} 1 & 0 \\ t & 1 \end{pmatrix}, \qquad (2.20)$$

where t is an arbitrary periodic function. Observe that these matrices form a group G. Let this group act on the symplectic manifold S according to

$$S' = TST^{-1} + T_x T^{-1}. \qquad (2.21)$$

Show that this is an action of G on S. Write this action in the component–wise form

$$\begin{aligned} p' &= p - t \\ q' &= q + 2pt - t^2 + t_x \\ r' &= r = 1, \end{aligned} \qquad (2.22)$$

and show that the orbits of this action are the leaves of the distribution E.

To learn how to reduce the bihamiltonian structure of $\mathcal{M}$ onto $\mathcal{N}$, we have to understand the process of *"lifting the covectors"* from $T^*\mathcal{N}$ into $T_S^*\mathcal{M}$. Let us denote a covector at the point u of $\mathcal{N}$ by v, and let us adopt the convention that the value of v on the tangent vector $\dot{u}$ is given by

$$\langle v, \dot{u} \rangle = \int_{S^1} v(x)\dot{u}(x)\,dx. \qquad (2.23)$$

The process of lifting allows us to associate with any covector $(u, v) \in T^*\mathcal{N}$ a covector $(S, V) \in T_S^*\mathcal{M}$, defined only at the points of the symplectic leaf S. The conditions characterizing the covector (S, V) are:

1. S is a point of the fiber over u,

$$u = p_x + p^2 + q.$$

2. V annihilates the distribution D,

$$\langle V, D \rangle = 0 \implies \int_{S^1} [-\alpha w + \beta(w_x + 2pw)]\,dx = 0,$$

for any periodic function w.

3. V and v are related by $\langle V, \dot{S} \rangle = \langle v, \dot{u} \rangle$, where $\dot{S}$ is any vector tangent at S to $\mathcal{S}$ and $\dot{u}$ is its projection onto u,

$$\int_{S^1} \mathrm{Tr}(V\dot{S}) \, dx = \int_{S^1} (\dot{p}_x + 2p\dot{p} + \dot{q})v \, dx.$$

By using these conditions one proves that the components (α, β, γ) of V are given by

$$\alpha = -v_x + 2pv \qquad \beta = v \qquad \gamma = \text{arbitrary.} \tag{2.24}$$

Exercise 2.3. Check the previous computations.

To obtain the reduced Poisson tensor $P_0^{\mathcal{N}}$ on $\mathcal{N}$, we now have merely to compute the projection onto $\mathcal{N}$ of the vector field

$$\dot{S} = [A, V]$$

associated by the Poisson tensor P_0 with the lifted covector V. By using the component–wise form (2.14) we readily obtain

$$\dot{u} = \dot{p}_x + 2p\dot{p} + \dot{q}$$

$$\overset{(2.14)}{=} -\beta_x - 2p\beta + \alpha \tag{2.25}$$

$$\overset{(2.24)}{=} -2v_x.$$

An analogous computation yields the reduction of the second Poisson tensor $P_1^{\mathcal{N}}$:

$$\dot{u} = \dot{p}_x + 2p\dot{p} + \dot{q}$$

$$= \tfrac{1}{2}\alpha_{xx} + (q\beta)_x - \gamma_x + 2p(\tfrac{1}{2}\alpha_x + q\beta - \gamma) + \gamma_x + 2p\gamma - q\alpha \tag{2.26}$$

$$\overset{(2.24)}{=} -\tfrac{1}{2}v_{xxx} + 2uv_x + u_x v.$$

We will refer to the manifold $\mathcal{N}$ endowed with these Poisson tensors as the *Gel'fand–Dickey* manifold (henceforth GD manifold) associated with $\mathfrak{g} = \mathfrak{sl}(2, \mathbb{C})$. This manifold is the phase–space of the KdV hierarchy as will be shown in the next lecture.

References

1. P. Casati, F. Magri, and M. Pedroni, "Bihamiltonian manifolds and τ–function", in: *Mathematical Aspects of Classical Field Theory 1991*, M.J. Gotay et al., eds., Contemporary Mathematics, vol. 132, 213–234 (American Mathematical Society, Providence 1992)
2. J.E. Marsden and T. Ratiu, "Reduction of Poisson manifolds", Lett. Math. Phys. **11** (1986) 161–169
3. G. Segal and G. Wilson, "Loop groups and equations of KdV type", Publ. Math. IHES **61** (1985) 5–65

3rd Lecture: Generalized Casimir Functions

In this lecture we shall begin the study of the structure of the GD manifolds. We consider the simplest case, which corresponds to $\mathfrak{g} = \mathfrak{sl}(2, \mathbb{C})$. In this case the GD manifold may be identified with the space of periodic functions $u(x)$ on the circle S^1. The bihamiltonian structure is defined by the pair of Poisson tensors

$$\dot{u} = -2v_x \tag{3.1}$$

$$\dot{u} = -\tfrac{1}{2} v_{xxx} + 2uv_x + u_x v, \tag{3.2}$$

where u is the point of the manifold, $\dot{u}$ is a tangent vector at u, and v is a covector. The value of v on the vector $\dot{u}$ is given by

$$\langle v, \dot{u} \rangle = \int_{S^1} v(x)\dot{u}(x)\,dx. \tag{3.3}$$

We remark that this manifold is another example of the class of bihamiltonian manifolds on duals of Lie algebras illustrated at the end of the first lecture. Indeed, as a manifold, $\mathcal{M}$ may be identified with the dual of the Virasoro algebra of the vector fields on the circle. If v_1 and v_2 are two periodic functions defining two vector fields on S^1, the commutator $[v_1, v_2]_{\mathrm{Vir}}$ is defined by

$$[v_1, v_2]_{\mathrm{Vir}} = v_1 v_{2x} - v_2 v_{1x}. \tag{3.4}$$

It is then easy to show that the linear mappings,

$$\begin{aligned} \Omega_1 v &= v_x \\ \Omega_2 v &= v_{xxx}, \end{aligned} \tag{3.5}$$

are 2–cocycles of the Virasoro algebra. Furthermore, by using the identity

$$\langle u, [v_1, v_2] \rangle = \langle v_1, \mathrm{ad}^*_{v_2} u \rangle,$$

it is possible to show that the coadjoint action of the Virasoro algebra is

$$\mathrm{ad}^*_v u = 2uv_x + u_x v. \tag{3.6}$$

Assembling this information, we recognize that the Poisson tensors (3.1) and (3.2) are particular instances of Poisson tensors on the dual of a Lie algebra shown in Lecture 1.

Exercise 3.1. Use condition (1.24) to show directly that the second Poisson tensor (3.2) of the KdV theory satisfies the Jacobi identity.

Our specific aim in this lecture is to study the *characteristic distributions* defined by the pair of Poisson tensors (3.1), (3.2). As explained in the first lecture, they are constructed by an iterative procedure. We observe that the vector fields (3.1) obey the condition

$$\int_{S^1} \dot{u}(x)\, dx = 0. \tag{3.7}$$

This means that the symplectic leaves of $P_0^{\mathcal{N}}$ are the affine hyperplanes whose equation is

$$H_1 = \int_{S^1} u\, dx = c_1. \tag{3.8}$$

We impose constraint (3.7) on vector fields (3.2), and we look for the covectors v which satisfy this constraint. We obtain

$$\int_{S^1} u v_x\, dx = 0. \tag{3.9}$$

Let us call L_1 the subspace in $T_u^* \mathcal{N}$ spanned by the solutions of this equation. We construct the image of L_1 by $P_0^{\mathcal{N}}$. We obtain the distribution C_1 spanned by the vector fields obeying the conditions:

$$\int_{S^1} \dot{u}\, dx = 0 \qquad \int_{S^1} u \dot{u}\, dx = 0. \tag{3.10}$$

This distribution is clearly integrable, and its integral leaves are the submanifolds of $\mathcal{N}$ whose equations are

$$H_1 = \int_{S^1} u\, dx = c_1 \qquad H_2 = \int_{S^1} u^2\, dx = c_2. \tag{3.11}$$

Now we repeat the process. We impose conditions (3.10) on vector fields (3.2), we determine the subspace L_2 of the 1–forms v satisfying these constraints, and we construct the image $C_2 = P_0^{\mathcal{N}}(L_2)$. The integral leaves of the distribution C_2 are defined by the equations $H_1 = c_1$, and $H_2 = c_2$, and by

$$H_3 = \int_{S^1} (\tfrac{1}{2} u_x^2 + u^3)\, dx = c_3. \tag{3.12}$$

In this way we iteratively construct a sequence of functions $(H_1, H_2, H_3, \dots)$ on $\mathcal{N}$, which are called the *generalized Casimir functions* of the bihamiltonian structure. They are the Hamiltonians of the KdV hierarchy.

A more efficient way of computing the functions H_k is to study the Casimir functions of the Poisson pencil defined by P_0 and P_1. To determine these functions we study the kernel of the Poisson pencil by looking for the 1–forms v on $\mathcal{N}$ which solve the equation

$$-\tfrac{1}{2} v_{xxx} + 2(u + \lambda) v_x + u_x v = 0, \tag{3.13}$$

where λ is an arbitrary parameter. We integrate this equation immediately by observing that

$$v(-\tfrac{1}{2}v_{xxx} + 2(u+\lambda)v_x + u_x v) = \frac{d}{dx}(\tfrac{1}{4}v_x^2 - \tfrac{1}{2}vv_{xx} + (u+\lambda)v^2), \qquad (3.14)$$

and therefore we consider the equation

$$\tfrac{1}{4}v_x^2 - \tfrac{1}{2}vv_{xx} + (u+\lambda)v^2 = a(\lambda), \qquad (3.15)$$

where $a(\lambda)$ does not depend on x. If v is a solution of this equation, then v belongs to the kernel of the Poisson pencil. We are going to show that if $a(\lambda)$ is also independent of u, then v is an exact 1-form, i.e., the differential of a Casimir of the Poisson pencil.

It is easily checked that one can put $a(\lambda) = \lambda$ without loss of generality. Indeed, another choice would simply recombine the coefficients in λ of the solution $v(\lambda)$. We set $\lambda = z^2$, and we observe that equation (3.15), with $a(\lambda) = \lambda$, can be written in the form of a *Riccati equation*,

$$\left(\frac{z}{v} + \frac{1}{2}\frac{v_x}{v}\right)_x + \left(\frac{z}{v} + \frac{1}{2}\frac{v_x}{v}\right)^2 = u + z^2. \qquad (3.16)$$

For this reason, we set

$$h(z) = \frac{z}{v} + \frac{1}{2}\frac{v_x}{v}. \qquad (3.17)$$

Then we prove that all the solutions of (3.15) are exact 1–forms on $\mathcal{N}$. To this end we consider a curve $u(t)$ in $\mathcal{N}$, and we denote the solution of (3.15) at the points of this curve by $v(t)$. By differentiating this equation with respect to time t, we obtain

$$\tfrac{1}{2}v_x\dot{v}_x - \tfrac{1}{2}\dot{v}v_{xx} - \tfrac{1}{2}v\dot{v}_{xx} + 2(u+z^2)v\dot{v} + \dot{u}v^2 = 0, \qquad (3.18)$$

or

$$\begin{aligned}
\dot{u}v &= \frac{1}{2}\dot{v}\frac{v_{xx}}{v} + \frac{1}{2}\dot{v}_{xx} - \frac{1}{2}\frac{v_x}{v}\dot{v}_x - 2(u+z^2)\dot{v} \\
&= \frac{1}{2}\dot{v}\frac{v_{xx}}{v} + \frac{1}{2}\dot{v}_{xx} - \frac{1}{2}\frac{v_x}{v}\dot{v}_x + 2\dot{v}(-\frac{1}{2}\frac{v_{xx}}{v} + \frac{1}{4}\frac{v_x^2}{v^2} - \frac{z^2}{v^2}) \\
&= \frac{\partial}{\partial x}(-\frac{1}{2}\dot{v}\frac{v_x}{v} + \frac{1}{2}\dot{v}_x) + \frac{\partial}{\partial t}(\frac{2z^2}{v}) \\
&= \frac{\partial}{\partial x}(-\frac{1}{2}\dot{v}\frac{v_x}{v} + \frac{1}{2}\dot{v}_x + z\frac{v_x}{v}) + \frac{\partial}{\partial t}(2zh). \qquad (3.19)
\end{aligned}$$

Therefore, by integrating on S^1 we obtain

$$\langle v, \dot{u} \rangle = \int_{S^1} v(x)\dot{u}(x)\,dx = \frac{d}{dt}2z\int_{S^1} h\,dx. \qquad (3.20)$$

This relation shows that the 1–form v is the differential of the function

$$H = 2z \int_{S^1} h\,dx \tag{3.21}$$

which, therefore, is a Casimir function of the Poisson pencil. We have thus shown that the Casimir functions of the Poisson pencil may be computed by solving the Riccati equation,

$$h_x + h^2 = u + z^2. \tag{3.22}$$

To show the relation with the generalized Casimir functions, we expand the solution of this equation in powers of z,

$$h(z) = z + \sum_{j \geq 1} h_j z^{-j}. \tag{3.23}$$

The coefficients h_j may be computed by recurrence, by inserting this expansion in the Riccati equation.

Exercise 3.2. Compute the first coefficients of the series (3.23), and show that $h_1 = \frac{1}{2}u$, $h_2 = -\frac{1}{4}u_x$, $h_3 = \frac{1}{8}(u_{xx} - u^2)$.

We can then define the functions

$$H_k = \int_{S^1} h_k\,dx \tag{3.24}$$

and we observe that the even functions, H_{2k}, vanish, while the odd functions, H_{2k+1}, are the generalized Casimir functions of the GD manifold. This observation provides an efficient method for computing the Hamiltonians of the KdV equations and a proof that they are in involution. Indeed, since function (3.21) is a Casimir function of the Poisson pencil, it obeys the equation

$$z^2\{\cdot, H\}_0 = \{\cdot, H\}_1. \tag{3.25}$$

Therefore its coefficients, H_k, obey the recursion relations

$$\{\cdot, H_1\}_0 = 0 \tag{3.27}$$
$$\{\cdot, H_{2k+1}\}_0 = \{\cdot, H_{2k-1}\}_1. \tag{3.28}$$

These relations are called the *Lenard recursion relations*.

Proposition 3.1. The functions H_{2k+1} which obey the Lenard recursion relations are in involution with respect to both Poisson brackets.

Proof. By using repeatedly the recursion relation we get, for $j > k$,

$$
\begin{aligned}
\{H_{2j+1}, H_{2k+1}\}_0 &= \{H_{2j+1}, H_{2k-1}\}_1 \\
&= -\{H_{2k-1}, H_{2j+1}\}_1 \\
&= -\{H_{2k-1}, H_{2j+3}\}_0 \\
&= \{H_{2j+3}, H_{2k-1}\}_0 \\
&\quad\ \vdots \\
&= \{H_{2j+2k+1}, H_1, \}_0 \\
&= 0.
\end{aligned}
$$

This proves that the bracket $\{H_{2j+1}, H_{2k+1}\}_0$ vanishes. The same computation shows also that $\{H_{2j+1}, H_{2k+1}\}_1 = 0$.

The KdV equations are usually written in the form

$$
\frac{\partial u}{\partial t_j} = -2 \left(\frac{\delta h_{2j+1}}{\delta u} \right)_x
\tag{3.28}
$$

by using the reduced Poisson bracket and the generalized Casimir functions on $\mathcal{N}$. However, they can also be written in a more classical form by returning to the symplectic leaf $\mathcal{S}$ sitting above $\mathcal{N}$.

Exercise 3.3. Show that the symplectic 2–form of $\mathcal{S}$ is defined by

$$
\omega((\dot{p}_1, \dot{q}_1), (\dot{p}_2, \dot{q}_2)) = \int_{S^1} (\dot{q}_1 \dot{p}_2 - \dot{q}_2 \dot{p}_1) \, dx.
\tag{3.29}
$$

Then we compose the functions H_k with the projection $\pi : \mathcal{S} \to \mathcal{N}$ to obtain functions defined on $\mathcal{S}$, and we write the KdV equations in the Hamiltonian form,

$$
\frac{\partial q}{\partial t_j} = \frac{\delta H_{2j+1}}{\delta p} \qquad \frac{\partial p}{\partial t_j} = -\frac{\delta H_{2j+1}}{\delta q}
\tag{3.30}
$$

by using the symplectic structure on $\mathcal{S}$. These equations are projectable equations onto $\mathcal{N}$, and the projected equations coincide with the standard KdV equations (3.28).

Exercise 3.4. Compute the Hamiltonian H_3 on $\mathcal{S}$ and the corresponding equation (3.30). Then use the relation $u = p_x + p^2 + q$ to project this equation onto $\mathcal{N}$. Show that this equation is the KdV on $\mathcal{N}$.

4th Lecture: Gel'fand–Dickey Manifolds

This is the first of a set of three lectures devoted to the general theory of soliton equations. The purpose is to give a glimpse of the techniques which allow us to perform the analysis of a bihamiltonian structure defined over a general loop algebra. In particular this lecture aims to define the MR reduction of the loop algebras constructed over $\mathfrak{g} = \mathfrak{sl}(n+1)$. The quotient spaces $\mathcal{N}$ produced by this reduction are the Gel'fand–Dickey manifolds, which are the phase–spaces of soliton equations. This reduction process can be performed for an arbitrary Lie algebra $\mathfrak{g}$, and is equivalent to (although different from) the well-known Drinfeld–Sokolov reduction [4], as shown in [3, 5].

We recall that the bihamiltonian structure over the space of C^∞ maps from S^1 into $\mathfrak{sl}(n+1)$ is defined by the pair of Poisson tensors,

$$\dot{S} = [A, V] \tag{4.1}$$

$$\dot{S} = V_x + [V, S], \tag{4.2}$$

where $(V, S, \dot{S})$ are traceless matrices with entries which are periodic functions of x, and A is the matrix

$$A = \begin{pmatrix} 0 & 0 & \dots & 0 \\ \dots & \dots & \dots & \dots \\ \dots & \dots & \dots & \dots \\ 0 & 0 & \dots & 0 \\ 1 & 0 & \dots & 0 \end{pmatrix}. \tag{4.3}$$

The symbols have the same meaning as in the second lecture, S is a point, V is a covector, and $\dot{S}$ is a tangent vector at S. The elements entering into the reduction process are: a symplectic leaf, $\mathcal{S}$, of (4.1), a foliation, E of $\mathcal{S}$, and the quotient space $\mathcal{N} = \mathcal{S}/E$. We begin by describing $\mathcal{S}$. We observe that the symplectic leaves of the Poisson tensor (4.1) are affine hyperplanes, modelled on the vector space of the traceless matrices which have entries differing from zero only in the first column and in the last row. We choose the submanifold

$$S = \begin{pmatrix} p_0 & 1 & \dots & \dots & 0 \\ p_1 & 0 & 1 & \dots & 0 \\ \dots & \dots & \dots & \dots & \dots \\ p_{n-1} & \dots & \dots & \dots & 1 \\ q_0 & \dots & \dots & q_{n-1} & -p_0 \end{pmatrix}, \tag{4.4}$$

and we observe that functions $p_a(x)$ and $q_a(x)$ play the role of canonical coordinates on $\mathcal{S}$. Indeed, it is not hard to see that the Poisson bracket induced by the Poisson tensor (4.1) on $\mathcal{S}$ has the canonical form

$$\{F, G\}_{\mathcal{S}} = \sum_{a=0}^{n-1} \int_{S^1} \left(\frac{\delta F}{\delta q_a} \frac{\delta G}{\delta p_a} - \frac{\delta F}{\delta p_a} \frac{\delta G}{\delta q_a} \right) dx. \tag{4.5}$$

This is proved by observing that the derivative of a function

$$F = \int_{S^1} f(q, p, q_x, p_x, \dots) \, dx \tag{4.6}$$

along any vector $\dot{S}$ tangent to $\mathcal{S}$ can be written in the form

$$\frac{dF}{dt} = \sum_{a=0}^{n-1} \int_{S^1} \left(\frac{\delta F}{\delta p_a} \dot{p}_a + \frac{\delta F}{\delta q_a} \dot{q}_a \right) dx = \langle V_F, \dot{S} \rangle, \tag{4.7}$$

where the matrix V_F is

$$V_F = \begin{pmatrix} \frac{1}{2} \frac{\delta F}{\delta p_0} & \frac{\delta F}{\delta p_1} & \cdots & \cdots & \frac{\delta F}{\delta q_0} \\ \cdots & \cdots & \cdots & \cdots & \cdots \\ \cdots & \cdots & \cdots & \cdots & \cdots \\ \cdots & \cdots & \cdots & \cdots & \frac{\delta F}{\delta q_{n-1}} \\ \cdots & \cdots & \cdots & \cdots & -\frac{1}{2} \frac{\delta F}{\delta p_0} \end{pmatrix}, \tag{4.8}$$

the remaining elements being arbitrary. Then the Poisson bracket (4.5) is simply computed according to

$$\{F, G\}_S = \langle A, [V_F, V_G] \rangle. \tag{4.9}$$

The next step is to study the distribution D spanned by the vector fields (4.2), evaluated on the 1–forms V which obey the condition

$$[A, V] = 0. \tag{4.10}$$

This kernel is easily computed. The difficult problem is to find the intersection E of D with $\mathcal{S}$, and to compute the integral leaves of E. The distribution E is spanned by the vector fields (4.2) associated with the covectors V which obey a supplementary condition,

$$\langle V, W_x \rangle + \langle S, [V, W] \rangle = 0, \tag{4.11}$$

for every matrix W solution of (4.10). The last condition, indeed, is the condition for vector field (4.2) to be tangent to the symplectic leaf $\mathcal{S}$.

Exercise 4.1. Deduce condition (4.11).

To understand the properties of the solutions of these two conditions, it is convenient to study the distribution E at the special point B in $\mathcal{S}$, where

$$B = \begin{pmatrix} 0 & 1 & \cdots & \cdots & 0 \\ 0 & 0 & 1 & \cdots & 0 \\ \cdots & \cdots & \cdots & \cdots & \cdots \\ 0 & \cdots & \cdots & \cdots & 1 \\ 0 & \cdots & \cdots & 0 & 0 \end{pmatrix}. \tag{4.12}$$

It is a general principle of the geometry of the symplectic leaf (4.4) (formalized in the method of "dressing transformation" discussed in the next lecture) that the point B dominates the geometry of $\mathcal{S}$. What is true at B is true, in general, in $\mathcal{S}$. Then one can remark that the solutions of conditions (4.10) and (4.11) at B form a subalgebra $\mathfrak{t}$ of lower triangular matrices. This subalgebra can be exponentiated. We denote by T the corresponding group.

Proposition 4.1. The integral leaves of E are the orbits of the gauge action of T on $\mathcal{S}$ defined by

$$S' = TST^{-1} + T_x T^{-1}. \tag{4.13}$$

We shall not prove this property in this lecture. We shall simply make a remark which will explain the appearance of the gauge action (4.13). Suppose $T = I + \epsilon V$ is a group element near the identity. If we set $S' - S = \epsilon \dot{S}$, it is readily seen that the gauge action becomes the vector fields (4.2) at first order in ϵ. Thus the vector fields (4.2) are the infinitesimal generators of the gauge action (4.13).

Exercise 4.2. Assume $n = 1$. Show that the algebra $\mathfrak{t}$ is formed by the matrices

$$V = \begin{pmatrix} 0 & 0 \\ \gamma & 0 \end{pmatrix}. \tag{4.14}$$

Exponentiate this algebra and find the group T studied in Exercise 2.2. This remark proves Prop. 4.1 for the case $n = 1$. Do the same for $n = 2$.

The description of the leaves of E so far obtained is not sufficiently detailed to perform the reduction. We need the equations of the integral leaves of E. The basic remark is that (4.13) can be written in an equivalent form,

$$(-\partial + S' + \lambda A) \circ T = T \circ (-\partial + S + \lambda A). \tag{4.15}$$

This equation means that the points S and S' belong to the same orbit if and only if the associated matrix first-order differential operators $(-\partial + S + \lambda A)$ and $(-\partial + S' + \lambda A)$ are conjugate with respect to an element of the group T. So, our problem is to extract from the differential operator $(-\partial + S + \lambda A)$ a set of functions which are invariant with respect to conjugation by T. This problem is similar to the problem of constructing the invariants of the adjoint action in a Lie algebra, and it can be treated by a suitable extension of the Frobenius technique familiar from linear algebra.

Let

$$\langle e^k| = \langle 0, \ldots, 0, 1, 0, \ldots, 0| \tag{4.16}$$

be the standard basis of $\mathbb{R}^{n+1}$. We use the first vector $\langle e^0|$ as a starting point for the recursion relation

$$\langle g^{(k+1)}| = \langle g^{(k)}|_x + \langle g^{(k)}|(S + \lambda A), \qquad \langle g^{(0)}| = \langle e^0|. \tag{4.17}$$

One can show that the first $n+1$ vectors of this sequence form a basis of $\mathbb{R}^{n+1}$. This basis changes with the point S of the symplectic leaf. By developing the next vector $\langle g^{n+1}|$ on this basis, we obtain the equation

$$\langle g^{(n+1)}| = \sum_{j=0}^{n-1} u_j(S)\langle g^{(j)}| + \lambda\langle g^{(0)}|, \tag{4.18}$$

where the components $u_j(S)$ are differential polynomials of the coordinates (q_a, p_a) of the point S. The important observation is that these functions form a complete set of invariants characterizing the leaves of the foliation E.

Proposition 4.2. Two points S and S' in $\mathcal{S}$ belong to the same integral leaf of the foliation E if and only if

$$u_j(S) = u_j(S'). \tag{4.19}$$

To give a simple example of this construction, we again consider the case $n = 1$. A simple computation shows that, at the points of the symplectic leaf (2.16),

$$\langle g^{(0)}| = \langle 1, 0|$$
$$\langle g^{(1)}| = \langle p, 1| \tag{4.20}$$
$$\langle g^{(2)}| = \langle p_x + p^2 + q + \lambda, 0|.$$

Therefore the characteristic equation is

$$\langle g^{(2)}| - (p_x + p^2 + q)\langle g^{(0)}| = \lambda\langle g^{(0)}|. \tag{4.22}$$

In this case we have a unique invariant,

$$u = p_x + p^2 + q, \tag{4.22}$$

defining the projection from $\mathcal{S}$ onto the quotient space $\mathcal{N}$.

Exercise 4.3. Perform the same computation for $n = 2$, and find the pair of invariants $u_0 = q_0 - p_0 q_1 + p_0 p_1 + p_{1x} + p_{0xx}$ and $u_1 = q_1 - p_1 + 2p_{0x} + p_0^2$. Compare with the result of Exercise 4.2.

Definition 4.3. The Gel'fand–Dickey manifold associated with the Lie algebra $\mathfrak{sl}(n+1)$ is the quotient space $\mathcal{N}$ of the symplectic leaf $\mathcal{S}$ with respect to equivalence relation (4.19).

Since it is obtained by a MR reduction of a bihamiltonian manifold, the GD manifold is itself a bihamiltonian manifold. The reduced Poisson tensors may be computed by improving the techniques shown in the second lecture. The final results are the Adler–Gel'fand–Dickey brackets (henceforth, AGD brackets) for $\mathfrak{sl}(n+1)$ [1,2]. However, we can avoid this computation since we shall not need the explicit form of these brackets in what follows. Indeed, an interesting feature of the formulation of the KP theory we want to present in the next two lectures is that we can avoid a detailed study of the quotient space $\mathcal{N}$ by a suitable analysis of the symplectic leaf $\mathcal{S}$.

References

1. P. Casati, G. Falqui, F. Magri, M. Pedroni, "The KP theory revisited. II. Adler–Gel'fand–Dickey brackets", Preprint SISSA/3/96/FM (1996)
2. P. Casati, G. Falqui, F. Magri, and M. Pedroni, "Bihamiltonian Reductions and $\mathcal{W}_n$-Algebras", J. Geom. Phys. **26** (1998) 291–310
3. P. Casati and M. Pedroni, "Drinfeld–Sokolov Reduction on a simple Lie algebra from the bihamiltonian point of view", Lett. Math. Phys. **25** (1992) 89–101
4. V.G. Drinfeld and V.V. Sokolov, "Lie Algebras and Equations of Korteweg–de Vries Type", J. Sov. Math. **30** (1985) 1975–2036
5. M. Pedroni, "Equivalence of the Drinfeld-Sokolov reduction to a bihamiltonian reduction", Lett. Math. Phys. **35** (1995) 291–302

5th Lecture: Gel'fand–Dickey Equations

In this lecture we compute the generalized Casimir functions of the bihamiltonian structure of the GD manifolds by the method of dressing transformation [2,3]. We use these functions to define the GD equations and we study their conservation laws. We show that the local densities of the conserved quantities obey the reduced KP equations. By this result we construct a bridge between the GD and the KP theories.

The geometrical setting of our approach has been already explained in the previous lecture. It consists of the space $\mathcal{M}$ of $C^\infty-$ maps from S^1 into $\mathfrak{sl}(n+1)$, of the submanifold $\mathcal{S}$ given by (4.4), and of the quotient space $\mathcal{N}$ defined by equivalence relation (4.19). The space $\mathcal{M}$ is a bihamiltonian manifold equipped with the Poisson pencil

$$(P_\lambda)_s V = V_x + [V, S + \lambda A], \tag{5.1}$$

the submanifold $\mathcal{S}$ is a symplectic manifold, and the quotient space $\mathcal{N}$ is again a bihamiltonian manifold equipped with the AGD brackets. We want to compute the generalized Casimir functions on $\mathcal{N}$. A possible method would be to study the kernel of the Poisson pencil on $\mathcal{N}$. (This method has been already used in the third lecture for the particular example of the KdV hierarchy). Our strategy is, however, to avoid doing the computations on $\mathcal{N}$ since the AGD brackets are too complicated. We prefer to work on the symplectic leaf $\mathcal{S}$, and to compute the Casimir functions of the Poisson pencil (5.1) on $\mathcal{S}$, rather than the Casimir functions of the AGD brackets on $\mathcal{N}$. The problem we have to solve may be formulated as follows, how can we find a formal series of 1–forms, $V(\lambda) = \sum_{k \geq -1} V_k \lambda^k$, which solves the matrix equation

$$V_x + [V, S + \lambda A] = 0 \tag{5.2}$$

at the points of the symplectic leaf (4.4), and which is exact when restricted to $\mathcal{S}$. This means that there must exist a formal series of functions, $F(\lambda) = \sum_{k \geq -1} F_k \lambda^k$ on $\mathcal{S}$, such that

$$\langle V(\lambda), \dot{S} \rangle = \frac{d}{dt} F(\lambda) \tag{5.3}$$

for every vector field $\dot{S}$ tangent to $\mathcal{S}$. This series of functions solves our original problem. Indeed, one can show that the function $F(\lambda)$ and its components, F_k, are constant along the foliation E. Therefore they define functions on $\mathcal{N}$. The projection of $F(\lambda)$ onto $\mathcal{N}$ is a Casimir function of the AGD bracket; the projections of the components F_k are the generalized Casimir functions we are looking for.

5.1. The Spectral Analysis of $V(\lambda)$

The basic strategy for solving (5.2) is to look at the eigenvalues and eigenvectors of the matrix V. The eigenvalues of V may be chosen arbitrarily, provided that they are independent of x. The eigenvectors must solve an auxiliary linear problem. The argument rests on the observation that, by (5.2), the matrix V has to commute with the first–order matrix differential operator $-\partial_x + (S + \lambda A)$. Suppose that this differential operator has a set of "eigenvectors" $|\psi_a\rangle$, obeying the equation

$$-|\psi_a\rangle_x + (S + \lambda A)|\psi_a\rangle = h_a|\psi_a\rangle, \tag{5.4}$$

for some suitable functions, h_a. Suppose they form a basis in $\mathbb{R}^{n+1}$. Then any matrix V having constant eigenvalues and eigenvectors $|\psi_a\rangle$,

$$V|\psi_a\rangle = c_a|\psi_a\rangle,$$

solves (5.2).

5.2. The Auxiliary Eigenvalue Problem

The idea is to study the auxiliary linear problem,

$$-|\psi\rangle_x + (S + \lambda A)|\psi\rangle = h|\psi\rangle \tag{5.5}$$

in order to characterize the eigenvectors of V. We project this equation on the Frobenius basis $\langle g^{(k)}|$ at the point S introduced in the fourth lecture, to find

$$-\langle g^{(k)}|\psi_x\rangle + \langle g^{(k)}|S + \lambda A|\psi\rangle = h\langle g^{(k)}|\psi\rangle.$$

After an integration by parts, we find

$$\langle g^{(k+1)}|\psi\rangle = \langle g^{(k)}|\psi\rangle_x + h\langle g^{(k)}|\psi\rangle.$$

Then, by imposing the normalization condition,

$$\langle g^{(0)}|\psi\rangle = 1, \tag{5.6}$$

we finally obtain

$$\langle g^{(k)}|\psi\rangle = h^{(k)}, \tag{5.7}$$

where $h^{(k)}$ is the Faà di Bruno polynomial of order k of the eigenvalue h. These differential polynomials are defined by recurrence according to

$$h^{(k+1)} = h_x^{(k)} + h h^{(k)}, \tag{5.8}$$

starting from $h^{(0)} = 1$. Equation (5.7) completely characterizes the normalized eigenvector associated with the eigenvalue h. To find h, we recall that the Frobenius basis satisfies the characteristic equation (4.18). By projecting on $|\psi\rangle$, we obtain

$$\langle g^{(n+1)}|\psi\rangle - \sum_{j=0}^{n-1} u_j \langle g^{(j)}|\psi\rangle = \lambda \langle g^{(0)}|\psi\rangle,$$

or

$$h^{(n+1)} - \sum_{j=0}^{n-1} u_j h^{(j)} = \lambda. \tag{5.9}$$

This is the generalized Riccati equation for the eigenvalues of the auxiliary linear condition (5.5). Set $\lambda = z^{n+1}$. It is easy to see that this equation admits a unique solution of the form

$$h(z) = z + \sum_{j \geq 1} h_j z^{-j}. \tag{5.10}$$

Its coefficients can be computed algebraically by recurrence. We then conclude that the eigenvalue problem (5.5) admits $(n+1)$ distinct eigenvalues,

$$h_a(z) = h(\omega^a z), \tag{5.11}$$

where $a = 0, 1, \ldots, n$ and ω is the $(n+1)$-th root of unit

$$\omega = \exp\left(\frac{2\pi i}{n+1}\right). \tag{5.12}$$

The corresponding eigenvectors $|\psi_a\rangle$ are linearly independent and, consequently, they form a basis in $\mathbb{R}^{n+1}$ associated with the point S. We denote by $|l_a\rangle$ the basis at the point B, and we note that they are the eigenvectors of the linear problem

$$\Lambda|l_a\rangle = \omega^a z|l_a\rangle, \tag{5.13}$$

where Λ is the matrix

$$\Lambda = B + \lambda A.$$

5.3 Dressing Transformations

We introduce the matrices C, J, and K,

$$C = \sum_{q=1}^{n} c_q \Lambda^q$$

$$\text{(5.14)}$$

$$J|l_a\rangle = h_a|l_a\rangle$$
$$K|l_a\rangle = |\psi_a\rangle.$$

We assume that the coefficients c_q are constant (i.e., independent of x and z). We call C the *generator* of the dressing transformation, K the *dressing matrix*, and J the *momentum map*. Furthermore we define the scalar function

$$H_C = \langle J, C \rangle, \tag{5.15}$$

on $\mathcal{S}$, the 1–form

$$V_C = KCK^{-1}, \tag{5.16}$$

and the vector field

$$\dot{S}_C = [A, V_C]. \tag{5.17}$$

Proposition 5.1.

i)   The matrix V_C admits the expansion $\sum_{k \le d} V_k \lambda^k$ in powers of λ.
ii) It is a solution of equation (5.2) for any choice of the generator C.
iii) It defines an exact 1–form on $\mathcal{S}$. Its Hamiltonian is the function H_C.
iv) The Hamiltonian functions commute with respect to the symplectic form on $\mathcal{S}$,

$$\{H_{C_1}, H_{C_2}\}^{\mathcal{S}} = 0. \tag{5.18}$$

To compute the generalized Casimir functions on $\mathcal{N}$ explicitly, we have to compute the coefficients of the expansion of the Hamiltonian function H_C in powers of λ. It is sufficient to consider the Hamiltonians H_q corresponding to the elements of the basis $C_q = \Lambda^q$, $q = 1, \ldots, n$.

Proposition 5.2. The Hamiltonians H_q admit an expansion of the form

$$H_q = \delta_q^n(n+1)\lambda + \sum_{p \ge 0} H_{pq} \lambda^{-p}. \tag{5.19}$$

The coefficients H_{pq} are related to the coefficients h_j of solution (5.10) of the generalized Riccati equation according to

$$H_{pq} = (n+1) \int_{S^1} h_{(n+1)p+q} \, dx. \tag{5.20}$$

This proposition explains how the generalized Casimir functions on the GD manifold $\mathcal{N}$ may be computed by solving the generalized Riccati equation (5.9). The following proposition summarizes the properties of these functions (the proof proceeds exactly as in the case of the KdV hierarchy treated in the third lecture). Henceforth we will write $H_j = H_{pq}$ when $n + 1 + j = (n+1)p + q$, so that

$$H_j = (n + 1) \int_{S^1} h_{n+1+j} \, dx. \tag{5.21}$$

Proposition 5.3. The first $(n + 1)$ GD Hamiltonians $(H_{-n}^{\mathcal{N}}, \ldots, H_0^{\mathcal{N}})$ are Casimir functions of the first Poisson bracket defined on the GD manifold $\mathcal{N}$,

$$\{\cdot, H_q^{\mathcal{N}}\}_0^{\mathcal{N}} = 0 \qquad q = -n, \ldots, 0. \tag{5.22}$$

The others satisfy the Lenard recursion relations,

$$\{\cdot, H_j^{\mathcal{N}}\}_1^{\mathcal{N}} = \{\cdot, H_{n+1+j}^{\mathcal{N}}\}_0^{\mathcal{N}}, \qquad j \geq -1, \tag{5.23}$$

with respect to the pair of AGD brackets defined on $\mathcal{N}$. Therefore they are in involution with respect to both Poisson brackets. The corresponding bihamiltonian equations

$$\dot{u}_a = \{u_a, H_j^{\mathcal{N}}\}_1^{\mathcal{N}} = \{u_a, H_{n+1+j}^{\mathcal{N}}\}_0^{\mathcal{N}}, \tag{5.24}$$

are the GD equations on $\mathcal{N}$.

Exercise 5.1. Compute the first few Hamiltonians of the Boussinesq hierarchy (corresponding to $n = 2$) by solving the generalized Riccati equation $h_{xx} + 3h_x h^2 + h^3 - u_1 h - u_0 = z^3$ by expansion in Laurent series. Compute the Hamiltonians on $\mathcal{S}$ by using the projection $\pi : \mathcal{S} \to \mathcal{N}$ found in Exercise 4.3.

The form (5.24) of the GD equations on $\mathcal{N}$ is purely nominal, since we do not know the form of the AGD brackets explicitly. To pursue the study, therefore, we need some alternative way of writing the GD equations. According to our strategy, we go back to the symplectic leaf, and we write the GD equations in their canonical form,

$$\frac{\partial q_a}{\partial t_j} = \frac{\delta H_j}{\delta p_a}, \qquad \frac{\partial p_a}{\partial t_j} = -\frac{\delta H_j}{\delta q_a}. \tag{5.25}$$

We denote the vector field on $\mathcal{S}$ defined by these equations by $\dot{S}_j$. The crucial observation is that this vector field admits a special representation in terms of the bihamiltonian pencil (5.1) which was at the origin of the whole story. Indeed, let us denote the particular 1–form V_C corresponding to $C = \Lambda$ by

$$V := K\Lambda K^{-1}. \tag{5.26}$$

Furthermore, let us denote the residue in λ of the power V^j of V by

$$V_j := \mathrm{res}_\lambda V^j, \tag{5.27}$$

and the positive part of the expansion of V^j in powers of λ by $(V^j)_+$.

Proposition 5.4. The vector fields $\dot{S}_j$ of the GD equations on the symplectic leaf S admit the bihamiltonian representation

$$\dot{S}_j = [A, V_j] = ((V^j)_+)_x + [(V^j)_+, S + \lambda A]. \tag{5.28}$$

This is the basic representation of the GD equations, the one which allows us to study in a concise way the conservation laws of these equations. Since the Hamiltonians H_k commute in pairs according to Proposition 5.1, we conclude that each Hamiltonian H_k is constant along the trajectories of the GD equations. Thus we can state that

$$\frac{\partial H_k}{\partial t_j} = 0. \tag{5.29}$$

By passing to the local densities h_k of these Hamiltonians, we then claim that they obey local conservation laws of the form

$$\frac{\partial h_k}{\partial t_j} = \partial_x H_k^{(j)}, \tag{5.30}$$

where the current densities $H_k^{(j)}$ have to be computed. Before doing that, we assemble (5.30) into

$$\frac{\partial h}{\partial t_j} = \partial_x H^{(j)} \tag{5.31}$$

by introducing the Laurent series $h(z)$ and $H^{(j)}(z)$ whose components are h_k and $H_k^{(j)}$, respectively. The current densities $H^{(j)}$ are the most interesting objects of the theory. The following preliminary result connects the current density $H^{(j)}$ to the matrix $(V^j)_+$ which generates the vector field $\dot{S}_j$ of the GD hierarchy.

Proposition 5.5. The current density $H^{(j)}$ is the element

$$H^{(j)} = \langle g^{(0)} | (V^j)_+ | \psi \rangle \tag{5.32}$$

of the matrix $(V^j)_+$ relative to the first row vector $\langle g^{(0)}| = \langle 1, 0, \ldots, 0|$ of the Frobenius basis at the point $S \in \mathcal{S}$, and to the eigenvector $|\psi\rangle$ associated with the solution $h(z)$ of the generalized Riccati equation (5.9).

This proposition is the bridge with the KP theory as will be shown in the next lecture.

References

1. P. Casati, G. Falqui, F. Magri, and M. Pedroni "The KP theory revisited. III. Gel'fand–Dickey equations", Preprint SISSA/4/96/FM (1996)
2. V.G. Drinfeld and V.V. Sokolov, "Lie Algebras and Equations of Korteweg–de Vries Type", J. Sov. Math. **30** (1985) 1975–2036
3. V.E. Zakharov and A.B. Shabat, "A scheme for integrating the nonlinear equations of mathematical physics by the method of the inverse scattering problem", Funct. Anal. Appl. **8** (1974) 226–235

6th Lecture: KP Equations

This is the final lecture on the theory of soliton equations from the point of view of Poisson manifolds. The GD equations have been defined as classical canonical systems

$$\frac{\partial q_a}{\partial t_j} = \frac{\delta H_j}{\delta p_a}, \qquad \frac{\partial p_a}{\partial t_j} = -\frac{\delta H_j}{\delta q_a}, \tag{6.1}$$

on $2n$ periodic functions $(q_a(x), p_a(x))$ playing the role of coordinates on a symplectic manifold. The Hamiltonian functions

$$H_j = \int_{S^1} h_j(q, p, q_x, p_x, \dots) \, dx \tag{6.2}$$

are computed by solving a generalized Riccati equation

$$h^{(n+1)} - \sum_{j=0}^{n-1} u_j(q, p) h^{(j)} = z^{n+1} \tag{6.3}$$

whose coefficients, $u_j(q, p)$, are obtained according to the MR reduction scheme explained in the fourth lecture. These equations admit an infinite set of conservation laws,

$$\frac{\partial H_k}{\partial t_j} = 0, \tag{6.4}$$

which may be written in the local form,

$$\frac{\partial h}{\partial t_j} = \partial_x H^{(j)}. \tag{6.5}$$

The current densities $H^{(j)}$ are given by

$$H^{(j)} = \langle g^{(0)} | (V^j)_+ | \psi \rangle \tag{6.6}$$

by the method illustrated in the last lecture.

In this lecture we shall explain how the theory of KP equations naturally evolves from the previous results. The point is to focus attention on the current densities $H^{(j)}$. They may be computed directly in terms of the solution $h(z)$ of the generalized Riccati equation, without passing through a computation of the matrix $(V^j)_+$. One has to pay attention to two consequences of the definition (6.6). The first concerns the expansion of the current density $H^{(j)}$ in powers of z. We claim that the Laurent expansion of $H^{(j)}$ has the form

$$H^{(j)} = z^j + \sum_{k \geq 1} H_k^j z^{-k}. \tag{6.7}$$

This is shown by writing $(V^j)_+ = V^j - (V^j)_-$, by observing that $V|\psi\rangle = z|\psi\rangle$ and by further observing that $(V^j)_-|\psi\rangle = O(z^{-1})$, since $(V^j)_- = O(\lambda^{-1})$, $|\psi\rangle = O(z^n)$ and $\lambda = z^{n+1}$. The second consequence concerns the expansion of $H^{(j)}$ on the Faà di Bruno polynomials $h^{(j)}$ associated with $h(z)$. We claim that the current densities $H^{(j)}$ admit an expansion of the form

$$H^{(j)} = h^{(j)} + \sum_{k=0}^{j-1} c_k^j h^{(k)}, \tag{6.8}$$

where the coefficients c_k^j are independent of z. This is proved by developing the row vector $\langle g^{(0)}|(V^j)_+$ on the Frobenius basis attached to the point S. We obtain

$$\langle g^{(0)}|(V^j)_+ = \sum_{i=0}^{n} a_i^j \langle g^{(i)}|, \tag{6.9}$$

and therefore

$$H^{(j)} = \sum_{i=0}^{j} a_i^j h^{(i)}, \tag{6.10}$$

since $\langle g^{(i)}|\psi\rangle = h^{(i)}$. Furthermore, the coefficients a_i^j are polynomials in λ. To know what is the effect of multiplying a Faà di Bruno polynomial $h^{(j)}$ by a power of λ, we act on the Riccati equation (6.3) by the operator $\partial_x + h$. We obtain

$$\lambda h = h^{(n+2)} - \sum_{j=0}^{n-2} u_j^1 h^{(j)}. \tag{6.11}$$

By recurrence, this result entails that $\lambda^k h$ admits a linear expansion on the Faà di Bruno polynomials, with coefficients which are independent of z. This proves our claim.

To show the practical use of (6.8), let us consider the current density

$$H^{(2)} = h^{(2)} + c_1 h^{(1)} + c_0 h^{(0)}. \tag{6.12}$$

Since

$$h^{(1)} = h = z + h_1 z^{-1} + h_2 z^{-2} + \ldots$$
$$h^{(2)} = h_x + h^2 = z^2 + 2h_1 + (h_{1x} + 2h_2)z^{-1} + (h_{2x} + h_1^2 + 2h_3)z^{-2} + \ldots \tag{6.13}$$

we obtain $H^{(2)} = z^2 + c_1 z + (2h_1 + c_0) + \dots$ Therefore, using (6.7), we obtain

$$c_1 = 0, \qquad c_0 = -2h_1 \tag{6.14}$$

and

$$H^{(2)} = z^2 + (h_{1x} + 2h_2)z^{-1} + (h_{2x} + h_1^2 + 2h_3)z^{-2} + \dots \tag{6.15}$$

To give the general form of the currents $H^{(j)}$, we introduce the infinite matrix

$$\mathbb{H} = \begin{bmatrix} h^{(0)} & h^{(1)} & h^{(2)} & \dots \\[2mm] \operatorname{res}\dfrac{h^{(0)}}{z} & \operatorname{res}\dfrac{h^{(1)}}{z} & \operatorname{res}\dfrac{h^{(2)}}{z} & \dots \\[2mm] 0 & \operatorname{res}\dfrac{h^{(1)}}{z^2} & \operatorname{res}\dfrac{h^{(2)}}{z^2} & \dots \\[2mm] & 0 & \operatorname{res}\dfrac{h^{(2)}}{z^3} & \dots \\[2mm] & & 0 & \dots \\[2mm] \dots & \dots & \dots & \dots \end{bmatrix}. \tag{6.16}$$

Proposition 6.1. The currents $H^{(j)}$ are, up to a sign, the principal minors of the matrix $\mathbb{H}$.

The proof of this statement is left to the reader. We observe that the latter expression of the current densities $H^{(j)}$ actually generalizes the definition given in equation (6.6). Indeed, it is independent of the requirement that $h(z)$ be a solution of the Riccati equation. Therefore we can define $H^{(j)}$ for a generic Laurent series $h(z)$ of the form (5.10), and then equations (6.5) become an infinite system of partial differential equations in an infinite number of periodic functions, $(h_1(x), h_2(x), \dots)$. These equations are the celebrated *KP equations*. To obtain an intuition about these equations, let us consider the first currents:

$$H^{(1)} = h = z + \frac{h_1}{z} + \frac{h_2}{z^2} + \dots$$

$$H^{(2)} = z^2 + \frac{2h_2 + h_{1x}}{z} + \frac{h_1^2 + 2h_3 + h_{2x}}{z^2} + \dots \tag{6.17}$$

$$H^{(3)} = z^3 + \frac{3h_3 + 3h_{2x} + h_{1xx}}{z} + \dots .$$

By expanding (6.5) in powers of z, we immediately see that the first three KP equations are:

$$\frac{\partial h_1}{\partial t_1} = h_{1x} \qquad\qquad \frac{\partial h_1}{\partial t_2} = \partial_x(h_{1x} + 2h_2)$$

$$\frac{\partial h_2}{\partial t_1} = h_{2x} \qquad\qquad \frac{\partial h_2}{\partial t_2} = \partial_x(h_{2x} + h_1^2 + 2h_3) \qquad (6.18)$$

$$\frac{\partial h_3}{\partial t_1} = h_{3x} \qquad\qquad \frac{\partial h_3}{\partial t_2} = \partial_x(h_{3x} + 2h_4 + 2h_1h_2)$$

and

$$\frac{\partial h_1}{\partial t_3} = \partial_x(h_{1xx} + 3h_{2x} + 3h_3)$$

$$\frac{\partial h_2}{\partial t_3} = \partial_x(h_{2xx} + 3h_{3x} + 3h_1h_{1x} + 3h_4 + 3h_1h_2) \qquad (6.19)$$

$$\frac{\partial h_1}{\partial t_3} = \partial_x(h_{3xx} + 3h_{4x} + 3h_1h_{2x} + 3h_{1x}h_2 + 3h_5 + 3h_1h_3 + 3h_2^2 + h_1^3)$$

This is not the usual form of the KP equations. The usual equations do not have the form of local conservation laws, but they can be given this form by a suitable transformation defined at the end of this lecture. To recover the GD equations we need only to require that the series $h(z)$ be a solution of the generalized Riccati equation. So, for example, for $n = 1$, this equation reads

$$h_x + h^2 = u + z^2, \qquad (6.20)$$

and its solution is

$$h(z) = z + \tfrac{1}{2}uz^{-1} - \tfrac{1}{4}u_x z^{-2} + \tfrac{1}{8}(u_{xx} - u^2)z^{-3} + \dots . \qquad (6.21)$$

The coefficients of this expansion are easily computed by recurrence from equation (6.20). By inserting them into the KP equations, we immediately see that the equations relative to h_1 yield the first three KdV equations,

$$\frac{\partial u}{\partial t_1} = u_x \qquad \frac{\partial u}{\partial t_2} = 0 \qquad \frac{\partial u}{\partial t_3} = \tfrac{1}{8}(u_{xxx} - 6uu_x). \qquad (6.22)$$

The equations relative to $(h_2, h_3, \dots)$ are differential consequences of the KdV equations. The same is true for $n = 2$: the equations relative to (h_1, h_2) give the Boussinesq hierarchy, while the remaining equations are differential consequences of the previous ones, and so on.

To go further into the KP theory, we have to extend the definition of the Faà di Bruno polynomials to the negative integers. This is done by solving backwards the recurrence relation

$$h_x^{(-j+1)} + hh^{(-j+1)} = h^{(-j)}, \qquad (6.23)$$

starting from $h^{(0)} = 1$. A simple computation shows that

$$
\begin{aligned}
h^{(-1)} &= z^{-1} - h_1 z^{-3} + (-h_2 + h_{1x})z^{-4} + \dots \\
h^{(-2)} &= z^{-2} - 2h_1 z^{-4} + \dots .
\end{aligned}
\tag{6.24}
$$

In general,

$$
h^{(j)} = z^j + O(z^{j-2}), \qquad j \in \mathbb{Z}.
\tag{6.25}
$$

Therefore the "polynomials" $h^{(j)}$ form a basis of the space L of Laurent series in z. Along with the basis $h^{(j)}$, we consider the dual basis $h^*_{(l)}$ defined by

$$
\langle h^{(j)}, h^*_{(l)} \rangle := \operatorname{res}_z h^{(j)} h^*_{(l)} = \delta^{j+1}_{-l}.
\tag{6.27}
$$

We denote the element $h^*_{(0)}$ of this basis by h^*, and we call it the dual Hamiltonian of the KP theory,

$$
h^* = h^*_{(0)}.
\tag{6.28}
$$

We observe that the relation between h and h^* is invertible. If we call h^*_j the coefficients of the expansion of h^*,

$$
h^* = 1 - \sum_{j \geq 1} h^*_j z^{-(j+1)},
\tag{6.29}
$$

after some nontrivial calculation we obtain

$$
\begin{aligned}
h^*_1 &= h_1 \\
h^*_2 &= 2h_2 + h_{1x} \\
h^*_3 &= 3h_3 + 3h_{2x} + h_{1xx}
\end{aligned}
\tag{6.30}
$$

$$
\dots\dots ,
$$

showing that the components of $h^*(z)$ can be expressed as differential polynomials of the components of $h(z)$, and vice-versa. This means that the two series convey the same set of information. Then we can write the KP equations as equations on the dual Hamiltonian h^*. One can prove that h^* obeys a new set of local conservation laws,

$$
\frac{\partial h^*}{\partial t_j} = \partial_x H^*_{(j)},
\tag{6.31}
$$

which are called the *dual KP equations*. These equations do not seem to have been previously considered in the literature. Their current densities, $H^*_{(j)}$, are the main object of the theory. They have two special properties. The first one is of an algebraic type and concerns the coefficient H^*_{jk} of the expansion

$$
H^*_{(j)} = z^{j-1} + \sum_{k \geq 1} H^*_{jk} z^{-(k+1)}
\tag{6.32}
$$

of $H^*_{(j)}$ in powers of z. It claims that these coefficients obey the symmetry condition

$$
H^*_{jk} = H^*_{kj}.
\tag{6.33}
$$

The second property is differential in nature. It concerns the time evolution of the currents $H^*_{(j)}$ along the trajectories of the dual KP equations (6.31) or, what is the same thing, of the KP equations (6.5). We claim that the dual currents $H^*_{(j)}$ obey the "zero–curvature conditions",

$$\frac{\partial H^*_{(j)}}{\partial t_k} - \frac{\partial H^*_{(k)}}{\partial t_j} = 0. \tag{6.34}$$

The two conditions together mean that the second–order tensor field,

$$g = \sum H^*_{jk} dt_j dt_k, \tag{6.35}$$

satisfies the conditions which characterize a metric of Hessian type [3]. Therefore, we can introduce a Kähler potential τ satisfying

$$H^*_{jk} = \frac{\partial^2}{\partial t_j \partial t_k} \log \tau. \tag{6.36}$$

This potential is the τ–function of Hirota's approach to the KP equations.

We hope that this quick survey may convey the flavor of the geometrical approach to soliton equations and to the KP equations. We started from a very simple bihamiltonian structure, constructed over the space of C^∞–maps from S^1 into $\mathfrak{sl}(n+1)$. By a systematic use of the MR reduction scheme, we discovered the GD manifolds with their associated AGD brackets. Then the study of the generalized Casimir functions of these brackets led us to construct the Hamiltonian functions of the GD hierarchies. By Noether's theorem, these Hamiltonians are constant along the trajectories of the GD flows. We have studied these conservation laws, and we have put them in the local form (6.5). Furthermore, we have proved the existence of a second set of conjugate local conservation laws, and we have computed their associated dual current densities $H^*_{(j)}$. They convey the principal information about the GD and the KP hierarchies. They may be seen as defining a Hessian metric on each solution $h(t_1, t_2, \dots)$ of the KP equations, and the corresponding Kähler potential is the τ–function of the KP theory.

As previously noticed, the present form of the KP equations is different from that currently adopted in the literature. We end this lecture by showing the link between the two formulations. The key is the expansion

$$z = h - \sum_{j \geq 1} u_j h^{(-j)}, \tag{6.38}$$

which naturally extends the generalized Riccati equations (6.3) of the GD case. The new variables u_j are the components of the vector z on the Faà di Bruno basis $h^{(j)}$ associated with the point h in the space L of Laurent series in z. We note that expansion (6.38) uniquely defines the coefficients u_j. For instance, the first coefficients are

$$u_1 = h_1$$
$$u_2 = h_2 \tag{6.39}$$
$$u_3 = h_3 - h_1^2.$$

These relations define the change of variables which transforms our KP equations into the standard ones. The resulting equations are the KP equations in Lax representation (see [2] for more details).

References

1. P. Casati, G. Falqui, F. Magri, and M. Pedroni, "The KP theory revisited. IV. Baker–Akhiezer and τ functions", Preprint SISSA/5/96/FM (1996)
2. G. Falqui, F. Magri, and M. Pedroni, "Bihamiltonian geometry, Darboux coverings, and linearization of the KP hierarchy", Commun. Math. Phys. **197** (1998) 303–324
3. H. Shima and K. Yagi, "Geometry of Hessian manifolds", Diff. Geom. Appl. **7** (1997) 277–290

7th Lecture: Poisson–Nijenhuis Manifolds

We leave the field of soliton equations to return to the more geometrical setting of Poisson manifolds. These last two lectures are devoted to identifying the properties of a particular class of bihamiltonian manifolds, the Poisson–Nijenhuis manifolds.

Consider a *symplectic* manifold $\mathcal{M}$ endowed with a second compatible Poisson bracket. It is a special bihamiltonian manifold where the first Poisson tensor P_0 is invertible. Thus we may define a (1,1) tensor field,

$$N = P_1 P_0^{-1}, \tag{7.1}$$

as a linear map $N : T\mathcal{M} \to T\mathcal{M}$ acting on the tangent bundle. The adjoint map, acting on the cotangent bundle, will be denoted by

$$N^* = P_0^{-1} P_1. \tag{7.2}$$

This tensor field enjoys several special properties which are the subject of this lecture.

The first property concerns the Nijenhuis torsion of N. This torsion is a vector–valued 2–form on $\mathcal{M}$ constructed according to

$$T_N(X, Y) := [NX, NY] - N[NX, Y] - N[X, NY] + N^2[X, Y], \tag{7.3}$$

where (X, Y) is a pair of vector fields and the bracket $[X, Y]$ denotes the commutator of these fields. The symbol NX denotes the iterated vector field obtained from X by the action of N.

Proposition 7.1. The torsion of N vanishes as a consequence of the assumption that P_0 and P_1 are a pair of compatible Poisson tensors,

$$[NX, NY] - N[NX, Y] - N[X, NY] + N^2[X, Y] = 0. \tag{7.4}$$

Proof. We denote the vector fields which are Hamiltonian with respect to P_0 and P_1 by $X_f = P_0 df$ and by $Y_f = P_1 df$, respectively. Then we observe that $N X_f = Y_f$, and we evaluate the torsion of N on the vector fields X_f, using (1.26),

$$
\begin{aligned}
T_N(X_f, X_g) :&= [Y_f, Y_g] - N\left([Y_f, X_g] + [X_f, Y_g] - N[X_f, X_g]\right) \\
&= [Y_f, Y_g] - N\left([Y_f, X_g] + [X_f, Y_g] - Y_{\{f,g\}_0}\right) \\
&= [Y_f, Y_g] - N X_{\{f,g\}_1} \\
&= [Y_f, Y_g] - Y_{\{f,g\}_1} \\
&= 0.
\end{aligned}
$$

Let now P_0 be a Poisson tensor, which might be degenerate. By reversing the previous proof we see that the vanishing of the torsion of N implies that $P_1 = N P_0$ is a Poisson tensor if it is skew–symmetric,

$$ N P_0 = P_0 N^*, \tag{7.5} $$

and satisfies the compatibility conditions (1.26) with P_0. The iterated bivector of P_1,

$$ P_2 = N P_1 = N^2 P_0, \tag{7.6} $$

is skew–symmetric but, in general, not compatible with P_0 and P_1. Therefore it is not, in general, a Poisson tensor. We can reasonably ask if there exists any condition on N which implies that all the iterated bivectors $(P_1, P_2, P_3, \dots)$ are compatible with P_0 and pairwise. This condition is in fact

$$ L_X(N P_0) = N L_X(P_0) + L_X(P_0) N^* - L_{NX}(P_0). \tag{7.7} $$

We call a tensor field N with vanishing torsion a Nijenhuis tensor field, and we say that it is compatible with the Poisson tensor P_0 if it obeys conditions (7.5) and (7.7). With these conventions, we adopt the following definition.

Definition 7.2. A Poisson–Nijenhuis manifold is a Poisson manifold endowed with a compatible Nijenhuis tensor field.

Almost by definition, a sequence of compatible Poisson tensors,

$$ P_k = N^k P_0, \tag{7.8} $$

is defined on such a manifold. Let us call C_k the characteristic distribution of P_k,

$$ C_k = \mathrm{Im}\, P_k. \tag{7.9} $$

We can pass from C_k to C_{k+1} in two ways: by using N, $C_{k+1} = N(C_k)$, or by using P_0 and P_1. In this second case, we first have to construct the inverse image $L_k = P_0^{-1}(C_k)$, and then the direct image $C_{k+1} = P_1(L_k)$ of L_k. This is exactly the same construction used, in the third lecture, to introduce the generalized Casimir functions on a bihamiltonian manifold. If the manifold is a Poisson–Nijenhuis manifold, the generalized Casimir functions are Casimir functions of the iterated Poisson tensors P_k. This observation justifies the name of generalized Casimir functions used in that occasion.

The geometry of Poisson–Nijenhuis manifolds is rich in interesting properties [4, 5]. They mainly concern the study of invariants with respect to the action of the Nijenhuis tensor on vector fields and 1–forms. This study leads us to identify several classes of special geometrical objects (functions, vector fields, 1–forms, and distributions) canonically defined on such a manifold. In this lecture we shall limit ourselves to give a simple example.

Consider the functions

$$I_k = \frac{1}{k} \operatorname{Tr} N^k. \tag{7.10}$$

They are a set of privileged functions whose property is that they are in involution with respect to all the Poisson brackets defined on $\mathcal{M}$. To prove this property, we consider the differentials dI_j, and we observe that they verify the following important recursion relation,

$$N^* dI_k = dI_{k+1}. \tag{7.11}$$

This relation follows from the computation

$$
\begin{aligned}
\langle dI_{k+1}, X \rangle &= \frac{1}{k+1} L_X \operatorname{Tr} N^{k+1} \\
&= \operatorname{Tr} \left(N^{*k-1} L_X(N^*) N^* \right) \\
&= \operatorname{Tr} \left(N^{*k-1} L_{NX}(N^*) \right) \\
&= \frac{1}{k} \operatorname{Tr} L_{NX}(N^{*k}) \\
&= \langle dI_k, NX \rangle \\
&= \langle N^* dI_k, X \rangle, \tag{7.12}
\end{aligned}
$$

where we have used the identity

$$L_{NX}(N^*) = L_X(N^*) N^*, \tag{7.13}$$

which is equivalent to the vanishing of the torsion of N.

Exercise 7.1. Show that condition (7.4) for the vanishing of the torsion of N means that

$$L_{NX}(N) = N L_X(N) \tag{7.14}$$

for any vector field on $\mathcal{M}$. Then deduce condition (7.13) on the adjoint mapping N^*.

We then conclude that the differentials dI_k satisfy the relation

$$P_1 dI_k = P_0 dI_{k+1} \tag{7.15}$$

and that the functions I_k form a Lenard sequence,

$$\{\cdot, I_k\}_1 = \{\cdot, I_{k+1}\}_0. \tag{7.16}$$

Their property of being in involution then follows then from the general argument explained in the third lecture. Of course, not all differentials dI_k will be linearly independent. For each point $m \in \mathcal{M}$ there exists an integer p (depending on m) such that

$$dI_{p+1} + c_1 dI_p + \cdots + c_p dI_1 = 0, \tag{7.17}$$

where the functions c_a are the coefficients of the minimal polynomial of N.

To investigate the structure of this polynomial we restrict somewhat the class of the manifold $\mathcal{M}$. Thus we suppose that $\mathcal{M}$ is a symplectic manifold of dimension $2n$, and that the first Poisson tensor P_0 is invertible. Then we observe that the eigenvalues of N are the roots of the equation

$$\det(P_1 - \lambda P_0) = 0. \tag{7.18}$$

Since any pencil of skew–symmetric matrices on an even–dimensional vector space has degenerate eigenvalues, we see that the most favorable case is the one in which N has only *double* eigenvalues, so that the index p in (7.17) is exactly n, half of the dimension of the phase space. This index, however, may vary with the point. So, we make a further assumption to restrict our study to an open neighborhood of *"maximal rank"*, where the index p is everywhere equal to n, and where the differentials dI_k are linearly independent.

We now show that it is possible to define a special system of coordinates on a neighborhood of maximal rank which gives P_0 and N simultaneously a canonical form. These coordinates were used as separation variables for the Hamilton–Jacobi equation in, e.g., [1–3].

Proposition 7.3 (Darboux theorem). In a neighborhood of a point of maximal rank there exist coordinates (λ_j, μ_j) which give P_0 the canonical form

$$P_0 = \sum_j \frac{\partial}{\partial \lambda_j} \wedge \frac{\partial}{\partial \mu_j} \tag{7.19}$$

and N the diagonal form

$$N^* d\lambda_j = \lambda_j d\lambda_j \qquad N^* d\mu_j = \lambda_j d\mu_j. \tag{7.20}$$

These coordinates may be constructed by quadrature.

We sketch the proof of this statement, by dividing it into four steps.

1st Step. First of all, we consider the eigenvalues of the Nijenhuis tensor N. In our neighborhood there are n distinct eigenvalues $(\lambda_1, \ldots, \lambda_n)$ which are in involution since the traces I_k are in involution, and their differentials are eigenvectors of N^*,

$$N^* d\lambda_k = \lambda_k d\lambda_k. \tag{7.21}$$

This property is a consequence of the Lenard recursion relation (7.11) on the traces I_k. To prove it, one has simply to write the traces I_k as functions of the eigenvalues $(\lambda_1, \ldots, \lambda_n)$, to insert them into the Lenard recursion relations, and to solve the system for the first n relations with respect to the vectors $N^* d\lambda_k$. By this property we have already got half of the eigenvectors of N.

2nd Step. We now use the Darboux theorem for symplectic manifolds. Since we know n coordinates which are in involution, by quadrature we can locally define n other coordinate functions $(\mu_1, \ldots, \mu_n)$ which give the Poisson bracket the canonical form,

$$\{\lambda_j, \lambda_k\} = 0, \qquad \{\lambda_j, \mu_k\} = \delta_{jk}, \qquad \{\mu_j, \mu_k\} = 0. \tag{7.22}$$

Of course these new coordinates are not uniquely defined. We are allowed to change them according to

$$\mu'_k = \mu_k + \frac{\partial S}{\partial \lambda_k}(\lambda_1, \ldots, \lambda_n). \tag{7.23}$$

3rd Step. We now evaluate the Nijenhuis tensor N in the new coordinates. We already know that

$$N^* d\lambda_j = \lambda_j d\lambda_j. \tag{7.24}$$

We claim that

$$N^* d\mu_j = \lambda_j d\mu_j + \sum_k a_{jk} d\lambda_k, \tag{7.25}$$

where the coefficients $a_{jk}(\lambda_1, \ldots, \lambda_n)$ are the components of a closed 2–form in the space of the coordinates λ_k. This follows from the compatibility conditions between N and P_0.

4th Step. We now use the arbitrariness (7.23) in the choice of the coordinates μ_k. By repeating the same computation as before we find

$$N^* d\mu'_j = \lambda_j d\mu'_j + \sum_k \left[a_{jk} - (\lambda_j - \lambda_k) \frac{\partial^2 S}{\partial \lambda_k \partial \lambda_j} \right] d\lambda_k. \tag{7.26}$$

Therefore, it is enough to choose the function $S(\lambda_1, \ldots, \lambda_n)$ in such a way as to solve the equations

$$(\lambda_j - \lambda_k) \frac{\partial^2 S}{\partial \lambda_k \partial \lambda_j} = a_{kj}. \tag{7.27}$$

Their integrability conditions are satisfied because the torsion of N vanishes.

The Darboux theorem has many consequences. One of them, related to the Lax formulation of integrable systems, will be dealt with in the next and final lecture. Before doing that, we want to point out a possible relation between the theory of Poisson–Nijenhuis manifolds and the KdV equation.

Exercise 7.2. Consider again the Poisson tensors (3.1) and (3.2) associated with the KdV hierarchy. Let us consider these operators as elements of the algebra of pseudodifferential operators on the circle. This means that we introduce a new symbol ∂^{-1} that we define to be the inverse of ∂,

$$\partial \circ \partial^{-1} = \partial^{-1} \circ \partial = 1, \tag{7.28}$$

and which must be handled according to the following rule, the commutator of ∂^{-1} and any scalar function $f(x)$ on S^1 is given by

$$[\partial^{-1}, f] = \sum_{k \geq 2} (-1)^{k+1} \frac{d^{k-1} f}{dx^{k-1}} \partial^k.$$

Then, we formally invert the operator P_0 and we introduce the operator

$$N = P_1 \circ P_0^{-1}$$
$$= \left(-\frac{1}{2}\partial^3 + 2u\partial + u_x \right) \circ \left(-\frac{1}{2}\partial^{-1} \right)$$
$$= \frac{1}{4}\partial^2 - u - \frac{1}{2}u_x\partial^{-1}$$

which we call the Nijenhuis operator associated with the KdV hierarchy. Finally, we compute the square root of this operator, which is the unique (up to sign) operator

$$N^{\frac{1}{2}} = \frac{1}{2}\partial - u\partial^{-1} - u^2\partial^{-3} + \ldots$$

whose square is N. We call the integral, over S^1, of the coefficient of ∂^{-1} the *Adler trace* of a pseudodifferential operator. Compute now the traces of $(N^{\frac{1}{2}}, N, N^{\frac{3}{2}}, \ldots)$ and show that, for the lowest orders, they coincide with the generalized Casimir functions of the KdV theory.

We have no explanation of this coincidence.

References

1. G. Falqui, F. Magri, and M. Pedroni, "Bihamiltonian geometry and separation of variables for Toda lattices", J. Nonlinear Math. Phys. Suppl. **8** (2001) 118–127

2. G. Falqui, F. Magri, M. Pedroni, and J.P. Zubelli, "A Bi-Hamiltonian theory for stationary KdV flows and their separability", Regular and Chaotic Dynamics **5** (2000) 33–52

3. G. Falqui, F. Magri, and G. Tondo, "Bi-Hamiltonian systems and separation of variables: an example from the Boussinesq hierarchy", Theor. Math. Phys. **122** (2000) 176–192

4. Y. Kosmann–Schwarzbach, and F. Magri, "Poisson–Nijenhuis structures", Ann. Inst. Henri Poincaré, Phys. Théor. **53** (1990) 35–81

5. I. Vaisman, *Lectures on the Geometry of Poisson Manifolds*, Progress in Math. 118 (Birkhäuser 1994)

8th Lecture: The Calogero System

We end these lectures with an example showing possible connections of the bihamiltonian scheme to finite–dimensional integrable systems. The Calogero system is an n–body problem on the line for particles repelling each other with a force proportional to the cube of the inverse of the distance. For three particles the Hamiltonian function is

$$H(x,y) = \frac{1}{2}(y_1^2 + y_2^2 + y_3^2) + \frac{1}{(x_1 - x_2)^2} + \frac{1}{(x_2 - x_3)^2} + \frac{1}{(x_3 - x_1)^2}, \quad (8.1)$$

where (x_1, x_2, x_3) are the coordinates of the particles and (y_1, y_2, y_3) are the corresponding momenta. Consequently the equations of motion are

$$\begin{aligned}
\dot{x}_1 &= y_1 & \dot{y}_1 &= 2(x_{12}^2 - x_{31}^2) \\
\dot{x}_2 &= y_2 & \dot{y}_2 &= 2(x_{23}^2 - x_{12}^2) \\
\dot{x}_3 &= y_3 & \dot{y}_3 &= 2(x_{31}^2 - x_{23}^2),
\end{aligned} \quad (8.2)$$

where we have used the shorthand notation

$$x_{ij} = \frac{1}{x_i - x_j}. \quad (8.3)$$

These equations admit a Lax representation with a Lax matrix

$$L = \begin{pmatrix} y_1 & ix_{12} & ix_{13} \\ ix_{21} & y_2 & ix_{23} \\ ix_{31} & ix_{32} & y_3 \end{pmatrix}. \quad (8.4)$$

Consider the functions

$$\begin{aligned}
I_1 &= \operatorname{Tr} L = y_1 + y_2 + y_3 \\
I_2 &= \frac{1}{2} \operatorname{Tr} L^2 = \frac{1}{2}(y_1^2 + y_2^2 + y_3^2) + x_{12}^2 + x_{23}^2 + x_{31}^2 \\
I_3 &= \frac{1}{3} \operatorname{Tr} L^3 = \frac{1}{3}(y_1^3 + y_2^3 + y_3^3) + x_{12}^2(y_1 + y_2) + x_{23}^2(y_2 + y_3) + x_{31}^2(y_3 + y_1)
\end{aligned} \quad (8.5)$$

to be new Hamiltonians, and compute the corresponding equations of motion. They form the first three equations of the *Calogero hierarchy*. Since $I_2 = H$, the Calogero system is the second member of this hierarchy. A characteristic property of this hierarchy is that it admits an *extended Lax representation*. Indeed, for $k = 1, 2, 3$, there exist matrices B_k such that the equations of the Calogero hierarchy may be written in the form

$$\frac{dL}{dt_k} = [L, B_k]$$
$$\frac{dX}{dt_k} = [X, B_k] + L^{k-1},$$
(8.6)

where

$$X = diag(x_1, x_2, \ldots, x_n)$$
(8.7)

is the diagonal matrix of the positions of the particles. We now want to explain the meaning of this representation and its connection with the theory of Poisson–Nijenhuis manifolds.

Let us return to the concept of *Darboux coordinates* (λ_j, μ_j) on a neighborhood of maximal rank of a symplectic Poisson–Nijenhuis manifold $\mathcal{M}$. We denote by L and X any pair of $n \times n$ matrices whose entries are functions on the manifold $\mathcal{M}$, such that

$$\operatorname{Tr} L^i = \lambda_1^i + \cdots + \lambda_n^i$$
(8.8)
$$\operatorname{Tr}(XL^{i-1}) = \mu_1 \lambda_1^{i-1} + \cdots + \mu_n \lambda_n^{i-1}$$
(8.9)

for $i = 1, 2, \ldots, n$. They will be referred to as a pair of *Darboux matrices* for the manifold $\mathcal{M}$. Presently, we study how these matrices evolve along the trajectories of the Hamiltonian vector fields defined by the functions I_k. In Darboux coordinates these vector fields are given by

$$\frac{d\lambda_j}{dt_k} = 0 \qquad \frac{d\mu_j}{dt_k} = \lambda_1^{k-1} + \cdots + \lambda_n^{k-1}.$$
(8.10)

Proposition 8.1. For $k = 1, 2, \ldots, n$, there exist pairs of matrices (B_k, C_k) such that the equations of motion of the Darboux matrices may be written in the form of an extended Lax representation,

$$\frac{dL}{dt_k} = [L, B_k]$$
(8.11)

$$\frac{dX}{dt_k} = [X, B_k] + [L, C_k] + L^{k-1}.$$
(8.12)

Matrices B_k and C_k depend on the choice of matrices L and X. However, the previous system is gauge–invariant with respect to this choice.

The proof of this theorem is easy, and is left to the reader. One simply has to differentiate equations (8.8) and (8.9) defining the Darboux matrices with respect to the time t_k, to keep in mind the equations of motion (8.10), and to recall that any matrix V such that $\mathrm{Tr}(L^iV) = 0$ for $i = 0, 1, \ldots, n-1$ is given by $V = [L, B]$ for a suitable matrix B.

The relation with Calogero systems is presently obvious. The representation (8.6) of the Calogero hierarchy is a particular instance of an extended Lax representation with $C_k = 0$. We use this observation to construct the bihamiltonian structure of the Calogero systems. For convenience, we introduce a new set of coordinates, (I_k, J_k), defined by

$$I_k = \frac{1}{k}(\lambda_1^k + \cdots + \lambda_n^k) \tag{8.13}$$

$$J_k = \mu_1 \lambda_1^{k-1} + \cdots + \mu_n \lambda_n^{k-1}, \tag{8.14}$$

so that $kI_k = \mathrm{Tr}\, L^k$ and $J_k = \mathrm{Tr}(XL^{k-1})$. We call these coordinates the *conformal coordinates* of the symplectic Poisson–Nijenhuis manifold $\mathcal{M}$. For the three–particle Calogero system, these coordinates are the functions (8.5) and

$$
\begin{aligned}
J_1 &= x_1 + x_2 + x_3 \\
J_2 &= x_1 y_1 + x_2 y_2 + x_3 y_3 \\
J_3 &= x_1 y_1^2 + x_2 y_2^2 + x_3 y_3^2 + x_{12}^2(x_1 + x_2) + x_{23}^2(x_2 + x_3) + x_{31}^2(x_3 + x_1).
\end{aligned}
\tag{8.15}
$$

Then we consider the Poisson tensors P_0 and $P_1 = NP_0$ instead of P_0 and N, and we compute their components in the (I, J) coordinates. As a simple exercise in the transformation laws of tensor fields, one can prove that these components are given by

$$\{I_j, I_k\}_0 = 0, \quad \{I_j, J_k\}_0 = (j + k - 2)I_{j+k-2}, \quad \{J_k, J_l\}_0 = (l - k)J_{l+k-2}, \tag{8.16}$$

$$\{I_j, I_k\}_1 = 0, \quad \{I_j, J_k\}_1 = (j + k - 1)I_{j+k-1}, \quad \{J_k, J_l\}_1 = (l - k)J_{l+k-1}, \tag{8.17}$$

using the canonical form of the tensor P_0 and N in Darboux coordinates, as shown in the previous lecture. These formulas yield the bihamiltonian structure associated with any extended Lax representation. We conclude that, in this context, to give a dynamical system an extended Lax representation is the same as to determine its bihamiltonian structure. We do this explicitly for the two–particle Calogero system. In this case the coordinates (I, J) are defined by:

$$
\begin{aligned}
I_1 &= y_1 + y_2 & J_1 &= x_1 + x_2 \\
I_2 &= \frac{1}{2}(y_1^2 + y_2^2) + x_{12} & J_2 &= x_1 y_1 + x_2 y_2.
\end{aligned}
\tag{8.18}
$$

By inserting these functions into definitions (8.16) and (8.17) of the pair of compatible Poisson tensors on $\mathcal{M}$, we obtain the fundamental Poisson brackets in (x, y) coordinates. The first Poisson tensor is

$$\{x_i, x_j\}_0 = 0 \quad \{x_i, y_j\}_0 = \delta_{ij} \quad \{y_i, y_j\}_0 = 0, \tag{8.19}$$

since positions and momenta are canonical coordinates. The second Poisson tensor is

$$\{x_1, x_2\}_1 = \frac{2x_{12}}{4x_{12}^2 + (y_1 - y_2)^2}$$

$$\{x_1, y_1\}_1 = y_1 + \frac{x_{12}^2(y_1 - y_2)}{4x_{12}^2 + (y_1 - y_2)^2}$$

$$\{x_1, y_2\}_1 = -\frac{x_{12}^2(y_1 - y_2)}{4x_{12}^2 + (y_1 - y_2)^2}$$

$$\{x_2, y_2\}_1 = y_2 - \frac{x_{12}^2(y_1 - y_2)}{4x_{12}^2 + (y_1 - y_2)^2}$$

$$\{y_1, y_2\}_1 = 2x_{12}^3. \tag{8.20}$$

By calling the coordinates (x_j, y_j) on the phase–space $\mathcal{M}$ collectively z_α, the Calogero equations (8.2) may then be given the bihamiltonian form

$$\dot{z}_\alpha = \{z_\alpha, I_2\}_0 = \{z_\alpha, I_1\}_1. \tag{8.21}$$

The first Hamiltonian is the energy $H = I_2$; the second Hamiltonian is the total momentum I_1.

References

1. D. Kazhdan, B. Kostant, and S. Sternberg, "Hamiltonian group actions and dynamical systems of Calogero type", Comm. Pure Appl. Math. **31** (1978) 481–508
2. F. Magri and T. Marsico, "Some developments of the concept of Poisson manifold in the sense of A. Lichnerowicz", in: *Gravitation, Electromagnetism, and Geometric Structures*, G. Ferrarese, ed. (Pitagora editrice, Bologna 1996) pp. 207–222
3. M. Olshanetsky and A. Perelomov, "Completely integrable Hamiltonian systems connected with semisimple Lie algebras", Invent. Math. **37** (1976) 93–108

Bilinear Formalism in Soliton Theory

J. Satsuma

Graduate School of Mathematical Sciences, University of Tokyo, Komaba, Meguro-ku, Tokyo 153-8914, Japan, satsuma@ms.u-tokyo.ac.jp

Abstract. A brief survey of the bilinear formalism discovered by Hirota is given. First, the procedure to obtain soliton solutions of nonlinear evolution equations is discussed. Then the algebraic structure of the equations in bilinear form is explained in a simple way. A few extensions of the formalism are also presented.

1 Introduction

The bilinear formalism, which was discovered by Hirota almost a quarter century ago, has played a crucial role in the study of integrable nonlinear systems. The formalism is perfectly suitable for obtaining not only multi-soliton solutions but also several types of special solutions of many nonlinear evolution equations. Moreover, it has been used for the study of the algebraic structure of evolution equations and extension of integrable systems.

In these lectures, we attempt to present a brief survey of the bilinear formalism and discuss several recent developments. Main emphasis is on the solutions of various classes of nonlinear evolution equations. Section 2 is devoted to the explanation of the procedure for obtaining soliton solutions. A few examples , which include the Korteweg-deVries (KdV) equation, the nonlinear Schrödinger (NLS) equation and the Toda equation, are given to show how we obtain the solutions. In this method, the transformation of variable is crucial and the transformed variable becomes a key function. We shall call it the τ function. For multi-soliton solutions, it is written in the form of a polynomial in exponential functions.

The τ function can also be expressed in terms of Wronskian, Pfaffian or Casorati determinants. In Sect. 3, by using this fact we show that the τ functions of soliton equations satisfy algebraic identities in the bilinear form. This result is a reflection of the richness of the algebraic structure common to the soliton equations. Some of the indications of the richness will also be briefly mentioned in this section.

In Sect. 4, we discuss a few extensions of the bilinear formalism. The first one is q-discrete soliton equations. It is shown that the Toda equation is naturally q-discretized in its bilinear form, retaining the soliton structure of the solutions. The second is the trilinear formalism which gives a multi-dimensional extension of the soliton equations. The last is an extension to the ultra-discrete systems. We show that the idea of bilinear formalism is also applied to cellular automata which are time evolution systems all of whose variables are discrete.

J. Satsuma, Bilinear Formalism in Soliton Theory, Lect. Notes Phys. **638**, 251–268 (2004)
http://www.springerlink.com/

Finally in Sect. 5, we give concluding remarks.

An introduction to Hirota's bilinear formalism by J. Hietarinta can be found in this volume.

2   Hirota's Method

The first article on the bilinear formalism by Hirota [1] considers the KdV equation,

$$u_t + 6uu_x + u_{xxx} = 0. \tag{2.1}$$

Following his idea, let us construct soliton solutions of (2.1). First we introduce the transformation of dependent variables,

$$u = 2(\log f)_{xx}. \tag{2.2}$$

Then, assuming suitable boundary condition, we obtain the bilinear form,

$$f_{xt}f - f_x f_t + f_{xxxx}f - 4f_{xxx}f_x + 3f_{xx}^2 = 0. \tag{2.3}$$

In order to write this equation in a compact form, we define an operater,

$$D_x^n D_t^m a \cdot b = (\frac{\partial}{\partial x} - \frac{\partial}{\partial x'})^n (\frac{\partial}{\partial t} - \frac{\partial}{\partial t'})^m a(x,t)b(x',t') \Big|_{x=x', t=t'}, \tag{2.4}$$

which is now called Hirota's operater. The followings are a few simple cases:

$$D_x a \cdot b = a_x b - ab_x,$$
$$D_x^2 a \cdot b = a_{xx}b - 2a_x b_x + ab_{xx},$$
$$D_x^3 a \cdot b = a_{xxx}b - 3a_{xx}b_x + 3a_x b_{xx} - ab_{xxx}.$$

By means of this operater, (2.3) is rewritten as

$$(D_x D_t + D_x^4)f \cdot f = 0. \tag{2.5}$$

In order to obtain soliton solutions, we employ a perturbational technique. Let us expand the variable f as

$$f = 1 + \epsilon f_1 + \epsilon^2 f_2 + \epsilon^3 f_3 + \cdots, \tag{2.6}$$

where ϵ is a formal parameter (we shall take $\epsilon = 1$ later on). Substituting (2.6) into (2.5) and equating terms with the same powers in ϵ, we have

$$\mathcal{O}(\epsilon) \qquad 2(\partial_x \partial_t + \partial_x^4)f_1 = \mathcal{L}f_1 = 0, \tag{2.7}$$

$$\mathcal{O}(\epsilon^2) \qquad \mathcal{L}f_2 = -(D_x D_t + D_x^4)f_1 \cdot f_1, \tag{2.8}$$

$$\mathcal{O}(\epsilon^3) \qquad \mathcal{L}f_3 = -2(D_x D_t + D_x^4)f_1 \cdot f_2, \tag{2.9}$$

$$\vdots$$

If we start with $f_1 = e^{\eta_1}$ in (2.7), then we find that η_1 should be given by $\eta_1 = p_1(x - p_1^2 t) + \eta_1^{(0)}$, where p_1 and $\eta_1^{(0)}$ are arbitrary parameters. Furthermore, taking into account the formula,

$$D_x^n e^{\alpha x} \cdot e^{\beta x} = (\alpha - \beta)^n e^{(\alpha + \beta)x}, \tag{2.10}$$

we see that if all the higher order terms of (2.6) vanish, (2.8), (2.9), $\cdots$ are satisfied. Hence $f = 1 + e^{\eta_1}$ is an exact solution of (2.5), which gives the one soliton solution of the KdV equation (2.1),

$$u = 2(\log f)_{xx} = \frac{p_1^2}{2} \text{sech}^2 \frac{1}{2} \{p_1(x - p_1^2 t) + \eta_1^{(0)}\}. \tag{2.11}$$

Since (2.7) is linear in f_1, we may take a linear sum of exponential functions as a starting function. Let us start with $f_1 = e^{\eta_1} + e^{\eta_2}$, where $\eta_j = p_j(x - p_j^2 t)$ $+\eta_j^{(0)}, p_j, \eta_j^{(0)} \in R$. For this function, (2.8) is satisfied by

$$f_2 = e^{\eta_1 + \eta_2 + A_{12}}, \qquad e^{A_{12}} = (\frac{p_1 - p_2}{p_1 + p_2})^2,$$

and again $f_3, f_4, \cdots$ can be taken zero. Thus we have an exact solution,

$$f = 1 + e^{\eta_1} + e^{\eta_2} + e^{\eta_1 + \eta_2 + A_{12}}. \tag{2.12}$$

In the physical variable u, this corresponds to the two soliton solution which describes a collision of two solitons. The parameter A_{12} relates to the phase shift after the collision.

In principe, we can obtain a solution describing collision of any number of solitons if we carry the perturbational calculation to higher orders. It is called the N-soliton solution and will be given in Sect. 3 in an elegant form.

The bilinear formalism has been successfully applied to various classes of nonlinear evolution equations. One of the important examples in one spatial dimension is the NLS equation [2],

$$i\frac{\partial \psi}{\partial t} + \frac{\partial^2 \psi}{\partial x^2} + 2|\psi|^2 \psi = 0. \tag{2.13}$$

For the complex variable ψ, we introduce the transformation of variables, $\psi = g/f$ with real f. Then we obtain

$$(iD_t + D_x^2)g \cdot f - \frac{g}{f}(D_x^2 f \cdot f - 2gg^*) = 0,$$

where the asterisk denotes complex conjugate. Since we have introduced two variables, f and g, for one variable ψ, we may decouple this equation to yield

$$(iD_t + D_x^2)g \cdot f = 0,$$
$$D_x^2 f \cdot f = 2gg^*. \tag{2.14}$$

Again by applying a perturbational technique,

$$f = 1 + \epsilon^2 f_2 + \epsilon^4 f_4 + \cdots, \quad g = \epsilon g_1 + \epsilon^3 g_3 + \cdots,$$

we get soliton solutions.

In particular, the one soliton solution is given by

$$g = e^\eta, \quad f = 1 + \frac{1}{(P + P^*)^2} e^{\eta + \eta^*},$$

where $\eta = Px + iP^2t + \eta^{(0)}, P, \eta^{(0)} \in C$. Rewriting $P = p + ik$ for $p, k \in R$ and using $\psi = g/f$, we have

$$\psi = p \operatorname{sech} p(x - 2kt - x_0) e^{i\{kx - (k^2 - p^2)t\}}, \tag{2.15}$$

where x_0 is an appropriate phase constant.

Another example is the Toda lattice equation [3],

$$\frac{d^2}{dt^2} \log(1 + V_n) = V_{n-1} - 2V_n + V_{n+1}. \tag{2.16}$$

According to Hirota, this is the first equation to which he applied the bilinear formalism in order to obtain soliton solutions, although the article describing it was published two years later than the one for the KdV equation.

Let us substitute

$$V_n = \frac{d^2}{dt^2} \log \tau_n \tag{2.17}$$

into (2.16). Then assuming a suitable boundary condition, we have

$$\frac{d^2 \tau_n}{dt^2} \tau_n - \left(\frac{d\tau_n}{dt}\right)^2 = \tau_{n+1}\tau_{n-1} - \tau_n^2. \tag{2.18}$$

It is to be observed that (2.18) may be rewritten by

$$(D_t^2 - 4\sinh^2 \frac{D_n}{2})\tau_n \cdot \tau_n = 0, \tag{2.19}$$

where we have introduced the difference operaters,

$$e^{D_n} f_n \cdot f_n = e^{\partial_n - \partial_{n'}} f_n f_{n'}\Big|_{n=n'} = f_{n+1}f_{n-1}, \tag{2.20}$$

with

$$e^{\epsilon \frac{\partial}{\partial x}} f(x) = f(x + \epsilon) \quad \text{or} \quad e^{\partial_n} f_n = f_{n+1}. \tag{2.21}$$

The lattice one-soliton solution is given by

$$\tau_n = 1 + e^{2\eta}, \quad \eta = P_n - \Omega t + \eta^{(0)}, \quad \Omega^2 = \sinh P^2, \tag{2.22}$$

or

$$V_n = \Omega^2 \text{sech}^2 \eta. \tag{2.23}$$

The above three examples are all in two dimensions. The equations extended to 3 dimensions, like the Kadomtsev-Petviashvili (KP), the Davey-Stuartson (DS) and the 2-dimensional Toda (2D Toda) equations, have also been successfully treated in the bilinear formalism [4–6]. The algebraic structure of soliton solutions becomes very clear in this formalism, as we shall see in the following section.

3 Algebraic Inentities

The 2D Toda equation,

$$\frac{\partial^2}{\partial x \partial y} \log(1 + V_n) = V_{n-1} - 2V_n + V_{n+1}, \tag{3.1}$$

was first presented by Darboux in 19th century. This is now well known as a generic semi-discrete soliton equation. Equation (3.1) is reduced to

$$D_x D_y \tau_n \cdot \tau_n = 2(\tau_{n+1}\tau_{n-1} - \tau_n^2), \tag{3.2}$$

by substituting

$$V_n = \frac{\partial^2}{\partial x \partial y} \log \tau_n, \tag{3.3}$$

and assuming an appropriate boundary condition.

We now show that (3.2) is nothing but an algebraic identity for determinants [7].

Proposition 3.1. Equation (3.2) is satisfied by the following Casorati determinant:

$$\tau_n(x, y) = \begin{vmatrix} f_n^{(1)} & f_{n+1}^{(1)} & \cdots & f_{n+N-1}^{(1)} \\ f_n^{(2)} & f_{n+1}^{(2)} & \cdots & f_{n+N-1}^{(2)} \\ \vdots & \vdots & & \vdots \\ f_n^{(N)} & f_{n+1}^{(N)} & \cdots & f_{n+N-1}^{(N)} \end{vmatrix}, \tag{3.4}$$

where

$$\frac{\partial}{\partial x} f_m^{(j)} = f_{m+1}^{(j)}, \qquad \frac{\partial}{\partial y} f_m^{(j)} = -f_{m-1}^{(j)}, \qquad j = 1, 2, 3, \cdots, N. \tag{3.5}$$

Let us give a rough proof. For $N = 1$, substituting $\tau_n = f_n^{(1)}$ into (3.2), we obtain

$$\begin{aligned} D_x D_y \tau_n \cdot \tau_n &= 2(\tau_{n,xy}\tau_n - \tau_{n,x}\tau_{n,y}) \\ &= 2(\frac{\partial^2 f_n^{(1)}}{\partial x \partial y} f_n^{(1)} - \frac{\partial f_n^{(1)}}{\partial x} \cdot \frac{\partial f_n^{(1)}}{\partial y}) \\ &= -2(f_n^{(1)} f_n^{(1)} - f_{n+1}^{(1)} f_{n-1}^{(1)}) \\ &= \text{RHS.} \end{aligned}$$

For $N = 2$, we first observe that the identity,

$$\begin{vmatrix} a_0 & a_1 & a_2 & a_3 \\ b_0 & b_1 & b_2 & b_3 \\ 0 & a_1 & a_2 & a_3 \\ 0 & b_1 & b_2 & b_3 \end{vmatrix} = 0,$$

holds for any entries a_j, b_j. Applying a Laplace expansion in 2×2 minors to the left-hand side, we get

$$\begin{vmatrix} a_0 & a_1 \\ b_0 & b_1 \end{vmatrix} \begin{vmatrix} a_2 & a_3 \\ b_2 & b_3 \end{vmatrix} - \begin{vmatrix} a_0 & a_2 \\ b_0 & b_2 \end{vmatrix} \begin{vmatrix} a_1 & a_3 \\ b_1 & b_3 \end{vmatrix} + \begin{vmatrix} a_0 & a_3 \\ b_0 & b_3 \end{vmatrix} \begin{vmatrix} a_1 & a_2 \\ b_1 & b_2 \end{vmatrix} = 0, \tag{3.6}$$

which is called the Plücker relation. If we simply write (3.6) by

$$(0, 1)(2, 3) - (0, 2)(1, 3) + (0, 3)(1, 2) = 0, \tag{3.7}$$

and have a correspondence,

$$\tau_n = \begin{vmatrix} f_n^{(1)} & f_{n+1}^{(1)} \\ f_n^{(2)} & f_{n+1}^{(2)} \end{vmatrix} \quad \Longleftrightarrow \quad (0, 1),$$

then, by observing the correspondences, $\tau_{n,x} \Longleftrightarrow (0, 2)$, $\tau_{n,y} \Longleftrightarrow -(-1, 1)$, $\tau_{n,xy} \Longleftrightarrow -(0, 1) - (-1, 2)$, $\tau_{n+1} \Longleftrightarrow (1, 2)$, $\tau_{n-x} \Longleftrightarrow (-1, 0)$, we easily find that (3.2) is equivalent to the identity (3.7). For $N \geq 3$, we may employ the same idea to show that (3.4) satisfies (3.2) or the equivalent indentity (3.7). For example, in the case of $N = 3$, we can start with the identity,

$$\begin{vmatrix} f & a_0 & a_1 & 0 & a_2 & a_3 \\ g & b_0 & b_1 & 0 & b_2 & b_3 \\ h & c_0 & c_1 & 0 & c_2 & c_3 \\ 0 & 0 & a_1 & f & a_2 & a_3 \\ 0 & 0 & b_1 & g & b_2 & b_3 \\ 0 & 0 & c_1 & h & c_2 & c_3 \end{vmatrix} = 0. \tag{3.8}$$

It is noted however that small modification is necessary to reduce (3.2) to the identity (3.7).

The soliton solutions of the 2D Toda equation are obtained from (3.4) by making a particular choice of the functions $f_n^{(j)}$. The size N of the determinant in (3.4) corresponds to the number of solitons. The one soliton solution is, for example, given by

$$\tau_n = f_n^{(1)} = p^n e^{px - \frac{1}{p}y} + q^n e^{qx - \frac{1}{q}y}, \tag{3.9}$$

where p and q are arbitrary parameters.

The Plücker relation (3.6) is a key identity for soliton equations. Actually Sato [8, 9] noticed that the bilinear form of the KP equation,

$$(4u_t - 12uu_x - u_{xxx})_x - 3u_{yy} = 0 \tag{3.10}$$

is nothing but the Plücker relation and he discovered that the totality of solutions of the KP equation as well as of its generalization constitutes an infinite-dimensional Grasmann manifold. The class of equations is now called the KP hierarchy.

Let us briefly sketch a part of his result. Through the transformation of variables, $u = (\log \tau)_{xx}$, we have the bilinear form of the KP equation,

$$(4D_x D_t - D_x^4 - 3D_y^2)\tau \cdot \tau = 0. \tag{3.11}$$

By applying the same tecknique as for the 2D Toda equation, it is shown that (3.11) is satisfied by the Wronski determinant,

$$\tau(x, y, t) = \begin{vmatrix} f^{(1)} & \partial_x f^{(1)} & \cdots & \partial_x^{N-1} f^{(1)} \\ f^{(2)} & \partial_x f^{(2)} & \cdots & \partial_x^{N-1} f^{(2)} \\ \vdots & \vdots & & \vdots \\ f^{(N)} & \partial_x f^{(N)} & \cdots & \partial_x^{N-1} f^{(N)} \end{vmatrix}, \tag{3.12}$$

where

$$\frac{\partial}{\partial y} f^{(j)} = \frac{\partial^2}{\partial x^2} f^{(j)}, \quad \frac{\partial}{\partial t} f^{(j)} = \frac{\partial^3}{\partial x^3} f^{(j)}. \tag{3.13}$$

For our purpose, it is convenient to introduce the notation [10], $\tau = (0, 1, 2, \cdots, N-1)$. Then observing that $\tau_x = (0, 1, 2, \cdots, N-2, N), \tau_y = -(0, 1, 2, \cdots, N-3, N-1, N) + (0, 1, 2, \cdots, N-3, N-2, N+1)$ and so on, we find that (3.11) is essentially the same as (3.7), which means (3.12) automatically satisfies the KP equation.

Shortly after Sato's discovery, Date, Jimbo, Kashiwara and Miwa [11] extended his idea and developed the theory of transformation groups for soliton equations. Moreover, the 2D Toda equation has been shown to belong to an extension of the KP hierarchy [12, 13]. All these results make it possible to understand the soliton theory from a unified point of view. For example, the relationship among the inverse sccattering transform, Hirota's method and the Bäcklund transformation is clearly explained by the infinite dimensional Lie algebra and its representation on a function space.

As we see from (3.4) and (3.12), the semi-discrete 2D Toda and the continuous KP equations possess solutions with a common structure. The Casorati determinant is the discrete version of Wronski determinant. Moreover, the Casorati determinant (3.4) itself is considered to be a Wronski determinant if we employ the linear relation (3.5) for the entries. Actually both equations are related because the KP is obtained by taking a proper continuous limit of the 2D Toda equaton.

Then a natural question is whether there exists a fully discrete equation which has the same type of solutions. One answer was given by Hirota[14]. The equation, which Hirota called the discrete analogue of generalized Toda equation, is written in bilinear form by

$$\tau_n(l+1, m+1)\tau_n(l, m) - \tau_n(l+1, m)\tau_n(l, m+1)$$

$$= ab\{\tau_{n+1}(l, m+1)\tau_{n-1}(l+1, m) - \tau_n(l+1, m+1)\tau_n(l, m)\}, \qquad (3.14)$$

where a and b are parameters related to the difference interval (see below). Since the algebraic structure of this equation was studied by Miwa [15] shortly after Hirota's discovery, we call (3.14) the Hirota-Miwa equation.

As expected, the solution of (3.14) is again given by the Casorati determinant. Its explicit form is exactly the same as (3.4). Only difference lies in the linear equations which should be satisfied by the entries. In this case they are given by

$$\Delta_l f_n^{(j)}(l, m) \equiv \frac{1}{a}\{f_n^{(j)}(l+1, m) - f_n^{(j)}(l, m)\} = f_{n+1}^{(j)}(l, m), \qquad (3.15)$$

$$\Delta_m f_n^{(j)}(l, m) \equiv \frac{1}{b}\{f_n^{(j)}(l, m+1) - f_n^{(j)}(l, m)\} = -f_{n-1}^{(j)}(l, m). \qquad (3.16)$$

If we read l, m for x, y, respectively, and take a continuous limit, then we obtain the 2D Toda equation (3.2).

The Hirota-Miwa equation may be considered as one of the master equations of soliton theory, since we recover many of the soliton equations by taking proper continuous limits [14].

Finally in this section, we comment on another class of solutions of the 2D Toda equation. The Casorati determinant solution (3.4) is obtained by assuming suitable boundary conditions in an infinite lattice. We may instead consider a finite lattice. If we impose the boundary condition $V_0 = V_M = 0$ for some positive integer M, the system is called the 2D Toda molecule equation. In this context, we call the infinite lattice system the 2D Toda lattice equation.

The Toda molecule equation is reduced to its bilinear form,

$$D_x D_y \tau_n \cdot \tau_n = 2\tau_{n+1}\tau_{n-1}, \qquad (3.17)$$

with $\tau_{-1} = \tau_{M+1} = 0$, by introducing the transformation of variables (3.3). It is known [7] that (3.17) admits the solution,

$$\tau_n(x,y) = \begin{vmatrix} f(x,y) & \partial_x f & \cdots & \partial_x^{n-1} f \\ \partial_y f & \partial_x \partial_y f & \cdots & \partial_x^{n-1} \partial_y f \\ \vdots & \vdots & & \vdots \\ \partial_y^{n-1} f & \partial_x \partial_y^{n-1} f & \cdots & \partial_x^{n-1} \partial_y^{n-1} f \end{vmatrix}, \tag{3.18}$$

for $n \geq 1$ and $\tau_0 = 1$, where the function $f(x,y)$ is obtained by

$$f(x,y) = \sum_{k=1}^{M} f_k(x) g_k(y), \tag{3.19}$$

for arbitrary f_k and g_k. Since this solution is a Wronskian with respect to x in the horizontal direction and with respect to y in the vertical direction, we call the determinant a two-directional Wronskian. The proof is given by using the Laplace expansion or the Jacobi identity for determinants. It should be remarked that the solution (3.18) is meaningful only for discrete system since the discrete variable n determines the size of determinant.

4 Extensions

The algebraic structure of the determinant solutions discussed in the preceding section is crucial when one considers extensions of nonlinear integrable systems. In this section we present a few examples which have been obtained based on this algebraic structure.

4.1 q-Discrete Toda Equation

As we have seen in Sect. 3, the solutions of the (continuous) KP, the (semi-discrete) 2D Toda and the (fully discrete) Hirota-Miwa has the same structure. The only difference lies in the linear equations satisfied by the entries of the determinant. This fact suggests that if we can generalize the linear equations we may have another integrable system. In this case, integrable means that the equation admits a similar type of determinant solutions. The q-difference version of the 2D Toda equation is just such a case [16].

Let us introduce an operator,

$$\delta_{q^\alpha, x} f(x,y) = \frac{f(x,y) - f(q^\alpha x, y)}{(1-q)x}, \tag{4.1}$$

which reduces to $\alpha \partial / \partial x$ at the limit of $q \to 1$. Note that this operator reduces to the original q-difference operator if α is taken to be 1.

The q-difference version of the 2D Toda equation is given by

$$\{\delta_{q^2,x} \delta_{q^2,y} \tau_n(x,y)\} \tau_n(x,y) - \{\delta_{q^2,x} \tau_n(x,y)\} \{\delta_{q^2,y} \tau_n(x,y)\}$$

$$= \tau_{n+1}(x, q^2 y) \tau_{n-1}(q^2 x, y) - \tau_n(q^2 x, q^2 y) \tau_n(x,y). \tag{4.2}$$

Again by using a Laplace expansion, we can show that (4.2) admits the solution of Casorati determinant type, (3.4). The linear equations (3.5) now become

$$\delta_{q^2,x} f_n^{(j)}(x,y) = f_{n+1}^{(j)}(x,y) \qquad (4.3)$$

and

$$\delta_{q^2,y} f_n^{(j)}(x,y) = -f_{n-1}^{(j)}(x,y). \qquad (4.4)$$

It is noted that the q-discrete version of the 2D Toda equation is considered to be an extension of Hirota-Miwa equation. The former equation is obtained by reading $l+1 \to q^2 x$, $m+1 \to q^2 y$, $a \to (q-1)x$, $b \to (q-1)y$ in the latter.

If we impose a restriction on the variables in (4.2), we are able to obtain a reduced system. Let us introduce a variable r where $xy = r^2$. Then, for example, we have

$$\{\delta_{q^2,x} \tau_n(x,y)\}\{\delta_{q^2,y} \tau_n(x,y)\} = \{\delta_{q,r} \tau_n(r)\}^2.$$

By using this kind of reduction, we obtain from (4.2),

$$(\frac{1}{r}\delta_{q,r} + q^2\delta_{q,r}^2)\tau_n(r) \cdot \tau_n(r) - \{\delta_{q,r}\tau_n(r)\}^2$$
$$= \tau_{n+1}(qr)\tau_{n-1}(qr) - \tau_n(q^2 r)\tau_n(r), \qquad (4.5)$$

which is considered to be the q-difference version of the cylindrical Toda equation. We find that the solution for (4.5) is given by the Casorati determinant whose entries are expressed by the q-Bessel function.

Finally in this subsection, we remark that a q-discrete version of the Toda molecule equation and its solution can also be constructed by extending (3.17) and (3.18) [17].

4.2   Trilinear Formalism

In order to prove Proposition 3.1, we have used identities for determinants. We have seen that the Plücker relation (3.6) is obtained by applying a Laplace expansion to the determinants and that the 2D Toda equation is equivalent to the relation. One extension of soliton equations is possible by following this simple idea. It is the trilinear formalism [18-20]. By this formalism, we can costruct four dimensional nonlinear equations which admit solutions expressed by Wronski or Casorati determinants. We here show the procedure for the semi-discrete case [19].

First we consider the following identities for $(3N+3) \times (3N+3)$ determinant:

$$
\begin{vmatrix}
A\,F_{n-1} & 0 & 0 & B \\
0 & A\,F_{n+N-1} & 0 & B \\
0 & 0 & A\,F_{n+N} & B
\end{vmatrix} = 0,
\tag{4.6}
$$

where F_n is a vector given by

$$
F_n = \begin{pmatrix}
f_{m,n} \\
f_{m+1,n} \\
\vdots \\
f_{m+N,n}
\end{pmatrix}.
\tag{4.7}
$$

and A, B are matrices given by

$$
A = (F_n \; F_{n+1} \; \cdots \; F_{n+N-2}),
\tag{4.8}
$$

and

$$
B = \begin{pmatrix}
1 & 0 & 0 \\
0 & 0 & 0 \\
\vdots & \vdots & \vdots \\
0 & 0 & 0 \\
0 & 1 & 0 \\
0 & 0 & 1
\end{pmatrix},
\tag{4.9}
$$

respectively. Applying a Laplace expansion in $(N+1) \times (N+1)$ minors to the left-hand side of (4.6), we have a trilinear form,

$$
\begin{vmatrix}
\partial_y \tau_{m,n-1} & \tau_{m,n-1} & \tau_{m+1,n-1} \\
\partial_y \tau_{m,n} & \tau_{m,n} & \tau_{m+1,n} \\
\partial_y \partial_x \tau_{m,n} & \partial_x \tau_{m,n} & \partial_x \tau_{m+1,n}
\end{vmatrix} = 0,
\tag{4.10}
$$

where

$$
\tau_{m,n} = \begin{vmatrix}
f_{m,n} & f_{m,n+1} & \cdots & f_{m,n+N-1} \\
f_{m+1,n} & f_{m+1,n+1} & \cdots & f_{m+1,n+N-1} \\
\vdots & \vdots & \ddots & \vdots \\
f_{m+N-1,n} & f_{m+N-1,n+1} & \cdots & f_{m+N-1,n+N-1}
\end{vmatrix},
\tag{4.11}
$$

and f satisfies

$$
\frac{\partial}{\partial x} f_{m,n} = f_{m,n+1}, \qquad \frac{\partial}{\partial y} f_{m,n} = f_{m+1,n}.
\tag{4.12}
$$

This result shows that the τ function (4.11) in the form of a two-directional Casorati determinant is a solution of the four (two discrete + two continuous) dimensional (4.10).

If we introduce the dependent variables ψ, ϕ by

$$\psi_{m,n} = \log\frac{\tau_{m,n}}{\tau_{m,n-1}}, \quad \phi_{m,n} = \log\frac{\tau_{m,n}}{\tau_{m+1,n}}, \tag{4.13}$$

then (4.10) is reduced to a coupled system,

$$\partial_x\partial_y\phi_{m,n} = \frac{\partial_x\phi_{m,n}\partial_y\psi_{m,n}}{e^{\phi_{m,n}-\phi_{m,n-1}}-1} - \frac{\partial_x\phi_{m+1,n}\partial_y\psi_{m+1,n}}{e^{\phi_{m+1,n}-\phi_{m+1,n-1}}-1}, \tag{4.14a}$$

$$\partial_x\partial_y\psi_{m,n} = \frac{\partial_x\phi_{m,n}\partial_y\psi_{m,n}}{e^{\phi_{m,n}-\phi_{m+1,n}}-1} - \frac{\partial_x\phi_{m,n-1}\partial_y\psi_{m,n-1}}{e^{\phi_{m,n-1}-\phi_{m+1,n-1}}-1}, \tag{4.14b}$$

with a constraint

$$\psi_{m+1,n} - \psi_{m,n} = \phi_{m,n-1} - \phi_{m,n}. \tag{4.15}$$

Furthermore, if the reduction, $\phi(x,y)_{m,n} = q(x+y)_{m+n}$, $\psi(x,y)_{m,n} = q(x+y)_{m+n-1}$, is imposed, then (4.12) reduce to

$$\partial_x^2 q_n = -\partial_x q_n \left(\frac{\partial_x q_{n-1}}{e^{q_n - q_{n-1}} - 1} - \frac{\partial_x q_{n+1}}{e^{q_{n+1} - q_n} - 1} \right), \tag{4.16}$$

which is nothing but the relativistic Toda equation proposed by Ruijsenaars. Therefore, (4.14) is considered to be a 2+2 dimensional extension of the relativistic Toda equation [21, 22].

In the continuous case, we have a hierarchy of trilinear equations [18],

$$\begin{vmatrix} p_i(\tilde\partial)p_l(-\tilde\partial')\tau & p_i(\tilde\partial)p_m(-\tilde\partial')\tau & p_i(\tilde\partial)p_n(-\tilde\partial')\tau \\ p_j(\tilde\partial)p_l(-\tilde\partial')\tau & p_j(\tilde\partial)p_m(-\tilde\partial')\tau & p_j(\tilde\partial)p_n(-\tilde\partial')\tau \\ p_k(\tilde\partial)p_l(-\tilde\partial')\tau & p_k(\tilde\partial)p_m(-\tilde\partial')\tau & p_k(\tilde\partial)p_n(-\tilde\partial')\tau \end{vmatrix} = 0, \tag{4.17}$$

for arbitrary nonnegative integers i, j, k, l, m, n, where τ is a function of $x_1, x_2, x_3, \cdots, y_1, y_2, y_3, \cdots, \tilde\partial, \tilde\partial'$ are defined by

$$\tilde\partial = (\frac{\partial}{\partial x_1}, \frac{1}{2}\frac{\partial}{\partial x_2}, \frac{1}{3}\frac{\partial}{\partial x_3}, \cdots), \tag{4.18a}$$

$$\tilde\partial' = (\frac{\partial}{\partial y_1}, \frac{1}{2}\frac{\partial}{\partial y_2}, \frac{1}{3}\frac{\partial}{\partial y_3}, \cdots), \tag{4.18b}$$

respectively and $p_j, j = 1, 2, \cdots$, are polynomials defined by

$$\exp\left(\sum_{n=1}^{\infty} x_n \lambda^n\right) = \sum_{j=0}^{\infty} p_j(x)\lambda^j. \tag{4.19}$$

The simplest case of (4.17) ($i = l = 0, j = m = 1, k = n = 2$) gives a 2+2 dimensional extension of the Brouer-Kaup system,

$$h_t = (h_x + 2hu)_x, \tag{4.20a}$$

$$u_t = (u^2 + 2h - u_x)_x, \tag{4.20b}$$

and the solution is again given by a two-directional Wronskian [18, 23].

4.3   Ultra-discrete Systems

As was mentioned in Sect. 3, the Hirota-Miwa equation is one of the master equations in the sense that it reduces to the KP equation via the 2D Toda equation in the continuous limit. Very recently we found a very interesting fact: there exists another limit, from which we can obtain cellular automata systems [24–26]. Since we obtain discrete systems in which all the variables, including the dependent ones, are discrete, we call it an ultra-discrete limit (the name is due to B. Grammaticos). In this subsection, we explain how to get a cellular automaton and its solutions starting from (3.14) [26].

The Hirota-Miwa equation may be written in a symmetric form [14],

$$\{Z_1 \exp(D_1) + Z_2 \exp(D_2) + Z_3 \exp(D_3)\} f \cdot f = 0, \tag{4.21}$$

where $Z_i(i = 1, 2, 3)$ are arbitrary parameters and $D_i(i = 1, 2, 3)$ stand for Hirota's operators with respect to variables of the unknown function f. We here consider a particular case of (4.21),

$$\{\exp(D_t) - \delta^2 \exp(D_x) - (1 - \delta^2) \exp(D_y)\} f \cdot f = 0, \tag{4.22}$$

or equivalently,

$$f(t - 1, x, y)f(t + 1, x, y) - \delta^2 f(t, x - 1, y)f(t, x + 1, y) -$$

$$(1 - \delta^2)f(t, x, y + 1)f(t, x, y - 1) = 0. \tag{4.23}$$

If we introduce a variable S by

$$f(t, x, y) = \exp[S(t, x, y)], \tag{4.24}$$

then (4.23) is reduced to

$$\exp[(\Delta_t^2 - \Delta_y^2)S(t, x, y)] = (1 - \delta^2) \left(1 + \frac{\delta^2}{1 - \delta^2} \exp[(\Delta_x^2 - \Delta_y^2)S(t, x, y)] \right), \tag{4.25}$$

where Δ_t^2, Δ_x^2 and Δ_y^2 represent central difference operators defined, for example, by

$$\Delta_t^2 S(t, x, y) = S(t + 1, x, y) - 2S(t, x, y) + S(t - 1, x, y). \tag{4.26}$$

Taking a logarithm of (4.25) and applying the operator $(\Delta_x^2 - \Delta_y^2)$, we have

$$(\Delta_t^2 - \Delta_y^2)u(t, x, y) = (\Delta_x^2 - \Delta_y^2) \log \left(1 + \frac{\delta^2}{1 - \delta^2} \exp[u(t, x, y)] \right), \tag{4.27}$$

where

$$u(t, x, y) = (\Delta_x^2 - \Delta_y^2)S(t, x, y). \tag{4.28}$$

Ultra-discretization is defined by the following formula:

$$\lim_{\varepsilon \to +0} \varepsilon \log(1 + e^{X/\varepsilon}) = F(X) = \max[0, X]. \tag{4.29}$$

It is noted that the function $F(x)$ maps positive integers to themselves. Let us take an ultra-discrete limit of (4.27). Putting

$$u(t,x,y) = \frac{v_\varepsilon(t,x,y)}{\varepsilon}, \quad \frac{\delta^2}{1-\delta^2} = e^{-\theta_0/\varepsilon}, \tag{4.30}$$

and taking the limit for small ε, we obtain the following equation:

$$(\Delta_t^2 - \Delta_y^2)v(t,x,y) = (\Delta_x^2 - \Delta_y^2)F(v(t,x,y) - \theta_0), \tag{4.31}$$

where we have rewritten $\lim_{\varepsilon \to +0} v_\varepsilon(t,x,y)$ as $v(t,x,y)$.

Equation (4.31) is considered to be an (extended) filter cellular automaton. This system is in 2 (spatial) and 1 (time) dimensions and may take only integer values. Since (4.31) is an ultra-discrete limit of the Hirota-Miwa equation, we expect that it admits soliton solutions. We here show that they are obtained also by taking an ultra-discrete limit of those for (4.23).

The one soliton solution of (4.23) is given by

$$f(t,x,y) = 1 + e^\eta, \quad \eta = px + qy + \omega t, \tag{4.32}$$

where the set of parameters (p,q,ω) satisfies

$$(e^{-\omega} + e^\omega) - \delta^2(e^{-p} + e^p) - (1-\delta^2)(e^{-q} + e^{-q}) = 0. \tag{4.33}$$

Then by means of (4.24) and (4.28), we have

$$u(t,x,y) = \log(1 + e^{\eta+p}) + \log(1 + e^{\eta-p}) - \log(1 + e^{\eta+q}) - \log(1 + e^{\eta-q}). \tag{4.34}$$

Introducing new parameters and variables by

$$P = \varepsilon p, \quad Q = \varepsilon q, \quad \Omega = \varepsilon\omega,$$

$$K = Px + Qy + \Omega t, \quad v_\varepsilon(t,x,y) = \varepsilon u(t,x,y),$$

and taking the limit $\varepsilon \to +0$, we obtain

$$v(t,x,y) = F(K+P) + F(K-P) - F(K+Q) - F(K-Q). \tag{4.35}$$

The dispersion relation (4.33) reduces, through the same limiting procedure, to

$$|\Omega| = \max[|P|,|Q| + \theta_0] - \max[0,\theta_0]. \tag{4.36}$$

This solution describes a solitary wave propagating in the xy plane at a constant speed without changing its shape.

The two-soliton solution describing a nonlinear interaction of two solitary wave is obtained starting from that of (4.23), which is expressed by

$$f(t, x, y) = 1 + e^{\eta_1} + e^{\eta_2} + e^{\eta_1 + \eta_2 + \theta_{12}}, \quad \eta_i = p_i x + q_i y + \omega_i t, \qquad (4.37)$$

$$(e^{-\omega_i} + e^{\omega_i}) - \delta^2(e^{-p_i} + e^{p_i}) - (1 - \delta^2)(e^{-q_i} + e^{-q_i}) = 0, \quad (i = 1, 2), \quad (4.39)$$

$$e^{\theta_{12}} = -\frac{(e^{-\omega_1 + \omega_2} + e^{\omega_1 - \omega_2}) - \delta^2(e^{-p_1 + p_2} + e^{p_1 - p_2}) - (1 - \delta^2)(e^{-q_1 + q_2} + e^{q_1 - q_2})}{(e^{\omega_1 + \omega_2} + e^{-\omega_1 - \omega_2}) - \delta^2(e^{p_1 + p_2} + e^{-p_1 - p_2}) - (1 - \delta^2)(e^{q_1 + q_2} + e^{-q_1 - q_2})}.$$

$$(4.37)$$

The variable θ_{12} stands for a phase shift. Again introducing new parameters and variables by

$$P_i = \varepsilon p_i, \ Q_i = \varepsilon q_i, \ \Omega_i = \varepsilon \omega_i,$$

$$K_i = P_i x + Q_i y + \Omega_i t, \ (i = 1, 2), \ v_\varepsilon(t, x, y) = \varepsilon u(t, x, y), \ \Theta_{12} = \varepsilon \theta_{12},$$

and taking the limit of $\varepsilon \to +0$, we have

$$\begin{aligned}
v(t, x, y) = \ &\max[0, K_1 + P_1, K_2 + P_2, K_1 + K_2 + P_1 + P_2 + \Theta_{12}] \\
&+ \max[0, K_1 - P_1, K_2 - P_2, K_1 + K_2 - P_1 - P_2 + \Theta_{12}] \\
&- \max[0, K_1 + Q_1, K_2 + Q_2, K_1 + K_2 + Q_1 + Q_2 + \Theta_{12}] \\
&- \max[0, K_1 - Q_1, K_2 - Q_2, K_1 + K_2 - Q_1 - Q_2 + \Theta_{12}],
\end{aligned}$$
$$(4.40)$$

where

$$|\Omega_i| = \max[|P_i|, |Q_i| + \theta_0] - \max[0, \theta_0] \quad (i = 1, 2), \qquad (4.41)$$

and

$$\max\left[\Theta_{12} + \max[0, \theta_0] + |\Omega_1 + \Omega_2|, \max[0, \theta_0] + |\Omega_1 - \Omega_2|\right]$$
$$= \max\left[\Theta_{12} + |P_1 + P_2|, \Theta_{12} + \theta_0 + |Q_1 + Q_2|, |P_1 - P_2|, \theta_0 + |Q_1 - Q_2|\right].$$
$$(4.42)$$

The following figure demonstrates a snapshot of the two-soliton solution (4.40) at $t = -4$ for $P_1 = 6, Q_1 = 1, P_2 = 6, Q_2 = 5$.

12	110000000000001500000000000000
11	011000000000000500000000000000
10	001100000000000510000000000000
9	000110000000000420000000000000
8	000010000000000330000000000000
7	000011000000000240000000000000
6	000001100000000150000000000000
5	000000110000000050000000000000
4	000000011000000051000000000000
3	000000001100000042000000000000
2	000000000100000033000000000000
1	000000000110000024000000000000

```
  0    0000000000011000015000000000000
 -1    0000000000011000050000000000000
 -2    0000000000001100051000000000000
 -3    0000000000000011004200000000000
 -4    0000000000000010033000000000000
 -5    0000000000000001102400000000000
 -6    0000000000000000111500000000000
 -7    0000000000000000011500000000000
 -8    0000000000000000001610000000000
 -9    0000000000000000000520000000000
-10    0000000000000000000431000000000
-11    0000000000000000000331000000000
-12    0000000000000000000241100000000
-13    0000000000000000000150110000000
-14    0000000000000000000050011000000
y/x    //////////////0123456789******
```

At the bottom of this figure, negative values of x coordinate are expressed as
"/" and values greater than 10 are expressed as "*" for convenience sake.

It is observed that a N-soliton solution is obtained by the same limiting
procedure. It is also remarked that we can costruct other types of integrable but
also nonintegrable cellular automata by using the ultra-discrete limit on several
fully discrete systems.

5   Concluding Remarks

In these lectures, we have given a brief survey of Hirota's bilinear formalism and
presented a few extensions. Because it yields explicit solutions and makes the
algebraic structure of equations clear, it is cleary powerful tool that lends itself
to many other applications. Here we mention only one example, the Painlevé
equations.

It has been shown by Okamoto [27] that the explicit solutions of Painlevé
equations are expressed in terms of the τ functions. For example, the Painlevé
II equation,

$$w_x x - 2w^3 + 2xw + \alpha = 0, \tag{5.1}$$

admits a solution for $\alpha = -(2N + 1)$,

$$w = \frac{d}{dx}\left(\log\frac{\tau_{N+1}}{\tau_N}\right), \tag{5.2}$$

where τ_N is given by an $N \times N$ two-directional Wronski determinant of the Airy
function. Recent finding of discrete analogue of the Painlevé equations [28] gives
rise to the question of whether there exist corresponding solutions for the discrete

case. An answer has been given by the bilinear formalism. For example, it has been shown through this formalism [29] that the discrete Painlevé II equation,

$$w_{n+1} + w_{n-1} = \frac{(\alpha n + \beta)w_n + \gamma}{1 - w_n^2}, \tag{5.3}$$

admits particular solutions written in terms of Casorati determinants whose entries are the discrete analogue of the Airy functions.

This example as well as the results in the preceding sections indicate that the bilinear formalism can be one of the most powerful tools to treat discrete problems, which the author believes to be an important subject for research in the 21st century.

References

1. R. Hirota, Phys. Rev. Lett. **27** (1971) 1192
2. R. Hirota, J. Math. Phys. **14** (1973) 805
3. R. Hirota, J. Phys. Soc. Jpn. **35** (1973) 289
4. J. Satsuma, J. Phys. Soc. Jpn. **40** (1976) 286
5. J. Hietarinta and R. Hirota, Phys. Lett. A **145** (1990) 237
6. R. Hirota, M. Ito and F. Kako, Prog. Theor. Phys. Suppl. **94** (1988) 42
7. R. Hirota, Y. Ohta and J. Satsuma, J. Phys. Soc. Jpn. **57** (1988) 1901; Prog. Theor. Phys. Suppl. **94** (1988) 59
8. M. Sato, RIMS Kokyuroku **439** (1981), 30
9. Y. Ohta, J. Satsuma, D. Takahashi, and T. Tokihiro, Prog. Theor. Phys. Suppl. **94** (1988) 210
10. N.C. Freeman, IMA J. Appl. Math. **32** (1984) 125
11. E. Date, M. Jimbo, M. Kashiwara, and T. Miwa, in *Non-linear Integrable Systems – Classical Theory and Quantum Theory*, ed. by M. Jimbo and T. Miwa (World Scientific, Singapore 1983) p. 39
12. K. Ueno and K. Takasaki, RIMS Kokyuroku **472** (1982) 62
13. M. Jimbo and T. Miwa, Publ. RIMS, Kyoto Univ. **19** (1983) 943
14. R. Hirota, J. Phys. Soc. Jpn. **50** (1981) 3785
15. T. Miwa, Proc. Jpn. Acad. A **58** (1982) 9
16. K. Kajiwara and J. Satsuma, J. Phys. Soc. Jpn. **60** (1991) 3986
17. K. Kajiwara, Y. Ohta, and J. Satsuma, Phys. Lett. A **180** (1993) 249
18. J. Matsukidaira, J. Satsuma, and W. Strampp, Phys. Lett. A. **147** (1990) 467
19. J. Matsukidaira and J.Satsuma, J. Phys. Soc. Jpn. **59** (1990) 3413
20. J. Matsukidaira and J. Satsuma, Phys. Lett. A. **158** (1991) 366
21. J. Hietarinta and J. Satsuma, Phys. Lett. A **161** (1991) 267
22. Y. Ohta, K. Kajiwara, J. Matsukidaira, and J. Satsuma, J. Math. Phys. **34** (1993) 5190
23. J. Satsuma, K. Kajiwara, J. Matsukidaira, and J. Hietarinta, J. Phys. Soc. Jpn. **61** (1992) 3096
24. T. Tokihiro, D. Takahashi, J. Matsukidaira, and J. Satsuma, Phys. Rev. Lett. **76** (1996) 3247
25. J. Matsukidaira, J. Satsuma, D. Takahashi, T. Tokihiro, and J. Satsuma, Phys. Lett. A **225** (1997) 287

26. S. Moriwaki, A. Nagai, J. Satsuma, T. Tokihiro, M. Torii, D. Takahashi, and J. Matsukidaira, London Math. Soc. Lecture Notes Series 255, ed. by P.A. Clarkson and F.W. Nijhoff, (Cambridge Univ. Press, Cambridge 1999) p. 334

27. K. Okamoto, Math. Ann. **275** (1986) 221; Japan J. Math. **13** (1987) 47; Ann. Mat. Pura. Appl. **146** (1987) 337; Funkcial. Ekvac. **30** (1987) 305

28. B. Grammaticos, A. Ramani, and J. Hietarinta, Phys. Rev. Lett. **67** (1991) 1825

29. K. Kajiwara, Y. Ohta, J. Satsuma, B. Grammaticos, and A. Ramani, J. Phys. A: Math. Gen. **27** (1994) 915. See also the article by B. Grammaticos and A. Ramani in this volume.

Quantum and Classical Integrable Systems

M.A. Semenov-Tian-Shansky

Laboratoire Gevrey de Mathématique physique, Université de Bourgogne, BP 47870, 21078 Dijon Cedex, France, `semenov@mail.u-bourgogne.fr`

Abstract. The key concept discussed in these lectures is the relation between the Hamiltonians of a quantum integrable system and the Casimir elements in the underlying hidden symmetry algebra. (In typical applications the latter is either the universal enveloping algebra of an affine Lie algebra or its q-deformation.) A similar relation also holds in the classical case. We discuss different guises of this very important relation and its implication for the description of the spectrum and the eigenfunctions of the quantum system. Parallels between the classical and the quantum cases are thoroughly discussed.

1 Introduction

The study of exactly solvable quantum mechanical models is at least as old as quantum mechanics itself. Over the last fifteen years there has been a major development aimed at a unified treatment of many examples known previously and, more importantly, at a systematic construction of new ones. The new method nicknamed the Quantum Inverse Scattering Method was largely created by L.D. Faddeev and his school in St. Petersburg as a quantum counterpart of the Classical Inverse Scattering Method, and has brought together many ideas believed to be unrelated. [Besides the Classical Inverse Scattering Method, one should mention the profound results of R. Baxter in Quantum Statistical Mechanics (which, in turn, go back to the work of L. Onsager, E. Lieb and many others) and the seminal article of H. Bethe on the ferromagnet model in which the now famous Bethe Ansatz was introduced.] It also allowed unraveling highly nontrivial algebraic structures, the Quantum Group Theory being one of its by-products.

The origins of QISM lie in the study of concrete examples; it is designed as a working machine which produces quantum systems together with their spectra, the quantum integrals of motion, and their joint eigenvectors. In the same spirit, the Classical Inverse Scattering Method (along with its ramifications) is a similar tool to produce examples of classical integrable systems together with their solutions. In these lectures it is virtually impossible to follow the history of the developement of this method, starting with the famous articles of [25], and [39]. (A good introduction close to the ideas of CISM may be found in [19].) I would like to comment only on one important turning point which gave the impetus to the invention of the Quantum Inverse Scattering Method. In 1979 L.D. Faddeev exposed in his seminar the draft article of B. Kostant on the quantization of

M.A. Semenov-Tian-Shansky, Quantum and Classical Integrable Systems, Lect. Notes Phys. **638**, 269–333 (2004)
`http://www.springerlink.com/`

the Toda lattice [36] which reduced the problem to the representation theory of semisimple Lie groups. This was an indication that completely integrable systems have an intimate relation to Lie groups and should also have exactly solvable quantum counterparts. In the same talk Faddeev introduced the now famous $RL_1L_2 = L_2L_1R$ commutation relations for quantum Lax matrices which could be extracted from R. Baxter's work on quantum transfer matrices. The research program outlined in that talk and implemented over the next few years was two-fold: On the one hand, the Yang-Baxter equation and the related algebra have led directly to exact solutions of several quantum models, such as the quantum sine-Gordon equation. On the other hand, connections with group theory and the orbit method resulted in a systematic treatment of numerous examples in the classical setting [45]. One major problem which has remained unsettled for more than a decade is how to fill the gap between the two approaches, and in particular, how to explain the group theoretical meaning of the Bethe Ansatz.

With hindsight we can now understand why this problem could not have been resolved immediately. First of all, while classical integrable systems are related to ordinary Lie groups, quantum systems quite often (though not always, cf. the discussion in Sect. 3) require the full machinery of Quantum Groups. The background took several years to prepare [12]. Second, even the simplest systems such as the open Toda lattice require a very advanced technique of representation theory [36, 47]. The Toda lattice is peculiar, since the underlying 'hidden symmetry' group is finite-dimensional. All the principal examples are related to *infinite-dimensional algebras,* mainly to classical or quantum affine Lie algebras. The representation theory of affine algebras which is an essential element in the study of integrable systems has been developed only in recent years (semi-infinite cohomologies, Wakimoto modules, critical level representations, cf. [21–23]).

The present lectures do not give a systematic overview of the Quantum Inverse Scattering Method (several good expositions are available, cf. [5], Faddeev [14–16, 18, 37, 51]). Instead, I shall try to explain the parallels between quantum and classical systems and the 'correspondence principles' which relate the quantum and the classical cases.

As already mentioned, the first key observation is that integrable systems always have an ample hidden symmetry. (Fixing this symmetry provides some rough classification of the associated examples. A nontrivial class of examples is related to loop algebras or, more generally, to their q-deformations. This is the class of examples I shall consider below.) Two other key points are the role of (classical or quantum) R-matrices and of the Casimir elements which give rise to the integrals of motion.

This picture appears in several different guises, depending on the type of examples in question. The simplest (so-called linear) case corresponds to classical systems which are modelled on coadjoint orbits of Lie algebras; a slightly more complicated group of examples is classical systems modelled on Poisson Lie groups or their Poisson submanifolds. In the quantum setting, the difference between these two cases is deeper: while in the former case the hidden symmetry algebra remains the same, quantization of the latter leads to Quantum Groups.

Still, for loop algebras, the quantum counterpart of the main construction is nontrivial even in the linear case; the point is that the universal enveloping algebra of a loop algebra has a trivial center which reappears only after a central extension at the critical value of the central charge. Thus to tackle the quantum case one needs the full machinery of the representation theory of loop algebras. (By contrast, in the classical case one mainly deals with the evaluation representations which allow reducing the solution of the equations of motion to a problem in algebraic geometry.)

The study of integrable models may be divided into two different parts. The first one is, so to say, kinematic: it consists in the choice of appropriate models together with their phase spaces or the algebra of observables and of their Hamiltonians. The second one is dynamical; it consists, classically, in the description of solutions or of the action-angle variables. The quantum counterpart consists in the description of the spectra, the eigenvectors, and of the various correlation functions. The algebraic scheme proved to be very efficient in the description of kinematics. (Quantum Group Theory may be regarded as a by-product of this kinematic problem.) The description of spectra at the present stage of the theory remains model-dependent. The standard tool for constructing the eigenvectors of the quantum Hamiltonians which was the starting point of QISM is the algebraic Bethe Ansatz. Until very recently, its interpretation in terms of representation theory was lacking. This problem has been finally settled by [22] for an important particular model with *linear* commutation relations (the *Gaudin model*); remarkably, their results follow the general pattern outlined above. A similar treatment of quantum models related to q-deformed affine algebras is also possible, although the results in this case are still incomplete. One should be warned that much of the 'experimental material' on Quantum Integrability still resists general explanations. I would like to mention in this respect the deep results of E.K. Sklyanin [51, 52] relating the Bethe Ansatz to the separation of variables; see also [29, 38].

2 Generalities

A quantum mechanical system is a triple consisting of an associative algebra with involution $\mathcal{A}$ called the algebra of observables, an irreducible $\star$-representation π of $\mathcal{A}$ in a Hilbert space $\mathfrak{H}$ and a distinguished self-adjoint observable $\mathcal{H}$ called the Hamiltonian. A typical question to study is the description of the spectrum and the eigenvectors of $\pi(\mathcal{H})$. Quantum mechanical systems usually appear with their classical counterparts. By definition, a classical mechanical system is again specified by its algebra of observables $\mathcal{A}_{cl}$ which is a commutative associative algebra equipped with a Poisson bracket (i. e., a Lie bracket which also satisfies the Leibniz rule

$$\{a, bc\} = \{a, b\}\, c + \{a, c\}\, b;$$

in other words, a Poisson bracket is a derivation of $\mathcal{A}_{cl}$ with respect to both its arguments), and a Hamiltonian $\mathcal{H} \in \mathcal{A}_{cl}$. A commutative algebra equipped with a Poisson bracket satisfying the Leibniz rule is called a *Poisson algebra*. Speaking informally, a quantum algebra of observables $\mathcal{A}$ arises as a deformation of the commutative algebra $\mathcal{A}_{cl}$ determined by the Poisson bracket. In these lectures we shall not be concerned with the quantization problem in its full generality (cf. [6, 57]). However, it will always be instructive to compare quantum systems with their classical counterparts.

The algebraic language which starts with Poisson algebras makes the gap between classical and quantum mechanics as narrow as possible; in practice, however, we also need the dual language based on the notion of the phase space. Roughly speaking, the phase space is the spectrum of the Poisson algebra. The accurate definition depends on the choice of a topology in the Poisson algebra. We shall not attempt to discuss these subtleties and shall always assume that the underlying phase space is a smooth manifold and that the Poisson algebra is realized as the algebra of functions on this manifold. In the examples that we have in mind, Poisson algebras always have an explicit geometric realization of this type. The Poisson bracket itself is then as usual represented by a bivector field on the phase space satisfying certain differential constraints which account for the Jacobi identity. This gives the definition of a *Poisson manifold* which is dual to the notion of a Poisson algebra. The geometry of Poisson manifolds has numerous obvious parallels with representation theory. Recall that the algebraic version of representation theory is based on the study of appropriate ideals in an associative algebra. For a Poisson algebra we have a natural notion of a *Poisson ideal* (i. e., a subalgebra which is an ideal with respect to both structures); the dual notion is that of a *Poisson submanifold* of a Poisson manifold. The classical counterpart of Hilbert space representations of an associative algebra is the restriction of functions to various Poisson submanifolds. Poisson submanifolds are partially ordered by inclusion; *minimal Poisson submanifolds* are those for which the induced Poisson structure is nondegenerate. (This means that the center of the Lie algebra of functions contains only constants.) Minimal Poisson submanifolds always carry a symplectic structure and form a stratification of the Poisson manifold; they are called *symplectic leaves*. The restriction of functions to symplectic leaves gives a classical counterpart of the irreducible representations of associative algebras.

Let $\mathcal{M}$ be a Poisson manifold, $\mathcal{H} \in C^{\infty}(\mathcal{M})$. A classical system $(\mathcal{M}, \mathcal{H})$ is called *integrable* if the commutant of the Hamiltonian $\mathcal{H}$ in $\mathcal{A}_{cl}$ contains an abelian algebra of maximal possible rank. (A technical definition is provided by the well known Liouville theorem.)

Let us recall that the key idea which has started the modern age in the study of classical integrable systems is how to bring them into *Lax form*. In the simplest case, the definition of a Lax representation may be given as follows. Let $(\mathcal{A}, \mathcal{M}, \mathcal{H})$ be a classical mechanical system. Let $\mathcal{F}_t : \mathcal{M} \to \mathcal{M}$ be the associated flow on $\mathcal{M}$ (defined at least locally). Suppose that $\mathfrak{g}$ is a Lie algebra. A mapping $L : \mathcal{M} \to \mathfrak{g}$ is called a *Lax representation* of $(\mathcal{A}, \mathcal{M}, \mathcal{H})$ if the following conditions are satisfied:

(i) The flow $\mathcal{F}_t$ factorizes over $\mathfrak{g}$, i. e., there exists a (local) flow $F_t : \mathfrak{g} \to \mathfrak{g}$ such that the following diagram is commutative.

$$
\begin{array}{ccc}
\mathcal{M} & \xrightarrow{\ \mathcal{F}_t\ } & \mathcal{M} \\
{\scriptstyle L}\big\downarrow & & \big\downarrow{\scriptstyle L} \\
\mathfrak{g} & \xrightarrow{\ F_t\ } & \mathfrak{g}
\end{array}
$$

(ii) The quotient flow F_t on $\mathfrak{g}$ is isospectral, i. e., it is tangent to the adjoint orbits in $\mathfrak{g}$.

Remark 1. In applications we have in mind, the Lie algebra $\mathfrak{g}$ is supposed to carry a nondegenerate invariant inner product and hence its adjoint and coadjoint representations are identical. In a more general way, we may assume that $\mathfrak{g}$ is arbitrary and replace the target space of the *generalized Lax representation* with its dual space $\mathfrak{g}^*$. (In that case 'isosectral flows' preserve coadjoint orbits in $\mathfrak{g}$.) The practical advantage of 'self-dual' Lie algebras is the possibility to use their finite-dimensional representations to construct spectral invariants or the integrals of motion, as discussed below.

Clearly, $L(x) \in \mathfrak{g}$ for any $x \in \mathcal{M}$; hence we may regard L as a 'matrix with coefficients in $\mathcal{A} = C^\infty(\mathcal{M})$, i. e., as an element of $\mathfrak{g} \otimes \mathcal{A}$; the Poisson bracket on $\mathcal{A}$ extends to $\mathfrak{g} \otimes \mathcal{A}$ by linearity. Property (ii) means that there exists an element $M \in \mathfrak{g} \otimes \mathcal{A}$ such that $\{\mathcal{H}, L\} = [L, M]$.

Let (ρ, V) be a (finite-dimensional) linear representation of $\mathfrak{g}$. Then $L_V = \rho \otimes id(L) \in \operatorname{End} V \otimes \mathcal{A}$ is a matrix-valued function on $\mathcal{M}$; the coefficients of its characteristic polynomial $P(\lambda) = \det(L_V - \lambda)$ are integrals of the motion.

One may replace in the above definition a Lie algebra $\mathfrak{g}$ with a Lie group G; in that case isospectrality means that the flow preserves conjugacy classes in G. In a more general way, the Lax operator may be a difference or a differential operator. The case of difference operator is of particular importance, since the quantization of difference Lax equations is the core of QISM and a natural source of Quantum Groups (cf. Sect. 3 below).

There is no general way to find a Lax representation for a given system (even if it is known to be completely integrable). However, there is a systematic way to produce *examples* of such representations. An ample source of such examples is provided by the general construction described in the next section.

2.1 Basic Theorem: Linear Case

The basic construction outlined in this section (summarized in Theorem 1 below) goes back to [36] and [1] in some crucial cases; its relation with the r-matrix method was established by [46]. We shall state it using the language of symmetric algebras which simplifies its generalization to the quantum case. Let $\mathfrak{g}$ be a Lie algebra over k (where $k = \mathbb{R}$ or $\mathbb{C}$). Let $\mathcal{S}(\mathfrak{g})$ be the symmetric algebra of $\mathfrak{g}$. Recall that there is a unique Poisson bracket on $\mathcal{S}(\mathfrak{g})$ (called the *Lie–Poisson bracket*) which extends the Lie bracket on $\mathfrak{g}$ (see, for example, [7]. This Poisson bracket is more frequently discussed from the 'spectral' point of view. Namely, let $\mathfrak{g}^*$ be the linear dual of $\mathfrak{g}$; the natural pairing $\mathfrak{g} \times \mathfrak{g}^* \to k$ extends to the 'evaluation map' $\mathcal{S}(\mathfrak{g}) \times \mathfrak{g}^* \to k$ which induces a canonical isomorphism of $\mathcal{S}(\mathfrak{g})$ with the space of polynomials $P(\mathfrak{g}^*)$; thus $\mathfrak{g}^*$ is a linear Poisson manifold and $\mathcal{S}(\mathfrak{g}) \simeq P(\mathfrak{g}^*)$ is the corresponding algebra of observables. Linear functions on $\mathfrak{g}^*$ form a subspace in $P(\mathfrak{g}^*)$ which may be identified with $\mathfrak{g}$. The Lie–Poisson bracket on $\mathfrak{g}^*$ is uniquely characterized by the following properties:

1. *The Poisson bracket of linear functions on $\mathfrak{g}^*$ is again a linear function.*
2. *The restriction of the Poisson bracket to $\mathfrak{g} \subset P(\mathfrak{g}^*)$ coincides with the Lie bracket in $\mathfrak{g}$.*

Besides the Lie–Poisson structure on $\mathcal{S}(\mathfrak{g})$ we shall need its Hopf structure. Further on we shall deal with other more complicated examples of Hopf algebras, so it is probably worth recalling the general definitions (though we shall not use them in full generality until Sect. 4). Recall that a Hopf algebra is a set $(\mathcal{A}, m, \Delta, \epsilon, S)$ consisting of an associative algebra $\mathcal{A}$ over k with multiplication $m : \mathcal{A} \otimes \mathcal{A} \to \mathcal{A}$ and the unit element 1, the coproduct $\Delta : \mathcal{A} \to \mathcal{A} \otimes \mathcal{A}$, the counit $\epsilon : \mathcal{A} \to k$ and the antipode $S : \mathcal{A} \to \mathcal{A}$ which satisfy the following axioms

- $m : \mathcal{A} \otimes \mathcal{A} \to \mathcal{A}$, $\Delta : \mathcal{A} \to \mathcal{A} \otimes \mathcal{A}$, $i : k \to \mathcal{A} : \alpha \longmapsto \alpha \cdot 1$, $\epsilon : \mathcal{A} \to k$ are *homomorphisms of algebras.*
- *The following diagrams are commutative:*

$$
\begin{array}{ccc}
\mathcal{A} \otimes \mathcal{A} \otimes \mathcal{A} & \xrightarrow{id \otimes m} & \mathcal{A} \otimes \mathcal{A} \\
{\scriptstyle m \otimes id}\downarrow & & \downarrow{\scriptstyle m} \\
\mathcal{A} \otimes \mathcal{A} & \xrightarrow{m} & \mathcal{A}
\end{array}
\quad , \quad
\begin{array}{ccc}
\mathcal{A} & \xrightarrow{\Delta} & \mathcal{A} \otimes \mathcal{A} \\
{\scriptstyle \Delta}\downarrow & & \downarrow{\scriptstyle \Delta \otimes id} \\
\mathcal{A} \otimes \mathcal{A} & \xrightarrow{id \otimes \Delta} & \mathcal{A} \otimes \mathcal{A} \otimes \mathcal{A}
\end{array}
$$

(these diagrams express the associativity of the product m and the coassociativity of the coproduct Δ, respectively),

$$
\begin{array}{ccc}
\mathcal{A} \otimes \mathcal{A} & \xleftarrow{id \otimes i} & \mathcal{A} \otimes k \\
{\scriptstyle m}\downarrow & & \downarrow{\scriptstyle m} \\
\mathcal{A} & \xleftarrow{id} & \mathcal{A}
\end{array}
\quad , \quad
\begin{array}{ccc}
\mathcal{A} \otimes \mathcal{A} & \xleftarrow{i \otimes id} & k \otimes \mathcal{A} \\
{\scriptstyle m}\downarrow & & \downarrow{\scriptstyle m} \\
\mathcal{A} & \xleftarrow{id} & \mathcal{A}
\end{array}
$$

$$A \otimes A \xrightarrow{id \otimes \epsilon} A \otimes k \qquad A \otimes A \xrightarrow{\epsilon \otimes id} k \otimes A$$

$$\Delta \uparrow \qquad \downarrow m \ , \qquad \Delta \uparrow \qquad \downarrow m$$

$$A \xrightarrow{\ id\ } A \qquad\qquad A \xrightarrow{\ id\ } A$$

(these diagrams express, respectively, the properties of the unit element $1 \in A$ and of the counit $\epsilon \in A^*$).

- *The antipode S is an antihomomorphism of algebras and the following diagrams are commutative:*

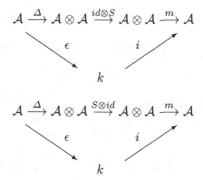

If $(A, m, 1, \Delta, \epsilon, S)$ is a Hopf algebra, its linear dual A^* is also a Hopf algebra; moreover, the coupling $A \otimes A^* \to k$ interchanges the roles of product and coproduct. Thus we have

$$\langle m\,(a \otimes b)\,,\,\varphi \rangle = \langle a \otimes b,\, \Delta\varphi \rangle\,.$$

The Hopf structure on the symmetric algebra $S\,(\mathfrak{g})$ is determined by the structure of the additive group on $\mathfrak{g}^*$. Namely, let

$$S\,(\mathfrak{g}) \times \mathfrak{g}^* \longrightarrow k : (a, X) \longmapsto a\,(X)$$

be the evaluation map; since $S\,(\mathfrak{g}) \otimes S\,(\mathfrak{g}) \simeq S\,(\mathfrak{g} \oplus \mathfrak{g}) \simeq P\,(\mathfrak{g}^* \oplus \mathfrak{g}^*)$, this map extends to $S\,(\mathfrak{g}) \otimes S\,(\mathfrak{g})$. We have

$$\Delta a\,(X, Y) = a\,(X + Y),\ \ \epsilon\,(a) = a\,(0),\ \ S\,(a)\,(X) = a\,(-X) \qquad (2.1)$$

(we shall sometimes write $S\,(a) = a'$ for brevity). It is important to notice that the Lie–Poisson bracket on $S(\mathfrak{g})$ is compatible with the Hopf structure:

$$\{\Delta\varphi,\, \Delta\psi\} = \Delta\,\{\varphi,\, \psi\}\,.$$

For future reference we recall also that the coproducts in the universal enveloping algebra $U\,(\mathfrak{g})$ and in $S\,(\mathfrak{g})$ coincide (in other words, $U\,(\mathfrak{g})$ is canonically isomorphic to $S\,(\mathfrak{g})$ as a coalgebra).

Let us now return to the discussion of the Poisson structure on $\mathcal{S}(\mathfrak{g})$, i. e., of the Lie–Poisson bracket of the Lie algebra $\mathfrak{g}$. It is well known that the Lie–Poisson bracket is always degenerate; its symplectic leaves are precisely the coadjoint orbits of the corresponding Lie group.

Proposition 1. *The center of $\mathcal{S}(\mathfrak{g})$ (regarded as a Poisson algebra) coincides with the subalgebra $I = \mathcal{S}(\mathfrak{g})^{\mathfrak{g}}$ of ad $\mathfrak{g}$-invariants in $\mathcal{S}(\mathfrak{g})$ (called Casimir elements).*

Restrictions of Casimir elements to symplectic leaves are constants; so fixing the values of Casimirs provides a rough classification of symplectic leaves, although it is not true in general that different symplectic leaves are separated by the Casimirs. (In the semisimple case there are enough Casimirs to separate generic orbits.)

Fix an element $\mathcal{H} \in \mathcal{S}(\mathfrak{g})$; it defines a derivation $D_{\mathcal{H}}$ of $\mathcal{S}(\mathfrak{g})$,

$$D_{\mathcal{H}}\varphi = \{\mathcal{H}, \varphi\}.$$

By duality, this derivation determines a (local) Hamiltonian flow on $\mathfrak{g}^*$. Since the Lie–Poisson bracket on $\mathfrak{g}^*$ is degenerate, this flow splits into a family of independent flows which are confined to Poisson submanifolds in $\mathfrak{g}^*$. Thus a Hamiltonian $\mathcal{H}$ defines Hamiltonian flows on the coadjoint orbits in $\mathfrak{g}^*$. In the special case when $\mathcal{H} \in I$ all these flows are trivial. However, one may still use Casimir elements to produce nontrivial equations of motion if the Poisson structure on $\mathcal{S}(\mathfrak{g})$ is properly modified, and this is the way in which Lax equations associated with $\mathfrak{g}$ do arise. The formal definition is as follows.

Let $r \in \operatorname{End}\mathfrak{g}$ be a linear operator; we shall say that r is a *classical r-matrix* if the *r-bracket*

$$[X,Y]_r = \frac{1}{2}\left([rX,Y] + [X,rY]\right), \quad X,Y \in \mathfrak{g}, \tag{2.2}$$

satisfies the Jacobi identity. In that case we get *two* different Lie brackets on the same underlying linear space $\mathfrak{g}$. We shall assume that r satisfies the following stronger condition called *the modified classical Yang–Baxter identity:*

$$[rX,rY] - r\left([rX,Y] + [X,rY]\right) = -[X,Y], X,Y \in \mathfrak{g}. \tag{2.3}$$

Proposition 2. *Identity (2.3) implies the Jacobi identity for (2.2).*

Remark 2. Identity (2.3) is of course only a sufficient condition; however, we prefer to impose it from the very beginning, since it assures a very important factorization property (see below).

Let $(\mathfrak{g}, r)$ be a Lie algebra equipped with a classical r-matrix $r \in \operatorname{End}\mathfrak{g}$ satisfying (2.3). Let $\mathfrak{g}_r$ be the corresponding Lie algebra (with the same underlying linear space). Put $r_{\pm} = \frac{1}{2}(r \pm id)$; then (2.3) implies that $r_{\pm} : \mathfrak{g}_r \to \mathfrak{g}$ are Lie algebra homomorphisms. Let us extend them to Poisson algebra morphisms

$S(\mathfrak{g}_r) \to S(\mathfrak{g})$ which we shall denote by the same letters. These morphisms also agree with the standard Hopf structure on $S(\mathfrak{g})$, $S(\mathfrak{g}_r)$. Define the action

$$S(\mathfrak{g}_r) \otimes S(\mathfrak{g}) \to S(\mathfrak{g})$$

by setting

$$x \cdot y = \sum r_+(x_i^{(1)}) \, y \, r_-(x_i^{(2)})', \quad x \in S(\mathfrak{g}_r), \ y \in S(\mathfrak{g}), \tag{2.4}$$

where $\Delta x = \sum x_i^{(1)} \otimes x_i^{(2)}$ is the coproduct and $a \mapsto a'$ is the antipode map.

Theorem 1. *(i) $S(\mathfrak{g})$ is a free graded $S(\mathfrak{g}_r)$-module generated by $1 \in S(\mathfrak{g})$. (ii) Let $i_r\colon S(\mathfrak{g}) \to S(\mathfrak{g}_r)$ be the induced isomorphism of graded linear spaces; its restriction to $I = S(\mathfrak{g})^{\mathfrak{g}}$ is a morphism of Poisson algebras. (iii) Assume, moreover, that $\mathfrak{g}$ is equipped with a nondegenerate invariant bilinear form; the induced mapping $\mathfrak{g}_r^* \to \mathfrak{g}$ defines a Lax representation for all Hamiltonians $\mathcal{H} = i_r(\widehat{\mathcal{H}})$, $\widehat{\mathcal{H}} \in S(\mathfrak{g})^{\mathfrak{g}}$.*

The geometric meaning of Theorem 1 is very simple. In fact, it becomes more transparent if one uses the language of Poisson manifolds rather than the dual language of Poisson algebras. (However, it is this more cumbersome language that may be generalized to the quantum case.) Since $\mathfrak{g} \simeq \mathfrak{g}_r$ as linear spaces, the polynomial algebras $P(\mathfrak{g}^*)$ and $P(\mathfrak{g}_r^*)$ are also isomorphic as graded linear spaces (though not as Poisson algebras). The subalgebra $I \subset P(\mathfrak{g}^*)$ remains commutative with respect to the Lie–Poisson bracket of $\mathfrak{g}_r$; moreover, the flows in $\mathfrak{g}^*$ which correspond to the Hamiltonians $\mathcal{H} \in I$ are tangent to *two* systems of coadjoint orbits in $\mathfrak{g}^*$, the orbits of $\mathfrak{g}$ and of $\mathfrak{g}_r$. The latter property is trivial, since *all* Hamiltonian flows in $\mathfrak{g}_r^*$ are tangent to coadjoint orbits; the former is an immediate corollary of the fact that $I = S(\mathfrak{g})^{\mathfrak{g}} \simeq P(\mathfrak{g}^*)^{\mathfrak{g}}$ is the subalgebra of coadjoint invariants. (As we mentioned, in the semisimple case generic orbits are level surfaces of the Casimirs.) Finally, if $\mathfrak{g}$ carries a nondegenerate invariant inner product, the coadjoint and the adjoint representations of $\mathfrak{g}$ are equivalent and I consists precisely of spectral invariants.

Let $\mathfrak{u} \subset \mathfrak{g}_r$ be an ideal, $\mathfrak{s} = \mathfrak{g}_r/\mathfrak{u}$ the quotient algebra, $p : S(\mathfrak{g}_r) \to S(\mathfrak{s})$ the canonical projection; restricting p to the subalgebra $I_r = i_r(S(\mathfrak{g})^{\mathfrak{g}})$ we get a Poisson commutative subalgebra in $S(\mathfrak{s})$; we shall say that the corresponding elements of $S(\mathfrak{s})$ are obtained by *specialization*.

The most common examples of classical r-matrices are associated with *decompositions of Lie algebras*. Let $\mathfrak{g}$ be a Lie algebra with a nondegenerate invariant inner product, $\mathfrak{g}_+, \mathfrak{g}_- \subset \mathfrak{g}$ two Lie subalgebras of $\mathfrak{g}$ such that $\mathfrak{g} = \mathfrak{g}_+ \dot{+} \mathfrak{g}_-$ as a linear space. Let P_+, P_- be the projection operators associated with this decomposition; the operator

$$r = P_+ - P_- \tag{2.5}$$

satisfies (2.3); moreover, we have:

$$[X, Y]_r = [X_+, Y_+] - [X_-, Y_-], \text{ where } X_\pm = P_\pm X, \ Y_\pm = P_\pm Y,$$

and hence the Lie algebra $\mathfrak{g}_r$ splits into two parts, $\mathfrak{g}_r \simeq \mathfrak{g}_+ \oplus \mathfrak{g}_-$.

Let us choose $\mathcal{A} = \mathcal{S}(\mathfrak{g}_+)$ as our algebra of observables. The construction of the Lax representation for the Hamiltonian equations of motion associated with $\mathcal{A}$ uses the notion of *canonical element*. Since this notion will be very important in the sequel, we shall briefly recall its definition.

Let V be a linear space, V^* its dual. Choose a linear basis $\{e_i\}$ in V and let $\{e^i\}$ be the dual basis in V^*. Set

$$C = \sum_i e_i \otimes e^i.$$

It is easy to see that C does not depend on the choice of $\{e_i\}$; under the canonical isomorphism $\operatorname{End} V \simeq V \otimes V^*$ it corresponds to the identity operator. If V is infinite dimensional, the canonical element does not belong to the algebraic tensor product $V \otimes V^*$. We shall always assume that in this case the space $V \otimes V^*$ is properly completed, so as to contain the canonical element. In different settings which we shall discuss below in these lectures canonical elements will appear in *different guises*: they are related to the construction of classical and quantum Lax representations (in the corresponding setting it is natural to denote it by L), as well as to classical and quantum R-matrices (in those cases letter R is more suggestive).

Let us now return to our example. Let $\mathfrak{g}_+^\perp, \mathfrak{g}_-^\perp$ be the orthogonal complements of $\mathfrak{g}_+, \mathfrak{g}_-$ in $\mathfrak{g}$. Then $\mathfrak{g} = \mathfrak{g}_+^\perp \oplus \mathfrak{g}_-^\perp$ and we may identify $\mathfrak{g}_+^* \simeq \mathfrak{g}_-^\perp, \mathfrak{g}_-^* \simeq \mathfrak{g}_+^\perp$. Let

$$L \in \mathfrak{g}_-^\perp \otimes \mathfrak{g}_+ \subset \mathfrak{g}_-^\perp \otimes \mathcal{S}(\mathfrak{g}_+)$$

be the canonical element; we may regard L as an embedding $\mathfrak{g}_+^* \hookrightarrow \mathfrak{g}_-^\perp$. Let $P : \mathcal{S}(\mathfrak{g}) \to \mathcal{S}(\mathfrak{g}_-)$ be the projection onto $\mathcal{S}(\mathfrak{g}_-)$ in the decomposition

$$\mathcal{S}(\mathfrak{g}) = \mathcal{S}(\mathfrak{g}_-) \oplus \mathfrak{g}_+ \mathcal{S}(\mathfrak{g}).$$

Corollary 1. *(i) The restriction of P to the subalgebra $I = \mathcal{S}(\mathfrak{g})^\mathfrak{g}$ of Casimir elements is a Poisson algebra homomorphism. (ii) L defines a Lax representation for all Hamiltonian equations of motion defined by the Hamiltonians $\overline{\mathcal{H}} = P(\mathcal{H})$, $\mathcal{H} \in I$. (iii) The corresponding Hamiltonian flows on $\mathfrak{g}_+^* \simeq \mathfrak{g}_-^\perp \subset \mathfrak{g}$ preserve the intersections of coadjoint orbits of $\mathfrak{g}_+$ with the adjoint orbits of $\mathfrak{g}$.*

One sometimes calls L *the universal Lax operator*; restricting the mapping $L : \mathfrak{g}_+^* \hookrightarrow \mathfrak{g}_-^\perp$ to various Poisson submanifolds in $\mathfrak{g}_+^*$ we get Lax representations for particular systems.

The situation is especially simple when $\mathfrak{g}_+, \mathfrak{g}_- \subset \mathfrak{g}$ are isotropic subspaces with respect to the inner product in $\mathfrak{g}$; then $\mathfrak{g}_+ = \mathfrak{g}_+^\perp$, $\mathfrak{g}_- = \mathfrak{g}_-^\perp$, $\mathfrak{g}_+ \simeq \mathfrak{g}_-^*$. In that case $(\mathfrak{g}_+, \mathfrak{g}_-, \mathfrak{g})$ is referred to as a *Manin triple*. This case is of particular importance since both $\mathfrak{g}$ and its Lie subalgebras $\mathfrak{g}_+, \mathfrak{g}_-$ then carry an additional structure of *Lie bialgebra*.

Definition 1. Let $\mathfrak{a}$ be a Lie algebra, $\mathfrak{a}^*$ its dual; assume that $\mathfrak{a}^*$ is equipped with a Lie bracket $[\,,]_* : \mathfrak{a}^* \wedge \mathfrak{a}^* \to \mathfrak{a}^*$. The brackets on $\mathfrak{a}$ and $\mathfrak{a}^*$ are said to be *compatible* if the dual map $\delta : \mathfrak{a} \to \mathfrak{a} \wedge \mathfrak{a}$ is a 1-cocycle on $\mathfrak{a}$ (with the values in the wedge square of the adjoint module). A pair $(\mathfrak{a}, \mathfrak{a}^*)$ with compatible Lie brackets is called a *Lie bialgebra*.

One can prove that this definition is actually symmetric with respect to $\mathfrak{a}, \mathfrak{a}^*$; in other words, if $(\mathfrak{a}, \mathfrak{a}^*)$ is a Lie bialgebra, so is $(\mathfrak{a}^*, \mathfrak{a})$.

Proposition 3. *Let $(\mathfrak{g}_+, \mathfrak{g}_-, g)$ be a Manin triple. Identify $\mathfrak{g}^*$ with $\mathfrak{g}$ by means of the inner product and equip it with the r-bracket associated with $r = P_+ - P_-$; then (i) $(\mathfrak{g}, \mathfrak{g}^*)$ is a Lie bialgebra. (ii) $(\mathfrak{g}_+, \mathfrak{g}_-)$ is a Lie sub-bialgebra of $(\mathfrak{g}, \mathfrak{g}^*)$. (iii) Conversely, if $(\mathfrak{a}, \mathfrak{a}^*)$ is a Lie bialgebra, there exists a unique Lie algebra $\mathfrak{d} = \mathfrak{d}(\mathfrak{a}, \mathfrak{a}^*)$ called the double of $(\mathfrak{a}, \mathfrak{a}^*)$ such that*

- $\mathfrak{d} = \mathfrak{a} \dotplus \mathfrak{a}^*$ *as a linear space,*
- $\mathfrak{a}, \mathfrak{a}^* \subset \mathfrak{d}$ *are Lie subalgebras,*
- *The canonical bilinear form on $\mathfrak{d}$ induced by the natural pairing between $\mathfrak{a}$ and $\mathfrak{a}^*$ is* ad $\mathfrak{d}$*-invariant.*

In a more general way, let $r \in \operatorname{End} \mathfrak{g}$ be a classical r-matrix on a Lie algebra $\mathfrak{g}$ which is equipped with a fixed nondegenerate invariant inner product. Identify $\mathfrak{g}_r$ with the dual of $\mathfrak{g}$ by means of the inner product. Assume that r is skew and satisfies the modified Yang-Baxter identity. Then $(\mathfrak{g}, \mathfrak{g}^*)$ is a Lie bialgebra; it is usually called *a factorizable Lie bialgebra*. Let us once more list the properties of factorizable Lie bialgebras for future reference:

- $\mathfrak{g}$ *is equipped with a nondegenerate invariant inner product.*
- $\mathfrak{g}^*$ *is identified with $\mathfrak{g}$ as a linear space by means of the inner product; the Lie bracket in $\mathfrak{g}^*$ is given by*

$$[X, Y]_* = \frac{1}{2} \left([rX, Y] + [X, rY]\right), X, Y \in \mathfrak{g},$$

- $r_\pm = \frac{1}{2}(r \pm id)$ *define Lie algebra homomorphisms $\mathfrak{g}^* \to \mathfrak{g}$; moreover, $r_+^* = -r_-$ and $r_+ - r_- = id$ (i. e., it coincides with the identification of $\mathfrak{g}^*$ and $\mathfrak{g}$ induced by the inner product).*

Lie bialgebras are particularly important, since the associated Lie–Poisson structures may be extended to Lie groups (cf. Sect. 3).

2.1.1 Hamiltonian Reduction.

Theorem 1 admits a very useful global version which survives quantization. To formulate this theorem we shall first recall some basic facts about Hamiltonian reduction.

Let G be a Lie group with Lie algebra $\mathfrak{g}$. Let T^*G be the cotangent bundle of G equipped with the canonical symplectic structure. Let $B \subset G$ be a Lie subgroup with Lie algebra $\mathfrak{b}$. The action of B on G by right translations extends canonically to a Hamiltonian action $B \times T^*G \to T^*G$. This means that the vector fields on T^*G generated by the corresponding infinitesimal action of $\mathfrak{b}$ are globally Hamiltonian; moreover, the mapping

$$\mathfrak{b} \to C^\infty (T^*G) : X \longmapsto h_X$$

which assigns to each $X \in \mathfrak{b}$ the Hamiltonian of the corresponding vector field is a homomorphism of Lie algebras. In particular, this map is linear and hence for each $X \in \mathfrak{g}, x \in T^*G$,

$$h_X(x) = \langle X, \mu(x) \rangle ,$$

where $\mu : T^*G \to \mathfrak{b}^*$ is the so-called *moment mapping*. Choose a trivialization $T^*G \simeq G \times \mathfrak{g}^*$ by means of left translations; then the action of B is given by

$$b : (g, \xi) \longmapsto \left(gb^{-1}, (\mathrm{Ad}^* b)^{-1} \xi \right).$$

The corresponding moment map $\mu : T^*G \to \mathfrak{b}^*$ is

$$\mu : (g, \xi) \longmapsto -\xi \mid_{\mathfrak{b}},$$

where $\xi \mid_{\mathfrak{b}}$ means the restriction of $\xi \in \mathfrak{g}^*$ to $\mathfrak{b} \subset \mathfrak{g}$. The moment mapping is *equivariant*; in other words, the following diagram is commutative

$$
\begin{array}{ccc}
B \times T^*G & \longrightarrow & T^*G \\
{\scriptstyle id \times \mu} \downarrow & & \downarrow {\scriptstyle \mu} \\
B \times \mathfrak{b}^* & \longrightarrow & \mathfrak{b}^*
\end{array}
$$

where $B \times \mathfrak{b}^* \to \mathfrak{b}^*$ is the ordinary coadjoint action.

Recall that the reduction procedure consists of two steps:

- *Fix the value of the moment map μ and consider the level surface*

$$\mathcal{M}_F = \{x \in T^*G; \mu(x) = F\}, F \in \mathfrak{b}^*.$$

Speaking in physical terms, we impose on T^*G *linear constraints* which fix the moment $\mu : T^*G \to \mathfrak{b}^*$.

- *Take the quotient of $\mathcal{M}_F$ over the action of the stabilizer of F in B.*
  Note that due to the equivariance of μ, the stabilizer $B^F \subset B$ preserves $\mathcal{M}_F$. In physical terms, the subgroup B^F plays the role of a *gauge group*.

The resulting manifold $\overline{\mathcal{M}}_F$ is symplectic; it is called the *reduced phase space* obtained by reduction over $F \in \mathfrak{b}^*$. A particularly simple case is the reduction over $0 \in \mathfrak{b}^*$; in that case the stabilizer coincides with B itself. We shall denote the reduced space $\mathcal{M}_0/B$ by $T^*G//B$. In a slightly more general fashion, we may start with an arbitrary Hamiltonian B-space $\mathcal{M}$ and perform the reduction over the zero value of the moment map getting the reduced space $\mathcal{M}//B$. Let S_G be the category of Hamiltonian G-spaces; recall that (up to a covering) homogeneous Hamiltonian G-spaces are precisely the coadjoint orbits of G. We shall define a functor (called *symplectic induction*, cf. [32]),

$$\mathrm{Ind}_B^G : \mathsf{S}_B \rightsquigarrow \mathsf{S}_G,$$

which assigns to each Hamiltonian B-space a Hamiltonian G-space. Namely, we set

$$\mathrm{Ind}_B^G(\mathcal{M}) = T^*G \times \mathcal{M}//B,$$

where the manifold $T^*G \times \mathcal{M}$ is equipped with the symplectic structure which is the *difference* of symplectic forms on T^*G and on $\mathcal{M}$ and B acts on $T^*G \times \mathcal{M}$ diagonally (hence the moment map on $T^*G \times \mathcal{M}$ is the *difference* of the moment maps on T^*G and on $\mathcal{M}$). The structure of the Hamiltonian G-space is induced on $T^*G \times \mathcal{M}//B$ by the action of G on T^*G by left translations.

The most simple case of symplectic induction is when the coadjoint orbit consists of a single point. In representation theory, this corresponds to representations induced by 1-dimensional representations of a subgroup. Let now $F \in \mathfrak{b}^*$ be an arbitrary element; let $\mathcal{O}_F$ be its coadjoint orbit, $\mathcal{K}_F = \mathrm{Ind}_B^G(\mathcal{O}_F)$. It is sometimes helpful to have a different construction of $\mathcal{K}_F$ (in much the same way as in representation theory it is helpful to have different realizations of a given representation). In many cases it is possible to induce $\mathcal{K}_F$ starting from a smaller subgroup $L \subset B$ and its single-point orbit (a similar trick in representation theory means that we wish to induce a given representation from a 1-dimensional one). The proper choice of L is the *Lagrangian subgroup* $L_F \subset B$ subordinate to F (whenever it exists). Recall that a Lie subalgebra $\mathfrak{l}_F \subset \mathfrak{b}$ is called a *Lagrangian subalgebra* subordinate to $F \in \mathfrak{b}^*$ if $ad^*\mathfrak{l}_F \cdot F \subset T_F\mathcal{O}_F$ is a Lagrangian subspace of the tangent space to the coadjoint orbit of F (with respect to the Kirillov form on $T_F\mathcal{O}_F$). This condition implies that $F\mid_{[\mathfrak{l}_F,\mathfrak{l}_F]} = 0$; in other words, F defines a character of $\mathfrak{l}_F$. This is of course tantamount to saying that F is a single-point orbit of $\mathfrak{l}_F$. To define the corresponding Lagrangian subgroup $L_F \subset B$ let us observe that $\mathfrak{l}_F$ always coincides with its own normalizer in $\mathfrak{b}$. Let us define $L_F \subset B$ as the normalizer of $\mathfrak{l}_F$ in B (this definition fixes the group of components of L_F which may be not necessarily connected).

Proposition 4. *Let* $L_F \subset B$ *be a Lagrangian subgroup subordinate to* F; *then* $T^*G \times \mathcal{O}_F//B \simeq T^*G \times \{F\}//L_F$.

In the next section we shall discuss an example of a coadjoint orbit with real polarization.

Let us now return to the setting of Theorem 1. Consider the action $G \times G \times G \to G$ by left and right translations:

$$(h, h') : x \longmapsto hxh'^{-1}; \tag{2.6}$$

this action may be lifted to the Hamiltonian action $G \times G \times T^*G \to T^*G$. Let us choose again a trivialization $T^*G \simeq G \times \mathfrak{g}^*$ by means of left translations; then

$$(h, h') : (g, \xi) \longmapsto \left(hgh'^{-1}, (\mathrm{Ad}^* h')^{-1} \xi \right). \tag{2.7}$$

The corresponding moment map $\mu : T^*G \to \mathfrak{g}^* \oplus g^*$ is

$$\mu : (g, \xi) \longmapsto \left(\xi, -(\mathrm{Ad}^* g)^{-1} \xi \right). \tag{2.8}$$

Using the left trivialization of T^*G we may extend polynomial functions $\varphi \in P(\mathfrak{g}^*) \simeq \mathcal{S}(\mathfrak{g})$ to left-invariant functions on T^*G which are polynomial on the fibers. The Casimir elements $\mathcal{H} \in \mathcal{S}(\mathfrak{g})^{\mathfrak{g}}$ give rise to *bi-invariant* functions; thus the corresponding Hamiltonian systems admit reduction with respect to any subgroup $S \subset G \times G$.

Lemma 1. *Let* $S \subset G \times G$ *be a Lie subgroup,* $\mathfrak{s} \subset \mathfrak{g} \oplus \mathfrak{g}$ *a Lie subalgebra,* $p : \mathfrak{g}^* \oplus \mathfrak{g}^* \to \mathfrak{s}^*$ *the canonical projection,* $S \times T^*G \to T^*G$ *the restriction to* H *of the action (2.7). This action is Hamiltonian and the corresponding moment map is* $\mu_S = p \circ \mu$.

Now let us assume again that $(\mathfrak{g}, r)$ is a Lie algebra equipped with a classical r-matrix $r \in \mathrm{End}\, \mathfrak{g}$ satisfying (2.3). Combining the Lie algebra homomorphisms $r_\pm : \mathfrak{g}_r \to \mathfrak{g}$ we get an embedding $i_r : \mathfrak{g}_r \to \mathfrak{g} \oplus \mathfrak{g}$; we may identify the Lie group which corresponds to $\mathfrak{g}_r$ with the subgroup $G_r \subset G \times G$ which corresponds to the Lie subalgebra $\mathfrak{g}_r \subset \mathfrak{g} \oplus \mathfrak{g}$. Fix $F \in \mathfrak{g}_r^*$ and let $\mathcal{O}_F \subset \mathfrak{g}_r^*$ be its coadjoint orbit. We want to apply to $\mathcal{O}_F$ the 'symplectic induction' procedure. Since half of G_r is acting by left translations and the other half by right translations, the action of G on the reduced space will be destroyed. However, the Casimir elements survive reduction.

Theorem 2. *(i) The reduced symplectic manifold* $T^*G \times \mathcal{O}_F // G_r$ *is canonically isomorphic to* $\mathcal{O}_F$. *(ii) Extend Casimir elements* $\mathcal{H} \in \mathcal{S}(\mathfrak{g})^{\mathfrak{g}} \simeq P(\mathfrak{g}^*)$ *to* $G \times G$-*invariant functions on* T^*G; *these functions admit reduction with respect to* G_r; *the reduced Hamiltonians coincide with those described in Theorem 1, namely,* $\mathcal{H}_{red} = i_r(\mathcal{H})|_{\mathcal{O}_F}$; *the quotient Hamiltonian flow on* $\mathcal{O}_F$ *which corresponds to* $\mathcal{H} \in \mathcal{S}(\mathfrak{g})^{\mathfrak{g}}$ *is described by the Lax equation.*

Theorem 2 allows to get explicit formulæ for the solutions of Lax equations in terms of the *factorization problem* in G. In these lectures we are mainly interested in its role for quantization.

The construction of Lax equations based on Theorem 1 may be applied to finite-dimensional semisimple Lie algebras [36]. However, its really important applications are connected with loop algebras, which possess sufficiently many Casimirs. We shall describe two examples both of which have interesting quantum counterparts: the *open Toda lattice* (this is the example first treated in [36]) and the so-called *generalized Gaudin model* (its quantum counterpart was originally proposed in [26]).

2.2   Two Examples

2.2.1 Generalized Toda Lattice.
The Toda Hamiltonian describes transverse oscillations of a 1-dimensional cyclic molecule with nearest-neighbour interaction around the equilibrium configuration. Thus the phase space is $\mathcal{M} = \mathbb{R}^{2n}$ with the canonical Poisson bracket; the Hamiltonian is

$$\mathcal{H} = \frac{1}{2}\sum_{i=1}^{n} p_i^2 + \sum_{i=1}^{n-1} \exp\left(q_i - q_{i+1}\right) + \exp\left(q_n - q_1\right).$$

Removing the last term in the potential energy gives the *open Toda lattice* whose behaviour is qualitatively different (all potentials are repulsive, and the Hamiltonian describes the scattering of particles in the conical valley

$$C_+ = \{q = (q_1, ..., q_n) \in \mathbb{R}^n;\ q_i - q_{i+1} \le 0,\ i = 1, ..., n-1\}$$

with steep walls). The *generalized open Toda lattice* may be associated with an arbitrary semisimple Lie algebra. (Generalized *periodic* Toda lattices correspond to affine Lie algebras.) I shall briefly recall the corresponding construction (cf. [27, 36, 45]).

Let $\mathfrak{g}$ be a real split semisimple Lie algebra, $\sigma : \mathfrak{g} \to \mathfrak{g}$ a Cartan involution, $\mathfrak{g} = \mathfrak{k} \oplus \mathfrak{p}$ the corresponding Cartan decomposition (i. e., $\sigma = id$ on $\mathfrak{k}$, $\sigma = -id$ on $\mathfrak{p}$). Fix a split Cartan subalgebra $\mathfrak{a} \subset \mathfrak{p}$; let $\Delta \subset \mathfrak{a}^*$ be the root system of $(\mathfrak{g}, a)$. For $\alpha \in \Delta$ let

$$\mathfrak{g}_\alpha = \{X \in \mathfrak{g};\ \mathrm{ad}\,H \cdot X = \alpha\left(H\right)X, H \in \mathfrak{a}\} \tag{2.9}$$

be the corresponding root space. Fix an order in the system of roots and let $\Delta_+ \subset \Delta$ be the set of positive roots, $P \subset \Delta_+$ the corresponding set of simple roots. Put $\mathfrak{n} = \oplus_{\alpha \in \Delta_+}\mathfrak{g}_\alpha$. Let $\mathfrak{b} = \mathfrak{a} + \mathfrak{n}$; recall that $\mathfrak{b}$ is a maximal solvable subalgebra in $\mathfrak{g}$ (Borel subalgebra). We have

$$\mathfrak{g} = \mathfrak{k}\dot{+}\mathfrak{a}\dot{+}\mathfrak{n} \tag{2.10}$$

(the Iwasawa decomposition). Equip $\mathfrak{g}$ with the standard inner product (the Killing form). Clearly, we have $\mathfrak{k}^\perp = \mathfrak{p}, \mathfrak{b}^\perp = \mathfrak{n}$; thus the dual of $\mathfrak{b}$ is modelled on $\mathfrak{p}$, and we may regard $\mathfrak{p}$ as a $\mathfrak{b}$-module with respect to the coadjoint representation. Let G be a connected split semisimple Lie group with finite center which corresponds to $\mathfrak{g}$, $G = ANK$ its Iwasawa decomposition which corresponds to the decomposition (2.10) of its Lie algebra.

Remark 3. The Iwasawa decomposition usually does *not* give rise to a Lie bialgebra structure on $\mathfrak{g}$; indeed, the subalgebras $\mathfrak{k}$ and $\mathfrak{b}$ are not isotropic (unless $\mathfrak{g}$ is a complex Lie algebra considered as an algebra over $\mathbb{R}$). In other words, the classical r-matrix associated with the decomposition (2.10) is not skew. We shall briefly recall the construction of the so-called *standard skew-symmetric classical r-matrix* on a semisimple Lie algebra in Section 2.2.2 below.

The Borel subalgebra admits a decreasing filtration $\mathfrak{b} \supset \mathfrak{b}^{(1)} \supset \mathfrak{b}^{(2)} \supset ...$, where $\mathfrak{b}^{(1)} = [\mathfrak{b}, \mathfrak{b}] = \mathfrak{n}, \mathfrak{b}^{(2)} = [\mathfrak{n}, \mathfrak{n}], ..., \mathfrak{b}^{(k)} = \left[\mathfrak{n}, \mathfrak{b}^{(k-1)}\right]$. By duality, there is an increasing filtration of $\mathfrak{p} \simeq \mathfrak{b}^*$ by $\mathrm{ad}^* \mathfrak{b}$ -invariant subspaces. To describe it let us introduce some notation. For $\alpha \in \Delta$ let

$$\alpha = \sum_{\alpha_i \in P} m_i \alpha_i \tag{2.11}$$

be its decomposition with respect to simple roots; set $d(\alpha) = \sum_{\alpha_i \in P} m_i$. Let

$$\mathfrak{d}_p = \bigoplus_{\{\alpha \in \Delta; d(\alpha) = p\}} \mathfrak{g}_\alpha, \ p \neq 0, \ \mathfrak{d}_0 = \mathfrak{a}, \ \mathfrak{n}^{(k)} = \bigoplus_{p \geq k} \mathfrak{d}_p. \tag{2.12}$$

Define the mapping $s : \mathfrak{g} \to \mathfrak{p} : X \longmapsto \frac{1}{2}(id - \sigma)X$, and set

$$\mathfrak{p}_k = \bigoplus_{0 \leq p \leq k} s(\mathfrak{d}_p), \ k = 0, 1, \ldots$$

It is easy to see that the subspaces $\mathfrak{p}_0 = \mathfrak{a} \subset \mathfrak{p}_1 \subset \ldots$ are invariant with respect to the coadjoint action of $\mathfrak{b}$; moreover, $\mathfrak{p}_k$ is the annihilator of $\mathfrak{b}^{(k+1)}$ in $\mathfrak{p}$. Thus $\mathfrak{p}_k$ is set in duality with $\mathfrak{b}/\mathfrak{b}^{(k+1)}$. Put $\mathfrak{s} = \mathfrak{b}/\mathfrak{b}^{(2)}$ and let $L \in \mathfrak{p}_1 \otimes \mathfrak{s} \subset \mathfrak{p}_1 \otimes \mathcal{S}(\mathfrak{s})$ be the corresponding canonical element. Let $I_\mathfrak{s}$ be the specialization to $\mathfrak{s}$ of the subalgebra $I = \mathcal{S}(\mathfrak{g})^\mathfrak{g}$ of Casimir elements. An example of a Hamiltonian $H \in I_\mathfrak{s}$ is constructed from the Killing form:

$$H = \frac{1}{2}(L, L), \tag{2.13}$$

in obvious notation. To describe all others we may use the following trick. Fix a faithful representation (ρ, V) of $\mathfrak{g}$ and put $L_V = (\rho \otimes id)L$. Then (ρ, V) is usually called the *auxiliary linear representation*.

Proposition 5. (i) $I_\mathfrak{s}$ is generated by $H_k = \mathrm{tr}_V L_V^k \in \mathcal{S}(\mathfrak{s}), k = 1, 2, \ldots$ (ii) L defines a Lax representation for any Hamiltonian equation on $\mathcal{S}(\mathfrak{s})$ with Hamiltonian $H \in I_\mathfrak{s}$.

We have $\mathfrak{s} = \mathfrak{a} \ltimes \mathfrak{u}$, where $\mathfrak{u} = \mathfrak{n}/[\mathfrak{n}, \mathfrak{n}]$; the corresponding Lie group is $S = A \ltimes U, U = N/N'$. Choose $e_\alpha \in \mathfrak{g}_\alpha, \alpha \in P$, in such a way that $(e_\alpha, e_{-\alpha}) = 1$, and put

$$\mathcal{O}_T = \left\{ p + \sum_{\alpha \in P} b_\alpha (e_\alpha + e_{-\alpha}) ; p \in \mathfrak{a}, b_\alpha > 0 \right\}; \tag{2.14}$$

by construction, $\mathcal{O}_T \subset \mathfrak{p}_1$. It is convenient to introduce the parametrization $b_\alpha = \exp \alpha(q), q \in \mathfrak{a}$.

Proposition 6. *(i)* $\mathcal{O}_T$ *is an open coadjoint orbit of S in* $\mathfrak{p}_1 \simeq \mathfrak{s}^*$ *(ii) Restriction to $\mathcal{O}_T$ of the Hamiltonian* (2.13) *is the generalized Toda Hamiltonian*

$$H_T = \frac{1}{2} (p,p) + \sum_{\alpha \in P} \exp 2\alpha (q) \,.$$

The Toda orbit has the natural structure of a polarized symplectic manifold, that is, $\mathcal{O}_T$ admits an S-invariant fibering whose fibers are Lagrangian submanifolds; the description of this Lagrangian structure is a standard part of the geometric quantization program [33]. More precisely, put $f = \sum_{\alpha \in P} (e_\alpha + e_{-\alpha})$; let us choose f as a marked point on $\mathcal{O}_T$.

Proposition 7. *(i) We have $S/U \simeq A$; the isomorphism $\mathcal{O}_T = S \cdot f$ induces a Lagrangian fibering $\mathcal{O}_T \to A$ with fiber U. (ii) $\mathcal{O}_T$ is isomorphic to T^*A as a polarized symplectic manifold. (iii) The Lagrangian subalgebra $\mathfrak{l}_f \subset \mathfrak{s}$ subordinate to f is $\mathfrak{l}_f = \mathfrak{u}$.*

Remark 4. We may of course regard $\mathcal{O}_T$ as a coadjoint orbit of the group $B = AN$ itself; indeed, there is an obvious projection $B \to S$ and its kernel lies in the stabilizer of f. The 'big' Lagrangian subgroup $\hat{L}_f \subset B$ subordinate to f coincides with N.

An alternative description of the open Toda lattice which provides a more detailed information on its behavior and survives quantization is based on Hamiltonian reduction. The following result is a version of Theorem 2. Let us consider again the Hamiltonian action of $G \times G$ on T^*G. To get the generalized Toda lattice we shall restrict this action to the subgroup $G_r = K \times B \subset G \times G$. We may regard $\mathcal{O}_T$ as a G_r-orbit, the action of K being trivial.

Theorem 3. *(i) $T^*G \times \mathcal{O}_T // G_r$ is isomorphic to $\mathcal{O}_T$ as a symplectic manifold. (ii) Let $\mathcal{H}_2$ be the quadratic Casimir element which corresponds to the Killing form on $\mathfrak{g}$. The Toda flow is the reduction of the Hamiltonian flow on T^*G generated by $\mathcal{H}_2$.*

Theorem 3 leads to explicit formulæ for the trajectories of the Toda lattice in terms of the Iwasawa decomposition of matrices [27, 40, 45].

Since $\mathcal{O}_T$ admits a real polarization, we may state the following version of proposition 4. Observe that if we regard $\mathcal{O}_T$ as a G_r-orbit, the Lagrangian subgroup subordinate to $f \in \mathcal{O}_T$ is $K \times N$.

Theorem 4. *(i) $T^*G \times \{f\} // K \times N$ is isomorphic to $\mathcal{O}_T$ as a symplectic manifold. (ii) The standard Lagrangian polarization on T^*G by the fibers of projection $T^*G \to G$ gives rise to the Lagrangian polarization on $\mathcal{O}_T$ described in proposition 7.*

It is easy to see that the Toda flow is again reproduced as the reduction of the Hamiltonian flow on T^*G generated by the quadratic Hamiltonian $\mathcal{H}_2$. As discussed in Sect. 4.1.2, Theorem 4 gives a very simple hint to the solution of the quantization problem.

2.2.2 More Examples: Standard r-Matrices on Semisimple Lie Algebras and on Loop Algebras.

As already noted, the classical r-matrix associated with the Iwasawa decomposition of a simple Lie algebra $\mathfrak{g}$ is not skew and hence does not give rise to a Lie bialgebra structure on $\mathfrak{g}$. While this is not a disadvantage for the study of the Toda lattice, the bialgebra structure is of course basic in the study of the q-deformed case. Let us briefly recall the so-called *standard* Lie bialgebra structure on $\mathfrak{g}$. We shall explicitly describe the corresponding Manin triple.

Let $\mathfrak{g}$ be a complex simple Lie algebra, $\mathfrak{a} \subset \mathfrak{g}$ a Cartan subalgebra, $\Delta \subset \mathfrak{a}^*$ the root system of $(\mathfrak{g}, a)$, $\mathfrak{b}_+ = \mathfrak{a} + \mathfrak{n}_+$ a positive Borel subalgebra which corresponds to some choice of order in Δ, $\mathfrak{b}_-$ the opposite Borel subalgebra. For $\alpha \in \Delta$, let e_α be the corresponding root space vector normalized in such a way that $(e_\alpha, e_{-\alpha}) = 1$. We may identify $\mathfrak{a}$ with the quotient algebra $\mathfrak{b}_\pm/\mathfrak{n}_\pm$; let $\pi_\pm : \mathfrak{b}_\pm \to \mathfrak{a}$ be the corresponding canonical projection. Put $\mathfrak{d} = \mathfrak{g} \oplus \mathfrak{g}$ (direct sum of two copies); we equip $\mathfrak{d}$ with the inner product which is the *difference* of the Killing forms on the first and the second copy. Let $\mathfrak{g}^\delta \subset \mathfrak{d}$ be the diagonal subalgebra,

$$\mathfrak{g}^* = \{(X_+, X_-) \in \mathfrak{b}_+ \oplus \mathfrak{b}_-; \ \pi_+ (X_+) = -\pi_- (X_-)\}.$$

(Observe the important minus sign in this definition!)

Proposition 8. *(i)* $(\mathfrak{d}, \mathfrak{g}^\delta, \mathfrak{g}^*)$ *is a Manin triple. (ii) The corresponding Lie bialgebra structure on* $\mathfrak{g}$ *is associated with the classical r-matrix*

$$r = \sum_{\alpha \in \Delta_+} e_\alpha \wedge e_{-\alpha}. \tag{2.15}$$

(iii) The Lie bialgebra $(\mathfrak{g}^\delta, \mathfrak{g}^*)$ *is factorizable.*

The construction described above admits a straightforward generalization for loop algebras. The associated r-matrix is called *trigonometric*. For future reference we shall recall its definition as well.

Let $\mathfrak{g}$ be a complex semisimple Lie algebra, let $L\mathfrak{g} = \mathfrak{g} \otimes \mathbb{C}[z, z^{-1}]$ be the *loop algebra* of $\mathfrak{g}$ consisting of rational functions with values in $\mathfrak{g}$ which are regular on $\mathbb{C}P_1 \smallsetminus \{0, \infty\}$. Let $L\mathfrak{g}_0, L\mathfrak{g}_\infty$ be its local completions at 0 and ∞; by definition, $L\mathfrak{g}_0, L\mathfrak{g}_\infty$ consist of formal Laurent series in local parameters z, z^{-1}, respectively. Put $\widehat{\mathfrak{d}} = L\mathfrak{g}_0 \oplus L\mathfrak{g}_\infty$ (direct sum of Lie algebras). Clearly, the diagonal embedding $L\mathfrak{g} \hookrightarrow \widehat{\mathfrak{d}} : X \longmapsto (X, X)$ is a homomorphism of Lie algebras. Let

$$L\mathfrak{g}_\pm = \mathfrak{g} \otimes \mathbb{C}[[z^{\pm 1}]],$$

$$\widehat{\mathfrak{b}}_+ = \{X \in L\mathfrak{g}_+; \ X(0) \in \mathfrak{b}_+\}, \quad \widehat{\mathfrak{b}}_- = \{X \in L\mathfrak{g}_-; \ X(\infty) \in \mathfrak{b}_-\}$$

(where $X(0), X(\infty)$ denote the constant term of the formal series). Combining the 'evaluation at 0' (respectively, at ∞) with the projections $\pi_\pm : \mathfrak{b}_\pm \to \mathfrak{a}$ we get two canonical projection maps $\widehat{\pi}_\pm : \widehat{\mathfrak{b}}_\pm \to \mathfrak{a}$. Put

$$(L\mathfrak{g})^* = \left\{(X_+, X_-) \in \widehat{\mathfrak{b}}_+ \oplus \widehat{\mathfrak{b}}_-; \ \widehat{\pi}_+ (X_+) = -\widehat{\pi}_- (X_-)\right\} \subset \widehat{\mathfrak{d}}.$$

Let us set $(L\mathfrak{g})^*$ and $L\mathfrak{g}$ in duality by means of the bilinear form on $\widehat{\mathfrak{d}}$

$$\langle (X_+, X_-), (Y_+, Y_-) \rangle =$$
$$\text{Res}_{z=0} \left(X_+(z), Y_+(z) \right) dz/z + \text{Res}_{z=\infty} \left(X_-(z), Y_-(z) \right) dz/z. \qquad (2.16)$$

Proposition 9. $\left(\widehat{\mathfrak{d}}, L\mathfrak{g}, (L\mathfrak{g})^* \right)$ *is a Manin triple.*

Remark 5. Properly speaking, with our choice of a completion for $(L\mathfrak{g})^*$ the Lie bialgebra $\left(L\mathfrak{g}, (L\mathfrak{g})^* \right)$ is *not* factorizable; indeed, our definition requires that the Lie algebra and its dual should be isomorphic as linear spaces. However, a slightly weaker assertion still holds true: the dual Lie algebra $(L\mathfrak{g})^*$ contains an *open dense subalgebra*

$$(L\mathfrak{g})^\circ = \left\{ (X_+, X_-) \in (L\mathfrak{g})^* ; X_\pm \in L\mathfrak{g} \right\}$$

such that the mappings $(L\mathfrak{g})^\circ \rightarrow L\mathfrak{g} : (X_+, X_-) \longmapsto X_\pm$ are Lie algebra homomorphisms and $(L\mathfrak{g})^\circ$ is isomorphic to $L\mathfrak{g}$ as a linear space. This subtlety is quite typical of infinite-dimensional Lie algebras.

Let us finally say a few words on the trigonometric r-matrix itself. Define the cobracket $\delta : L\mathfrak{g} \rightarrow L\mathfrak{g} \otimes L\mathfrak{g}$ by

$$\langle \delta X, Y \otimes Z \rangle = \left\langle X, [Y, Z]_{L\mathfrak{g}^*} \right\rangle, \quad X \in L\mathfrak{g},\ Y, Z \in (L\mathfrak{g})^*, \qquad (2.17)$$

i. e., as the dual of the Lie bracket in $(L\mathfrak{g})^*$ with respect to the pairing (2.16). The element $\delta(X) \in L\mathfrak{g} \otimes L\mathfrak{g}$ may be regarded as a Laurent polynomial in two variables with values in $\mathfrak{g}$; the cobracket (2.17) is then given by

$$\delta(X)(z, w) = \left[r_{trig} \left(\frac{z}{w} \right), X_1(z) + X_2(w) \right],$$

where $r_{trig}(z/w)$ is a rational function in z/w with values in $\mathfrak{g} \otimes \mathfrak{g}$. (We use the dummy indices $1, 2, \ldots$ to denote different copies of linear spaces; in other words,

$$X_1(z) := X(z) \otimes 1, \quad X_2(w) := 1 \otimes X(w);$$

below we shall frequently use this abridged tensor notation.) The explicit expression for r_{trig} is given by

$$r_{trig}(x) = t \cdot \frac{x+1}{x-1} + r, \qquad (2.18)$$

where $t \in \mathfrak{g} \otimes \mathfrak{g}$ is the canonical element (*tensor Casimir*) which represents the inner product $\langle\,,\rangle$ in $\mathfrak{g}^1$ and $r \in \mathfrak{g} \wedge \mathfrak{g}$ is the standard classical r-matrix in $\mathfrak{g}$ given by (2.15).

Remark 6. Note that the expression for r_{trig} which may be found in [3] is different from (2.18), since these authors use a different grading in the loop algebra (the so-called *principal grading*, as compared to the *standard grading* which we use in these lectures.)

---

[1] More precisely, the tensor Casimir $t \in \mathfrak{g} \otimes \mathfrak{g}$ is the image of the canonical element $C \in \mathfrak{g} \otimes \mathfrak{g}^*$ under the isomorphism $\mathfrak{g} \otimes \mathfrak{g}^* \simeq \mathfrak{g} \otimes \mathfrak{g}$ induced by the inner product.

2.2.3 Generalized Gaudin Model. We shall return to the bialgebras described in the previous section when we shall discuss the q-deformed algebras and the quadratic Poisson Lie groups. In order to deal with our next example of an integrable system we shall use another important Lie bialgebra associated with the Lie algebra of rational functions. The example to be discussed is a model of spin-spin interaction (the so-called *Gaudin model*) which may be deduced as a limiting case of lattice spin models (such as the Heisenberg XXX model) for small values of the coupling constant [26]. The original Gaudin model is related to the $sl\,(2)$ or $su\,(2)$ algebras; below we shall discuss its generalization to arbitrary simple Lie algebras. The generalized Gaudin Hamiltonians also arise naturally in the semiclassical approximation to the Knizhnik-Zamolodchikov equations [43]. The underlying r-matrix associated with the Gaudin model is the so called *rational r-matrix* which is described below.

Let $\mathfrak{g}$ be a complex simple Lie algebra. Fix a finite set $D = \{z_1, ..., z_N, \infty\}$ $\subset \mathbb{C}P_1$ and let $\mathfrak{g}(D)$ be the algebra of rational functions on $\mathbb{C}P_1$ with values in $\mathfrak{g}$ which are regular outside D. Set $\mathfrak{g}_{z_i} = \mathfrak{g} \otimes \mathbb{C}((z - z_i))$, $z_i \neq \infty$, and $\mathfrak{g}_\infty = \mathfrak{g} \otimes \mathbb{C}((z^{-1}))$. By definition, $\mathfrak{g}_{z_i}$ is the *localization* of $\mathfrak{g}(D)$ at $z_i \in D$. [As usual, we denote by $\mathbb{C}((z))$ the algebra of formal Laurent series in z, and by $\mathbb{C}\,[[z]]$ its subalgebra consisting of formal Taylor series.] Put $\mathfrak{g}_D = \oplus_{z_i \in D} \mathfrak{g}_{z_i}$. There is a natural embedding $\mathfrak{g}(D) \hookrightarrow \mathfrak{g}_D$ which assigns to a rational function the set of its Laurent expansions at each point $z_i \in D$. Put $\mathfrak{g}_{z_i}^+ = \mathfrak{g} \otimes \mathbb{C}\,[[z - z_i]]$ for $z_i \neq \infty$ and $\mathfrak{g}_\infty^+ = \mathfrak{g} \otimes z^{-1}\mathbb{C}\,[[z^{-1}]]$. Let $\mathfrak{g}_D^+ = \oplus_{z_i \in D}\,\mathfrak{g}_{z_i}$. Then

$$\mathfrak{g}_D = \mathfrak{g}_D^+ \dotplus \mathfrak{g}(D) \qquad (2.19)$$

as a linear space.

[This assertion has a very simple meaning. Fix an element $X = (X_i)_{z_i \in D} \in \mathfrak{g}_D$; truncating the formal series X_i, we get a finite set of Laurent polynomials which may be regarded as principal parts of a rational function. Let X^0 be the rational function defined by these principal parts; it is unique up to a normalization constant. In order to resolve this potential ambiguity we have modified the definition of the algebra $\mathfrak{g}_\infty^+$ (dropping the constant term in the formal Taylor series). Expand X^0 in its Laurent series at $z = z_i$; by construction, $X_i - X_i^0 \in \mathfrak{g}_{z_i}^+$; thus $P : X \longmapsto X_0$ is a projection operator from $\mathfrak{g}_D$ onto $\mathfrak{g}(D)$ parallel to $\mathfrak{g}_D^+$. In other words, the direct sum decomposition (2.19) is equivalent to the existence of a rational function with prescribed principal parts; this is the assertion of the well known Mittag-Leffler theorem in complex analysis.]

Fix an inner product on $\mathfrak{g}$ and extend it to $\mathfrak{g}_D$ by setting

$$\langle X, Y \rangle = \sum_{z_i \in D} \mathrm{Res}\,(X_i, Y_i)\,dz, \quad X = (X_i)_{z_i \in D}\ Y = (Y_i)_{z_i \in D}\,. \qquad (2.20)$$

(One may notice that the key difference from the formula (2.16) in the previous section is in the choice of the differential dz instead of dz/z.) Both subspaces $\mathfrak{g}(D), \mathfrak{g}_D^+$ are isotropic with respect to the inner product (2.20) which sets them in duality. Thus $\left(\mathfrak{g}_D, \mathfrak{g}_D^+, \mathfrak{g}(D)\right)$ is a Manin triple. The linear space $\mathfrak{g}(D)$ may be regarded as a $\mathfrak{g}_D^+$-module with respect to the coadjoint representation. The action of $\mathfrak{g}_D^+$ preserves the natural filtration of $\mathfrak{g}(D)$ by the order of poles. In particular, rational functions with simple poles at finite points $z_i \neq \infty$ form an invariant subspace $\mathfrak{g}(D)_1 \subset \mathfrak{g}(D)$. Let $\mathfrak{g}_{z_i}^{++} \subset \mathfrak{g}_{z_i}^+$ be the subalgebra consisting of formal series without constant terms. Put

$$\mathfrak{g}_D^{++} = \mathfrak{g}_\infty^+ \oplus \bigoplus_{z_i \neq \infty} \mathfrak{g}_{z_i}^{++};$$

clearly, $\mathfrak{g}_D^{++}$ is an ideal in $\mathfrak{g}_D^+$ and its action on $\mathfrak{g}(D)_1$ is trivial.[2] The quotient algebra $\mathfrak{g}_D^+/\mathfrak{g}_D^{++}$ is isomorphic to $\mathfrak{g}^N = \oplus_{z_i \neq \infty} \mathfrak{g}$. The inner product (2.20) sets the linear spaces $\mathfrak{g}(D)_1$ and $\mathfrak{g}^N$ into duality. Let $L(z) \in \mathfrak{g}(D)_1 \otimes \mathfrak{g}^N$ be the canonical element; we shall regard $L(z)$ as a matrix-valued rational function with coefficients in $\mathfrak{g}^N \subset S(\mathfrak{g}^N)$. Fix a faithful linear representation (ρ, V) of $\mathfrak{g}$ (the *auxiliary linear representation*); it extends canonically to a representation $(\rho, V(z))$ of the Lie algebra $\mathfrak{g}(z) = \mathfrak{g} \otimes \mathbb{C}(z)$ in the space $V(z) = V \otimes \mathbb{C}(z)$. (In Sect. 2.2.1 we have already used this auxiliary linear representation for a similar purpose, cf. proposition 5.) Put

$$L_V(z) = (\rho \otimes id)L(z). \tag{2.21}$$

The matrix coefficients of $L_V(z)$ generate the algebra of observables $S(\mathfrak{g}^N)$. The Poisson bracket relations in this algebra have a nice expression in 'tensor form', the brackets of the matrix coefficients of L_V forming a matrix in $\text{End } V \otimes \text{End } V$ with coefficients in $\mathbb{C}(u,v) \otimes S(\mathfrak{g}^N)$; the corresponding formula, suggested for the first time by [50], was the starting point of the whole theory of classical r-matrices. To describe it let us first introduce the *rational r-matrix*,

$$r_V(u,v) = \frac{t_V}{u - v}, \tag{2.22}$$

Here t is the *tensor Casimir* of $\mathfrak{g}$ which corresponds to the inner product in $\mathfrak{g}$ (i. e., $t = \sum e_a \otimes e_a$, where $\{e_a\}$ is an orthonormal basis in $\mathfrak{g}$) and $t_V = \rho_V \otimes \rho_V(t)$. Notice that $r_V(u,v)$ is essentially the Cauchy kernel solving the Mittag-Leffler problem on $\mathbb{C}P_1$ with which we started.

Proposition 10. *The matrix coefficients of $L_V(u)$ satisfy the Poisson bracket relations*

$$\{L_V(u) \overset{\otimes}{,} L_V(v)\} = [r_V(u,v), L_V(u) \otimes 1 + 1 \otimes L_V(v)]. \tag{2.23}$$

---

[2] Observe the special role of the point at infinity: since the local algebra $\mathfrak{g}_\infty^+$ consists of formal series without constant terms, its coadjoint orbits are different from those of other local algebras. For simplicity we consider only trivial orbits of this exceptional algebra; this corresponds to the subspace of rational functions (with values in $\mathfrak{g}$) which are regular at infinity. The obvious reason for which it is impossible to discard the contribution of infinity is the standard residue theorem: the sum of all residues of a rational function (including the residue at infinity) is zero.

We shall now specialize corollary 1 to the present setting; it shows that, in a way, the spectral invariants of the Lax operator $L(z)$ may be regarded as 'radial parts' of the Casimir elements.

Remark 7. One should use some caution, since Casimir elements do not lie in the symmetric algebra $\mathcal{S}(\mathfrak{g}_D)$ itself but rather in its appropriate local completion; however, their projections onto $\mathcal{S}(\mathfrak{g}^N)$ are well defined. More precisely, for each $z_i \in D$ consider the projective limit

$$\tilde{\mathcal{S}}(\mathfrak{g}_{z_i}) = \varprojlim_{n \to +\infty} \mathcal{S}(\mathfrak{g}_{z_i}) / \mathcal{S}(\mathfrak{g}_{z_i}) (\mathfrak{g} \otimes z_i^n \mathbb{C}[[z_i]])$$

and set $\tilde{\mathcal{S}}(\mathfrak{g}_D) = \otimes_i \tilde{\mathcal{S}}(\mathfrak{g}_{z_i})$. Passing to the dual language, let us observe that the inner product on $\mathfrak{g}_D$ induces the evaluation map $\mathfrak{g}(D) \times \mathcal{S}(\mathfrak{g}_D) \to \mathbb{C}$; the completion is chosen in such a way that this map makes sense for $\tilde{\mathcal{S}}(\mathfrak{g}_D)$. To relate Casimir elements lying in $\tilde{\mathcal{S}}(\mathfrak{g}_D)$ to the more conventional spectral invariants, let us notice that if $L(z) \in \mathfrak{g}(D)$ is a 'Lax matrix', its spectral invariants, e.g., the coefficients of its characteristic polynomial

$$P(z, \lambda) = \det(L(z) - \lambda) = \sum \sigma_k(z) \lambda^k, \qquad (2.24)$$

are rational functions in z. To get numerical invariants we may expand $\sigma_k(z)$ in a local parameter (at any point $\zeta \in \mathbb{C}P_1$) and take any coefficient of this expansion. The resulting functionals are well defined as polynomial mappings from $\mathfrak{g}(D)$ into $\mathbb{C}$. It is easy to see that any such functional is obtained by applying the evaluation map to an appropriate Casimir element $\zeta \in \tilde{\mathcal{S}}(\mathfrak{g}_D)^{\mathfrak{g}_D}$.

Clearly, we have $\mathfrak{g}_D = \mathfrak{g}(D) \dotplus \mathfrak{g}^N \dotplus \mathfrak{g}_D^{++}$, and hence

$$\mathcal{S}(\mathfrak{g}_D) = \mathcal{S}(\mathfrak{g}^N) \oplus (\mathfrak{g}(D)\,\mathcal{S}(\mathfrak{g}_D) + \mathcal{S}(\mathfrak{g}_D)\,\mathfrak{g}_D^{++}).$$

Let $\gamma : \mathcal{S}(\mathfrak{g}_D) \to \mathcal{S}(\mathfrak{g}^N)$ be the projection map associated with this decomposition.

Proposition 11. *The restriction of γ to the subalgebra $\tilde{\mathcal{S}}(\mathfrak{g}_D)^{\mathfrak{g}_D}$ of Casimir elements is a morphism of Poisson algebras; under the natural pairing $\mathcal{S}(\mathfrak{g}^N) \times \mathfrak{g}(D)_1 \to \mathbb{C}$ induced by the inner product (2.20) the restricted Casimirs coincide with the spectral invariants of $L(z)$.*

The mapping γ is an analogue of the *Harish-Chandra homomorphism* ([10]); its definition may be extended to the quantum case as well.

Corollary 2. *Spectral invariants of $L(z)$ are in involution with respect to the standard Lie–Poisson bracket on $\mathfrak{g}^N$; $L(z)$ defines a Lax representation for any of these invariants (regarded as a Hamiltonian on $(\mathfrak{g}^N)^* \simeq \mathfrak{g}^N$.)*

The generalized Gaudin Hamiltonians are, by definition, the quadratic Hamiltonians contained in this family; one may take, e.g.,

$$\mathcal{H}_V(u) = \frac{1}{2} \mathrm{tr}_V\, L(u)^2. \qquad (2.25)$$

(Physically, they describe, e.g., the bilinear interaction of several 'magnetic momenta'.)

Corollary 2 does not mention the 'global' Lie algebra $\mathfrak{g}_D$ (and may in fact be proved by elementary means). However, of the three Lie algebras involved, it is probably the most important one, as it is responsible for the dynamics of the system. It is this 'global' Lie algebra that may be called the *hidden symmetry algebra*.

Setting $P(z, \lambda) = 0$ in (2.24), we get an affine algebraic curve (more precisely, a *family of curves* parametrized by the values of commuting Hamiltonians); this switches on the powerful algebro-geometric machinery and makes the complete integrability of the generalized Gaudin Hamiltonians almost immediate [45].

Remark 8. In the exposition above we kept the divisor D fixed. A more invariant way to deal with these problems is of course to use adélic language. Thus one may replace the algebra $\mathfrak{g}_D$ with the global algebra $\mathfrak{g}_{\mathbf{A}}$ defined over the ring $\mathbf{A}$ of adèles of $\mathbb{C}(z)$; fixing a divisor amounts to fixing a Poisson subspace inside the Lie algebra $\mathfrak{g}(z)$ of rational functions (which is canonically identified with the dual space of the subalgebra $\mathfrak{g}_{\mathbf{A}}^{+}$). The use of adélic freedom allowing us to add new points to the divisor is essential in the treatment of the quantum Gaudin model.

Remark 9. Further generalizations consist in replacing the rational r-matrix with more complicated ones. One is of course tempted to repeat the construction above, replacing $\mathbb{C}P_1$ with an arbitrary algebraic curve. There are obvious obstructions which come from the Mittag-Leffler theorem for curves: the subalgebra $\mathfrak{g}(C)$ of rational functions on the curve C does not admit a complement in $\mathfrak{g}_{\mathbf{A}}$ which is again a Lie subalgebra; for elliptic curves (and $\mathfrak{g} = \mathfrak{sl}(n)$) this problem may be solved [3,45] by considering *quasiperiodic* functions on C, and in this way we recover elliptic r-matrices (which were originally the first example of classical r-matrices ever considered [50])!

3   Quadratic Case

The integrable systems which we have considered so far are modelled on Poisson submanifolds of linear spaces, namely of the dual spaces of appropriate Lie algebras equipped with the Lie–Poisson bracket. As we shall see, quantization of such systems (whenever possible) still leaves us in the realm of ordinary (i. e., non-quantum) Lie groups and algebras. Let us note in passing that, in the context of the Integrability phenomena, the name 'Quantum Groups' leads to some confusion: one is tempted to believe that quantization of integrable systems implies that the corresponding 'hidden symmetry groups' also become quantum. As a matter of fact, the real reason to introduce Quantum groups is *different*. (This is clear from the fact that the Planck constant and the deformation parameter q which enters the definition of Quantum groups are independent of each other!) The point is that we are usually interested not in *individual* integrable systems, but rather in *families* of such systems with an arbitrary number of 'particles'. (This is particularly natural for problems arising in Quantum Statistical Mechanics.) A good approximation is obtained if one assumes that the phase space of a multiparticle system is the direct product of phase spaces for single particles. However, the definition of *observables* for multiparticle systems is not straightforward and tacitly assumes the existence of a Hopf structure (or some of its substitutes) on the algebra $\mathcal{A}$ of observables, i. e., of a map $\Delta^{(N)} : \mathcal{A} \to \otimes^N \mathcal{A}$ which embeds the algebra of observables of a single particle into that of the multiparticle system and thus allows us to speak of individual particles inside the complex system. For systems which are modelled on dual spaces of Lie algebras, the underlying Hopf structure is that of $S(\mathfrak{g})$ or $\mathcal{U}(\mathfrak{g})$ which is derived from the additive structure on $\mathfrak{g}^*$ (cf. Sect. 2). There is an important class of integrable systems (*difference Lax equations*) for which the natural Hopf structure is different; it is derived from the multiplicative structure in a nonabelian Lie group. Let me briefly recall the corresponding construction.

As already mentioned, difference Lax equations arise when the 'auxiliary linear problem' which underlies the construction is associated with a difference equation. A typical example is a first order difference system,

$$\psi_{n+1} = L_n \psi_n, \ n \in \mathbb{Z}. \tag{3.1}$$

The discrete variable $n \in \mathbb{Z}$ labels the points of an infinite 1-dimensional lattice; if we impose the periodic boundary condition $L_n = L_{n+N}$, it is replaced with a finite periodic lattice parametrized by $\mathbb{Z}/N\mathbb{Z}$. It is natural to assume that the Lax matrices L_n belong to a matrix Lie group G. Difference Lax equations arise as compatibility conditions for the linear system,

$$\begin{aligned} \psi_{n+1} &= L_n \psi_n, \\ \partial_t \psi_n &= A_n \psi_n, \end{aligned} \tag{3.2}$$

with an appropriately chosen A_n. They have the form of *finite-difference zero-curvature equations*,

$$\partial_t L_n = L_n A_{n+1} - A_n L_n. \tag{3.3}$$

Let Ψ be the fundamental solution of the difference system (3.1) normalized by $\Psi_0 = I$. The value of Ψ at $n = N$ is called the *monodromy matrix* for the periodic problem and is denoted by M_L. Clearly,

$$\begin{aligned} \Psi_n &= \widehat{\prod}_{0 \leq k < n} L_k, \\ M_L &= \widehat{\prod}_{0 \leq k < N} L_k. \end{aligned}$$

Spectral invariants of the difference operator associated with the periodic difference system (3.1) are the eigenvalues of the monodromy matrix. Thus one expects

$$I_s = \operatorname{tr} M_L^s, s = 1, 2, ...,$$

to be the integrals of motion for any Lax equation associated with the linear problem (3.1). This will hold if the monodromy matrix itself evolves isospectrally, i. e.,

$$\partial_t M_L = [M_L, A(M_L)] \tag{3.4}$$

for some matrix $A(M_L)$. Let us now discuss possible choices of the Poisson structure on the phase space which will turn (3.3) into Hamiltonian equations of motion. Observe first of all that the phase space of our system is the product $\mathbb{G} = \prod^N G$; if we assume that the variables corresponding to different copies of G are independent of each other, we may equip $\mathbb{G}$ with the product Poisson structure. The monodromy may be regarded as a mapping $M : \mathbb{G} \to G$ which assigns to a sequence $(L_0, ..., L_{N-1})$ the ordered product $M_L = \overset{\frown}{\prod} L_k$. Let $\mathbf{F}_t : \mathbb{G} \to \mathbb{G}$ be the (local) dynamical flow associated with equation (3.3) and $F_t : G \to G$ the corresponding flow associated with equation (3.4) for the monodromy. We expect the following diagram to be commutative:

$$
\begin{array}{ccc}
\mathbb{G} & \xrightarrow{\ \mathbf{F}_t\ } & \mathbb{G} \\
{\scriptstyle M}\big\downarrow & & \big\downarrow{\scriptstyle M} \\
G & \xrightarrow{\ F_t\ } & G
\end{array}
$$

In other words, the dynamics in $\mathbb{G}$ factorizes over G. It is natural to demand that all maps in this diagram should be Poissonian; in particular, the monodromy map $M : \mathbb{G} \to G$ should be compatible with the product structure on $\mathbb{G} = \prod^N G$. By induction, this property is reduced to the following one:

Multiplication $m : G \times G \to G$ is a Poisson mapping.

This is precisely the axiom introduced by [11] as a definition of Poisson Lie groups. In brief, one can say that the multiplicativity property of the Poisson bracket on G matches perfectly with the kinematics of one-dimensional lattice systems.

3.1   Abstract Case: Poisson Lie Groups and Factorizable Lie Bialgebras

To construct a lattice Lax system one may start with an arbitrary Manin triple, or, still more generally, with a factorizable Lie bialgebra. Let us briefly describe this construction which is parallel to the linear case but involves the geometry of Poisson Lie groups. Although the main applications are connected with infinite-dimensional groups (e.g., loop groups) it is more instructive to start with the finite-dimensional case.

Let $(\mathfrak{g}, g^*)$ be a finite-dimensional factorizable Lie bialgebra. Recall that in this case $r_{\pm} = \frac{1}{2}(r \pm id)$ are Lie algebra homomorphisms from $\mathfrak{g}^* \simeq \mathfrak{g}_r$ into $\mathfrak{g}$. Let G, G^* be the connected simply connected Lie groups which correspond to $\mathfrak{g}, g^*$, respectively. Let us assume that G is a linear group; let (ρ_V, V) be a faithful linear representation of G. We extend $r_{\pm}$ to Lie group homomorphisms from G^* into G which we denote by the same letters; we shall also write $h_{\pm} = r_{\pm}(h)$, $h \in G^*$. Both G and G^* carry a natural Poisson structure which makes them Poisson Lie groups. The description of the Poisson bracket on G is particularly simple. For $\varphi \in F(G)$ let $D\varphi, D'\varphi \in \mathfrak{g}^*$ be its left and right differentials defined by

$$\langle D\varphi(L), X \rangle = \frac{d}{dt}\varphi(\exp tX \cdot L)\,|_{t=0}, \qquad (3.5)$$

$$\langle D'\varphi(L), X \rangle = \frac{d}{dt}\varphi(L \exp tX)\,|_{t=0}, \; L \in G, \; X \in \mathfrak{g}.$$

The *Sklyanin bracket* on G is defined by

$$\{\varphi, \psi\} = \langle r, D'\varphi \wedge D'\psi \rangle - \langle r, D\varphi \wedge D\psi \rangle. \qquad (3.6)$$

Here is an equivalent description which makes sense when G is a linear group:

For $L \in G$ put $L_V = \rho_V(L)$; the matrix coefficients of L_V generate the affine ring $\mathcal{F}[G]$ of G. It is convenient not to fix a representation (ρ_V, V) but to consider all finite-dimensional representations of G simultaneously. The reason is that the Hamiltonians which are used to produce integrable systems are spectral invariants of the Lax matrix. To get a complete set of invariants one may consider either $\operatorname{tr}_V L_V^n$ for a *fixed* representation and all $n \geq 1$ or, alternatively, $\operatorname{tr}_V L_V$ for *all* linear representations. The second version is more convenient in the quantum case. If (ρ_W, W) is another linear representation, we get from the definition (3.6):

$$\{L_V \overset{\otimes}{,} L_W\} = [r_{VW}, L_V \otimes L_W], \text{ where } r_{VW} = (\rho_V \otimes \rho_W)(r). \qquad (3.7)$$

To avoid excessive use of tensor product signs, one usually sets

$$L_V^1 = L_V \otimes I, L_W^2 = I \otimes L_W;$$

formula (3.7) is then condensed to

$$\{L_V^1, L_W^2\} = [r_{VW}, L_V^1 L_W^2]. \qquad (3.8)$$

Let $I \subset \mathcal{F}[G]$ be the subalgebra generated by $\operatorname{tr}_V L_V$, $V \in \operatorname{Rep} G$. Formula (3.8) immediately implies the following assertion.

Proposition 12. *$I \subset \mathcal{F}[G]$ is a commutative subalgebra with respect to the Poisson bracket (3.6).*

The dual Poisson bracket on G^* may be described in similar terms. Extend $r_\pm$ to Lie group homomorphisms $r_\pm : G^* \to G$; it is easy to see that

$$i_r : G^* \longrightarrow G \times G : T \longmapsto (r_+ (T), r_- (T))$$

is an injection; hence G^* may be identified with a subgroup in $G \times G$. We shall write $r_\pm (T) = T_\pm$ for short. Let again $(\rho_V, V), (\rho_W, W)$ be two linear representation of G; for $T \in G^*$ put

$$T_V^\pm = \rho_V(T_\pm), \ T_W^\pm = \rho_W(T_\pm).$$

The matrix coefficients of $T_V^\pm$, $\rho_V \in \mathsf{Rep}\, G$, generate the affine ring $\mathcal{F}\,[G^*]$ of G^*. The Poisson bracket relations for the matrix coefficients of $T_V^\pm$, $T_W^\pm$ are given by

$$\{T_V^{\pm 1}, T_W^{\pm 2}\} = \left[r_{VW}, T_V^{\pm 1}T_W^{\pm 2}\right], \tag{3.9}$$
$$\{T_V^{+1}, T_W^{-2}\} = \left[r_{VW} + t_{VW}, T_V^{+1}T_W^{-2}\right];$$

here $t \in \mathfrak{g} \otimes \mathfrak{g}$ is the tensor Casimir which corresponds to the inner product on $\mathfrak{g}$ and $t_{VW} = (\rho_V \otimes \rho_W)(t)$. Consider the action

$$G^* \times G \to G : (h, x) \longmapsto h_+ x h_-^{-1}. \tag{3.10}$$

Lemma 2. G^* *acts locally freely on an open cell in G containing the unit element.*

In other words, almost all elements $x \in G$ admit a decomposition $x = h_+ h_-^{-1}$, where $h_\pm = r_\pm (h)$ for some $h \in G^*$. (Moreover, this decomposition is unique if both h and (h_+, h_-) are sufficiently close to the unit element.) This explains the term 'factorizable group'.

Lemma 2 allows us to push forward the Poisson bracket from G^* to G. More precisely, we have the following result.

Proposition 13. *There exists a unique Poisson structure $\{\,,\}^*$ on G such that*

$$F : G^* \to G : (x_+, x_-) \longmapsto x_+ x_-^{-1}$$

is a Poisson mapping; it is given by

$$\{\varphi, \psi\}^* = \langle r, D\varphi \wedge D\psi \rangle + \langle r, D'\varphi \wedge D'\psi \rangle \tag{3.11}$$
$$- \langle 2r_+, D\varphi \wedge D'\psi \rangle - \langle 2r_-, D'\varphi \wedge D\psi \rangle, \ \varphi, \psi \in C^\infty (G).$$

There is an equivalent description of the bracket $\{\,,\}^*$ for linear groups. If $L_V = \rho_V (L), L_W = \rho_W (L)$, then

$$\{L_V^1, L_W^2\}^* = r_{VW} L_V^1 L_W^2 + L_V^1 L_W^2 r_{VW} \tag{3.12}$$
$$- L_V^1 (r_{VW} + t_{VW}) L_W^2 - L_W^2 (r_{VW} - t_{VW}) L_V^1.$$

Remark 10. (i) A priori the bracket $\{\,,\,\}^*$ is defined only on the big cell in G (i. e., on the image of G^*); however, formulae (3.12,3.13) show that it extends smoothly to all of G. (ii) Observe that the choice of F is actually rigid: this is essentially the only combination of x_+, x_- such that the bracket for $x = F(x_+, x_-)$ is expressed in terms of x (not of its factors).

Recall that $I \subset \mathcal{F}[G]$ consists of spectral invariants; thus we are again in the setting of the generalized Kostant-Adler theorem: there are two Poisson structures on the same underlying space and the Casimirs of the former are in involution with respect to the latter.

Theorem 5. *Hamiltonians $\mathcal{H} \in I$ generate Lax equations on G with respect to the Sklyanin bracket.*

Geometrically, this means that the Hamiltonian flows generated by $\mathcal{H} \in I$ preserve two systems of symplectic leaves in G, namely, the symplectic leaves of $\{\,,\,\}$ and $\{\,,\,\}^*$. The latter coincide with the *conjugacy classes* in G (for a proof see [48]). In order to include lattice Lax equations into our general framework we have to introduce the notion of *twisting* which is discussed in the next section.

Remark 11. The Hamiltonian reduction picture discussed in Sect. 2.1.1 may be fully generalized to the present case as well [48]. To define the symplectic induction functor we now need the *nonabelian moment map* (see [44] for a review). In the present exposition we shall not describe this construction.

3.2 Duality Theory for Poisson Lie Groups and Twisted Spectral Invariants

Our key observation so far has been the duality between the Hamiltonians of integrable systems on a Lie group and the Casimir functions of its Poisson dual. In Sect. 4.2 we shall see that a similar relation holds for quantized universal enveloping algebras. In order to keep the parallel between the two cases as close as possible we must introduce into the picture still another ingredient: *the twisted Poisson structure on the dual group*. This notion will also allow us to include lattice Lax equations into our general framework. Roughly speaking, twisting is possible whenever there is an automorphism $\tau \in \operatorname{Aut}\mathfrak{g}$ which preserves the r-matrix (i. e., $(\tau \otimes \tau)r = r$). Outer automorphisms are particularly interesting. However, even in the case of inner automorphisms the situation does not become completely trivial. The natural explanation of twisting requires the full duality theory which involves the notions of the double, the Heisenberg double and the twisted double of a Poisson Lie group [49]. In the present exposition we shall content ourselves by presenting the final formulae for the twisted Poisson brackets. The structure of these formulae is fairly uniform; we list them starting with the case of finite-dimensional simple Lie algebras (where all automorphisms are inner) and then pass to the lattice case (which accounts for the treatment of Lax equations on the lattice) and to the case of affine Lie algebras. In the quantum case which will be considered in the next section twisting will play an important role in the description of quantum Casimirs.

3.2.1 Finite-Dimensional Simple Lie Groups.

We begin with a finite-dimensional example and return to the setting of Sect. 2.2.2. Let again $\mathfrak{g}$ be a complex simple Lie algebra, $\mathfrak{b}_+ = \mathfrak{a} + \mathfrak{n}_+$ a Borel subalgebra, $\mathfrak{b}_-$ the opposite Borel subalgebra, and $\pi_\pm : \mathfrak{b}_\pm \to \mathfrak{a}$ the canonical projection,

$$\mathfrak{g}^* = \{(X_+, X_-) \in \mathfrak{b}_+ \oplus \mathfrak{b}_- \subset \mathfrak{g} \oplus \mathfrak{g};\ \pi_+(X_+) = -\pi_-(X_-)\}.$$

Let $B_\pm \subset G$ be the corresponding Borel subgroups, and $\pi_\pm : B_\pm \to A$ the canonical projections,

$$G^* = \left\{ (x_+, x_-) \in B_+ \times B_-;\ \pi_+(x_+) = \pi_-(x_-)^{-1} \right\}. \tag{3.13}$$

Define $r_d \in \bigwedge^2 (\mathfrak{g} \oplus \mathfrak{g})$ by

$$r_d = \begin{pmatrix} r & -r-t \\ -r+t & r \end{pmatrix}, \tag{3.14}$$

where $t \in \mathfrak{g} \otimes \mathfrak{g}$ is the tensor Casimir. (One may notice that r_d equips $\mathfrak{g} \oplus \mathfrak{g}$ with the structure of a Lie bialgebra which coincides, up to an isomorphism, with that of the *Drinfeld double* of $\mathfrak{g}$.) Our next assertion specializes the results stated in the previous section. Let us define the Poisson bracket on G^* by the formula

$$\{\varphi, \psi\} = \langle\langle r_d, D'\varphi \wedge D'\psi \rangle\rangle - \langle\langle r_d, D\varphi \wedge D\psi \rangle\rangle, \tag{3.15}$$

where

$$D\varphi = \begin{pmatrix} D_+\varphi \\ D_-\varphi \end{pmatrix},\ D'\varphi = \begin{pmatrix} D'_+\varphi \\ D'_-\varphi \end{pmatrix}$$

are left and right differentials of φ with respect to its two arguments and $\langle\langle\,,\rangle\rangle$ is the natural coupling between $\bigwedge^2 (\mathfrak{g} \oplus \mathfrak{g})$ and its dual.

Proposition 14. *(i) G^* is the dual Poisson Lie group of G which corresponds to the standard r-matrix described in (2.15). (ii) The mapping*

$$G^* \to G : (x_+, x_-) \longmapsto x_+ x_-^{-1}$$

is a bijection of G^* onto the 'big Schubert cell' in G. (iii) The dual Poisson bracket on G^* extends smoothly from the big cell to all of G; it is explicitly given by (3.12) and its Casimirs are central functions on G.

The next definition prepares the introduction of *twisting*.

Definition 2. Let $(\mathfrak{g}, \mathfrak{g}_r)$ be a factorizable Lie bialgebra with the classical r-matrix $r \in \mathfrak{g} \otimes \mathfrak{g}$. An automorphism of $(\mathfrak{g}, \mathfrak{g}_r)$ is an automorphism $\varphi \in \mathrm{Aut}\,\mathfrak{g}$ which preserves the r-matrix and the inner product on $\mathfrak{g}$.

We have the following simple result.

Proposition 15. *Let $\mathfrak{g}$ be a finite-dimensional simple Lie algebra with the standard classical r-matrix. Then* $\mathrm{Aut}\,(\mathfrak{g}, \mathfrak{g}_r)$ *coincides with the Cartan subgroup* $A \subset G$.

The action of A on $\bigwedge^2 \mathfrak{g}$ is the restriction to $A \subset G$ of the wedge square of the standard adjoint action.

Remark 12. If $\mathfrak{g}$ is only semisimple, $\mathrm{Aut}\,(\mathfrak{g}, \mathfrak{g}_r)$ may have a nontrivial group of components.

The Sklyanin bracket on G is invariant with respect to the action $A \times G \to G$ by right translations. By contrast, the dual bracket is *not* invariant, and in this way we get a family of Poisson structures on G^* with different Casimir functions. More precisely, define the action $A \times G^* \to G^*$ by

$$h : (x_+, x_-) \longmapsto \left(hx_+, h^{-1}x_- \right) \tag{3.16}$$

(Note that this action is compatible with (3.13); the flip $h \longmapsto h^{-1}$ is possible, since A is abelian.) We denote by λ_h the contragredient action of h on $\mathcal{F}\,[G^*]$: $\lambda_h \varphi\,(x_+, x_-) = \varphi\left(h^{-1}x_+, hx_- \right)$. Define the twisted Poisson bracket $\{\,,\,\}_h^*$ on G^* by

$$\{\varphi, \psi\}_h^* = \lambda_h^{-1} \left\{ \lambda_h \varphi, \lambda_h \psi \right\}.$$

Explicitly we get

$$\{\varphi, \psi\}_h^* = \langle\langle r_d^h D'\varphi, D'\psi \rangle\rangle - \langle\langle r_d D\varphi, D\psi \rangle\rangle, \tag{3.17}$$

where

$$r_d^h = \overset{2}{\bigwedge} \left(Ad\, h \oplus Ad\, h^{-1} \right) \cdot r_d = \begin{pmatrix} r & (-r-t)^h \\ (-r+t)_{h^{-1}} & r \end{pmatrix},$$

or, in matrix form,

$$\{T_V^{\pm 1}, T_W^{\pm 2}\}_h^* = [r_{VW}, T_V^{\pm 1} T_W^{\pm 2}], \tag{3.18}$$
$$\{T_V^{+1}, T_W^{-2}\}_h^* = (r_{VW}^h + t_{VW}^h) T_V^{+1} T_W^{-2} - T_V^{+1} T_W^{-2} (r_{VW} + t_{VW}),$$

where $r_{VW}^h = (Ad\, h \otimes Ad\, h^{-1})(r_{VW}) = (Ad\, h^2 \otimes id)(r_{VW})$ and similarly for t_{VW}^h. Let us denote by $I_h\,(G)$ the set of *twisted spectral invariants*; by definition, $\phi \in I_h\,(G)$ if

$$\phi\,(gx) = \phi\,(xgh), \text{ for any } x, g \in G.$$

Example 1. Let (ρ, V) be a linear representation of G; then $\phi : x \longmapsto \mathrm{tr}\,_V \rho(hx)$ is a twisted spectral invariant.

Proposition 16. *(i) There exists a unique Poisson structure on G such that the twisted factorization map $F_h : G^* \to G : (x_+, x_-) \longmapsto h^2 x_+ x_-^{-1}$ is a Poisson mapping. (ii) F_h transforms the Casimir functions of $\{\,,\}_h^*$ into the set of twisted spectral invariants $I_{h^2}(G)$. (ii) Twisted spectral invariants also commute with respect to the Sklyanin bracket on G and generate on G generalized Lax equations*

$$\frac{dL}{dt} = AL - LB, \ A = Ad \ h^2 \cdot B.$$

It is instructive to write an explicit formula for the bracket on G which is the push-forward of $\{\,,\}_h^*$ with respect to the factorization map. We get

$$\{\varphi, \psi\}_h^* = \langle r, D\varphi \wedge D\psi \rangle + \langle r, D'\varphi \wedge D'\psi \rangle - \langle r^h + t^h, D\varphi \wedge D'\psi \rangle$$
$$- \left\langle r^{h^{-1}} - t^{h^{-1}}, D'\varphi \wedge D\psi \right\rangle, \ \varphi, \psi \in C^\infty(G), \tag{3.19}$$

where t is the tensor Casimir, and we put $r^h = (Ad \ h \otimes Ad \ h^{-1})(r)$ and $t^h = (Ad \ h \otimes Ad \ h^{-1})(t)$. This bracket again extends smoothly to the whole of G (cf. (3.12)). Equivalently,

$$\{L_V^1, L_W^2\}_h^* = r_{VW} L_V^1 L_V^2 + L_V^1 L_W^2 r_{VW} \tag{3.20}$$
$$- L_V^1 (r_{VW} + t_{VW})^h L_V^2 - L_W^2 (r_{VW} - t_{VW})^{h^{-1}} L_V^1$$

The twisted bracket $\{\,,\}_h^*$ on G^* is *not* multiplicative with respect to the group structure on G^*; to explain its relation to the Poisson group theory we shall need some more work.

Let $\mathfrak{a} \ltimes \mathfrak{g}^*$ be the semidirect product of Lie algebras which corresponds to action (3.16); in other words, we set

$$[H, (X_+, X_-)] = ([H, X_+], -[H, X_-]), H \in \mathfrak{a}, X_\pm \in \mathfrak{b}_\pm.$$

Fix a basis $\{H_i\}$ of $\mathfrak{a}$ and let $\{H^i\}$ be the dual basis of $\mathfrak{a}^*$. We define a (trivial) 2-cocycle on $\mathfrak{g}$ with values in $\mathfrak{a}^*$ by

$$\omega(X, Y) = \sum_i (H_i, [X, Y]) H^i. \tag{3.21}$$

Let $\widehat{\mathfrak{g}} = \mathfrak{g} \oplus \mathfrak{a}^*$ be the central extension of $\mathfrak{g}$ by $\mathfrak{a}^*$ which corresponds to this cocycle.

Proposition 17. $(\mathfrak{a} \ltimes \mathfrak{g}^*, \widehat{\mathfrak{g}})$ *is a Lie bialgebra.*

The Poisson Lie group which corresponds to $\mathfrak{a} \ltimes \mathfrak{g}^*$ is the semidirect product $A \ltimes G^*$ associated with action (3.16). It is easy to see that the variable $h \in A$ is central with respect to this bracket; thus the bracket on $A \ltimes G^*$ splits into a family of brackets on G^* parametrized by $h \in A$. This is precisely the family $\{\,,\}_h^*$.

3.2.2 Twisting on the Lattice. One may notice that the previous example is in fact trivial, since the twisted bracket differs from the original one by a change of variables[3]. This is of course closely related to the fact that the central extension associated with the cocycle (3.21) is also trivial. Our next example is more interesting; it is adapted to the study of Lax equations on the lattice. Put $\Gamma = \mathbb{Z}/N\mathbb{Z}$. Let again $\mathfrak{g}$ be a finite dimensional simple Lie algebra equipped with a standard r-matrix. Let $\mathbf{g} = \mathfrak{g}^\Gamma$ be the Lie algebra of functions on Γ with values in $\mathfrak{g}$; obviously, $\mathbf{g}$ inherits from $\mathfrak{g}$ the natural Lie bialgebra structure; the corresponding r-matrix $\mathsf{r} \in \bigwedge^2 \mathbf{g}$ is given by $\mathsf{r}_{mn} = \delta_{mn} r$, $m, n \in \Gamma$. We define $\mathsf{r}_d \in \bigwedge^2(\mathbf{g} \oplus \mathbf{g})$ by

$$\mathsf{r}_d = \begin{pmatrix} \mathsf{r} & -\mathsf{r} - \mathsf{t} \\ -\mathsf{r} + \mathsf{t} & \mathsf{r} \end{pmatrix}, \tag{3.22}$$

where $\mathsf{t} \in \mathbf{g} \otimes \mathbf{g}$ is the tensor Casimir which corresponds to the standard inner product on $\mathbf{g}$,

$$\langle\langle X, Y \rangle\rangle = \sum_n \langle X_n, Y_n \rangle.$$

Let τ be the cyclic permutation on $\mathbf{g}$, $(\tau X)_n = X_{n+1 \bmod N}$.

Lemma 3. $\tau \in \mathrm{Aut}\,(\mathbf{g}, \mathbf{g}^*)$.

Let us extend the action of τ to $\mathbf{g} \oplus \mathbf{g}$ in the following way:

$$\tau \cdot (X, Y) = (\tau X, Y).$$

(By analogy with (3.16), one might rotate the second copy of $\mathbf{g}$ in the opposite sense, but then the total shift in formulæ below would be by two units, which is less natural on the lattice.) Put

$$\mathsf{r}_d^\tau = \bigwedge^2 (\tau \oplus id) \cdot \mathsf{r}_d = \begin{pmatrix} \mathsf{r} & -\mathsf{r}^\tau - \mathsf{t}^\tau \\ -\mathsf{r}^{\tau^{-1}} + \mathsf{t}^{\tau^{-1}} & \mathsf{r} \end{pmatrix},$$

$$\mathsf{r}^\tau = (\tau \otimes id) \cdot \mathsf{r}, \quad \mathsf{r}^{\tau^{-1}} = (\tau^{-1} \otimes id) \cdot \mathsf{r} = (id \otimes \tau) \cdot \mathsf{r}.$$

Put $\mathbb{G} = G^\Gamma$, $\mathbb{G}^* = G^{*\,\Gamma}$. The action of τ extends to $\mathbb{G}$ and to $\mathbb{G}^*$ in an obvious way. As in Sect. 3.2.1 we may embed $\mathbb{G}^*$ into the direct product $\mathbb{G} \times \mathbb{G}$. By definition, the twisted Poisson bracket on $\mathbb{G}^*$ is given by

$$\{\varphi, \psi\}_\tau^* = \langle\langle \mathsf{r}_d^\tau, D\varphi \wedge D\psi \rangle\rangle - \langle\langle \mathsf{r}_d D'\varphi \wedge D'\psi \rangle\rangle,$$

where $D\varphi, D'\varphi \in \mathbf{g} \oplus \mathbf{g}$ are the two-component left and right differentials of φ. We may push forward the Poisson bracket from $\mathbb{G}^*$ to $\mathbb{G}$ using the factorization map. More precisely:

---

[3] However, even in this case, twisted spectral invariants give rise to a *different* set of Hamiltonians, as compared with the non-twisted case.

Proposition 18. *There exists a unique Poisson structure on $\mathbb{G}$ (which we shall still denote by $\{\,,\}^*_\tau$) such that the* twisted factorization map $\mathbb{G}^* \to \mathbb{G} : (x_+, x_-)$ $\longmapsto x_+^\tau x_-^{-1}$ *becomes a Poisson mapping. This Poisson structure is given by*

$$\{\varphi, \psi\}^*_\tau = \langle r, D\varphi \wedge D\psi \rangle + \langle r, D'\varphi \wedge D'\psi \rangle$$
$$- \langle r^\tau, D\varphi \wedge D'\psi \rangle - \langle r^{\tau^{-1}}, D'\varphi \wedge D\psi \rangle \qquad (3.23)$$
$$- \langle t^\tau, D\varphi \wedge D'\psi \rangle + \langle t^{\tau^{-1}}, D'\varphi \wedge D\psi \rangle, \ \varphi, \psi \in C^\infty(\mathbb{G}).$$

We shall denote the Lie group $\mathbb{G}$ equipped with the Poisson structure (3.24) by $\mathbb{G}_\tau$, for short. There is an equivalent description of the bracket $\{\,,\}^*_\tau$ for linear groups. Let again $L_V = \rho_V(L)$; then

$$\{L_V^1, L_V^2\}^* = rL_V^1 L_V^2 + L_V^1 L_V^2 r \qquad (3.24)$$
$$- L_V^1 (r+t)^\tau L_V^2 - L_V^2 (r-t)^{\tau^{-1}} L_V^1.$$

Another obvious Poisson structure on $\mathbb{G}$ is the Sklyanin bracket which corresponds to the Lie bialgebra $(\mathbf{g}, \mathbf{g}^*)$. Let us consider the following action of $\mathbb{G}$ on $\mathbb{G}_\tau$ called the (lattice) gauge action:

$$x : T \longmapsto x^\tau T x^{-1}.$$

Theorem 6. *(i) The gauge action $\mathbb{G} \times \mathbb{G}_\tau \to \mathbb{G}_\tau$ is a Poisson group action. (ii) The Casimir functions of $\{\,,\}^*_\tau$ coincide with the gauge invariants. (iii) Let $M : \mathbb{G}_\tau \to \mathbb{G}$ be the monodromy map,*

$$M : T = (T_1, T_2, \dots, T_N) \longmapsto T_N \cdots T_1.$$

We equip the target space with Poisson structure (3.12). Then M is a Poisson map and the Casimirs of $\mathbb{G}_\tau$ coincide with the spectral invariants of the monodromy. (iv) Spectral invariants of the monodromy are in involution with respect to the Sklyanin bracket on $\mathbb{G}$ and generate lattice Lax equations on $\mathbb{G}$.

Parallels with Theorem 1 are completely obvious.

3.2.3 Twisting for Loop Algebras.

Our last example of twisting is based on the outer automorphisms of loop algebras. We shall use the notation and definitions introduced in Sect. 2.2.2. Let $\mathbf{g}$ be a simple Lie algebra, $L\mathbf{g}$ the corresponding loop algebra with the standard trigonometric r-matrix, and let $L\mathbf{g}_r = (L\mathbf{g})^*$ be the dual algebra. Let G be the connected simply connected complex Lie group associated with $\mathbf{g}$, $B_\pm = AN_\pm$ its opposite Borel subgroups. The Lie group which corresponds to $L\mathbf{g}$ is the group LG of polynomial loops with values in G; the elements of the dual Poisson group may be identified with pairs of formal series $\left(T^+(z), T^-(z^{-1})\right)$,

$$T^\pm(z) = \sum_{n \geq 0} T^\pm[\pm n] z^{\pm n}, \ T^\pm[0] \in B_\pm, \ \pi_+\left(T^+[0]\right) \cdot \pi_-\left(T^-[0]\right) = 1$$

(here $\pi_\pm : B_\pm \to A$ are the canonical projections). The group $(LG)^*$ is pro-algebraic and its affine ring is generated by the matrix coefficients of $T^\pm[n]$. The next proposition describes the Poisson structure on the affine ring of $(LG)^*$. Let $t \in \mathfrak{g} \otimes \mathfrak{g}$ be the canonical element (*tensor Casimir*) which represents the inner product $\langle\,,\,\rangle$ in $\mathfrak{g}$. Let $\delta(x)$ be the Dirac delta,

$$\delta(x) = \sum_{n=-\infty}^{\infty} x^n.$$

(Observe that $t\,\delta(z/w)$ represents the kernel of the identity operator acting in $L\mathfrak{g}$. To put it in a different way, $t\,\delta(z/w)$ is another guise of the canonical element in $L\mathfrak{g} \otimes L\mathfrak{g}$.)

Proposition 19. *The Poisson bracket on* $(LG)^*$ *which corresponds to the standard Lie bialgebra structure on* $L\mathfrak{g}$ *is given by*

$$\{T_1^\pm(z), T_2^\pm(w)\} = \left[r\left(\frac{z}{w}\right), T_1^\pm(z)T_2^\pm(w)\right], \tag{3.25}$$

$$\{T_1^+(z), T_2^-(w)\} = \left[r\left(\frac{z}{w}\right), T_1^+(z)T_2^-(w)\right] + \delta\left(\frac{z}{w}\right)\left[t, T_1^+(z)T_2^-(w)\right],$$

where $r(x)$ *is the trigonometric r-matrix.*

We pass to the description of the twisted structure on $(LG)^*$.

Lemma 4. $\mathrm{Aut}\,(L\mathfrak{g}, L\mathfrak{g}_r)$ *is the extended Cartan subgroup* $\hat{A} = A \times \mathbb{C}^\times$.

Recall that $G \supset A$ is canonically embedded into LG as the subgroup of constant loops. By definition, $p \in \mathbb{C}^\times$ acts on $L\mathfrak{g}$ by sending $X(z)$ to $X(pz)$; we shall denote the corresponding dilation operator by D_p. The action of A gives nothing new, as compared with our first example (Sect. 2.2.1). To describe the twisting of the Poisson structure on $(LG)^*$ by the action of $\mathbb{C}^\times$ we start with the r-matrix

$$r_d(z/w) = \begin{pmatrix} r(z/w) & -r(z/w) - t\delta(z/w) \\ -r(z/w) + t\delta(z/w) & r(z/w) \end{pmatrix}$$

(as usual, $r_d(z/w)$ does not belong to $\bigwedge^2(L\mathfrak{g} \oplus L\mathfrak{g})$, but is well defined as a kernel of a linear operator acting in $L\mathfrak{g} \oplus L\mathfrak{g}$). For $p \in \mathbb{C}^\times$ we define the twisted r-matrix by

$$r_d(z/w)^p = \bigwedge^2 (D_p \oplus D_p^{-1}) \cdot r_d(z/w) =$$
$$\begin{pmatrix} r(z/w) & -r(pz/w) - t\delta(pz/w) \\ -r(z/pw) + t\delta(z/pw) & r(z/w) \end{pmatrix}.$$

Definition 3. The twisted Poisson bracket on $(LG)^*$ is given by

$$\{T_1^{\pm}(z), T_2^{\pm}(w)\}_p^* = \left[r\left(\frac{z}{w}\right), T_1^{\pm}(z)T_2^{\pm}(w)\right],$$

$$\{T_1^+(z), T_2^-(w)\}_p^* = r\left(\frac{pz}{w}\right)T_1^+(z)T_2^-(w) - T_1^+(z)T_2^-(w)r\left(\frac{z}{w}\right) \qquad (3.26)$$

$$+ \delta\left(\frac{pz}{w}\right)t\, T_1^+(z)T_2^-(w) - T_1^+(z)T_2^-(w)\delta\left(\frac{z}{w}\right)t\,.$$

Define the new generating function ('full current') by

$$T(z) = T^+(pz)T^-(p^{-1}z)^{-1}; \qquad (3.27)$$

clearly, the coefficients of $T(z)$ generate the affine ring of $(LG)^*$.

Proposition 20. *The current $T(z)$ satisfies the following Poisson bracket relations:*

$$\{T_1(z), T_2(w)\}_p^* =$$
$$r\left(\frac{z}{w}\right)T_1(z)T_2(w) + T_1(z)T_2(w)r\left(\frac{z}{w}\right) \qquad (3.28)$$

$$- T_1(z)r\left(\frac{p^2z}{w}\right)T_2(w) - T_2(w)r\left(\frac{z}{p^2w}\right)T_1(z)$$

$$- T_1(z)\delta\left(\frac{p^2z}{w}\right)t\, T_2(w) + T_2(w)\delta\left(\frac{z}{p^2w}\right)t\, T_1(z).$$

The existence of twisting in the affine case is due to the existence of a nontrivial central extension of the loop algebra. (This should be compared with the situation in our first example, where the central extension is trivial because all automorphisms of simple Lie algebras are inner.) More precisely, we have the following result. Let ω be the 2-cocycle on $L\mathfrak{g}$ defined by

$$\omega(X, Y) = \operatorname{Res}_{z=0}\langle X(z), z\partial_z Y(z)\rangle\, dz/z. \qquad (3.29)$$

Let $\widehat{L\mathfrak{g}}$ be the central extension of $L\mathfrak{g}$ determined by this cocycle, and $\widehat{L\mathfrak{g}}^*$ the semidirect product of $(L\mathfrak{g})^*$ and $\mathbb{C}^\times$.

Proposition 21. $\left(\widehat{L\mathfrak{g}}, \widehat{L\mathfrak{g}}^*\right)$ *is a Lie bialgebra.*

The Poisson Lie group which corresponds to $\widehat{L\mathfrak{g}}^*$ is the semidirect product $\mathbb{C}^\times \ltimes (LG)^*$; it is easy to see that the affine coordinate on $\mathbb{C}^\times$ is central with respect to the Poisson structure on $\mathbb{C}^\times \ltimes (LG)^*$; fixing its value, we get the family of Poisson brackets $\{\,,\,\}_p^*$ on $(LG)^*$. In Sect. 4.2 we shall discuss a similar phenomenon for quantum systems associated with quasitriangular Hopf algebras.

3.3   Sklyanin Bracket on $G(z)$

In our previous example we have described the Poisson structure associated with the loop algebra, namely, the bracket on the dual group $(LG)^*$; applications to lattice systems involve the Poisson structure on the loop group itself. As a matter of fact, this was the first example of a Poisson Lie group which appeared prior to their formal definition [50]. To put it more precisely, let $\mathfrak{g}$ be a complex semisimple Lie algebra, G a linear complex Lie group with the Lie algebra $\mathfrak{g}$. Let $G(z)$ be the associated loop group, i. e., the group of rational functions with values in G. We may regard $G(z)$ as an affine algebraic group defined over $\mathbb{C}(z)$. Let us again fix a faithful representation (ρ, V) of G; the affine ring $\mathcal{A}_{aff}$ of $G(z)$ is generated by 'tautological functions' which assign to $L \in G(z)$ the matrix coefficients of $L_V(z) \in \operatorname{End} V \otimes \mathbb{C}(z)$. The Poisson bracket on the affine ring of $G(z)$ is defined by the following formula [50],

$$\left\{ L_V^1(u),\, L_V^2(v) \right\} = \left[r_V(u,v),\, L_V^1(u) L_V^2(v) \right]. \tag{3.30}$$

(By an abuse of notation, we do not distinguish between the generators of $\mathcal{A}_{aff}$ and their values at $L \in G(z)$ in the left-hand side of this formula.) Bracket (3.30) defines the structure of a Poisson-Lie group on $G(z)$. There is an equivalent formulation:

Let $\Delta : \mathcal{A}_{aff} \to \mathcal{A}_{aff} \otimes \mathcal{A}_{aff}$ be the coproduct in $\mathcal{A}_{aff}$ induced by the group multiplication in $G(z)$:

$$\Delta\left(L_V(z) \right) = L_V(z) \,\dot{\otimes}\, L_V(z),$$

or, in a less condensed notation,

$$\Delta\left(L_V^{ij}(z) \right) = \sum_k L_V^{ik}(z) \otimes L_V^{kj}(z). \tag{3.31}$$

Then Δ is a morphism of Poisson algebras (in other words,

$$\{\Delta\varphi, \Delta\psi\} = \Delta\{\varphi, \psi\}$$

for any $\varphi, \psi \in \mathcal{A}_{aff}$.

The Lax matrix $L_V(u)$ in this context is basically a tautological mapping which assigns to an element $L \in G(z)$ the matrix $L_V(u) \in \operatorname{End} V \otimes \mathbb{C}(u)$. Alternatively, for any Poisson submanifold $\mathcal{M} \subset G(z)$ the Lax matrix may be regarded as an embedding map $\mathcal{M} \hookrightarrow G(z) \to \operatorname{End} V \otimes \mathbb{C}(u)$. An explicit description of all Poisson submanifolds in $G(z)$ follows from the theory of *dressing transformations* [48]; for our present goals it suffices to know that rational functions with a prescribed divisor of poles form a finite-dimensional Poisson submanifold in $G(z)$; moreover, since the Poisson structure is multiplicative, the product of Poisson submanifolds is again a Poisson submanifold. Generic Poisson submanifolds correspond to functions with simple poles which may be written in multiplicative form

$$L_V(z) = \overset{\frown}{\prod_i} (I - \frac{X_i}{z - z_i}).$$ (3.32)

In other words, the description of Poisson submanifolds is tantamount to the choice of an *Ansatz* for the Lax matrix; the free parameters in this Ansatz (e.g., the residues X_i) become dynamical variables. Spectral invariants of 'Lax matrices' of this type may be used to generate completely integrable lattice Lax equations. More precisely, to get such a system one may proceed as follows:

- Pick a Poisson submanifold $\mathcal{M} \subset G(z)$.
- Consider the product space $\mathcal{M}^N \subset G(z)^{\mathbb{Z}/N\mathbb{Z}}$ and the monodromy map $m : G(z) \times ... \times G(z) \to G(z)$ (more precisely, its restriction to $\mathcal{M}^N$).
- Choose any central function $\mathcal{H}$ on $G(z)$ as a Hamiltonian; its pullback to $\mathcal{M}^N$ defines a lattice Lax equation.

Under some mild assumptions on the choice of $\mathcal{M}$, $\mathcal{H}$ defines a completely integrable system on $m\left(\mathcal{M}^N\right)$; its pullback to $\mathcal{M}^N$ remains completely integrable; this may be regarded as the main content of the Inverse Scattering Method (in this slightly simplified setting).

The study of Lax systems on the lattice thus breaks naturally into two parts: *(a) Solve the Lax equation for the monodromy. (b) Lift the solutions back to* $G(z)^N$. The second stage (by no means trivial) is the inverse problem, *stricto sensu*.

4   Quantization

4.1   Linear Case

Speaking informally, quantization consists in replacing Poisson bracket relations with commutation relations in an associative algebra. We shall have to distinguish between the linear and the quadratic case (as explained in the previous section, this difference stems from the different Hopf structures on the algebras of observables). The linear case is easier, since we remain in the conventional setting of Lie groups and Lie algebras. Quantization of quadratic Poisson bracket relations (3.7) leads to *quasitriangular Hopf algebras*.

The standard way to quantize the algebra $\mathcal{A}_{cl} = \mathcal{S}(\mathfrak{g})$ (equipped, as usual, with the Lie–Poisson bracket of $\mathfrak{g}$) is to replace it with the universal enveloping algebra $\mathcal{U}(\mathfrak{g})$. Let me briefly recall the corresponding construction.

Let $\mathcal{U}(\mathfrak{g}) = \bigcup_{n \geq 0} \mathcal{U}_n$ be the canonical filtration of $\mathcal{U}(\mathfrak{g})$, and $\mathcal{S}(\mathfrak{g}) = \oplus_{n \geq 0} \mathcal{S}_n$ the canonical grading of $\mathcal{S}(\mathfrak{g})$. By the classical Poincaré–Birkhoff–Witt theorem, $\mathcal{S}_n \simeq \mathcal{U}_n / \mathcal{U}_{n-1}$. Let $\mathrm{gr}_n : \mathcal{U}_n \to \mathcal{S}_n$ be the canonical projection. If $u \in \mathcal{U}_n, v \in \mathcal{U}_m$, their commutator $[u, v] \in \mathcal{U}_{n+m-1}$. It is easy to see that $\mathrm{gr}_{n+m-1}([u, v]) = \{\mathrm{gr}_n u, \mathrm{gr}_m v\}$ is precisely the Lie–Poisson bracket; hence $\mathcal{U}(\mathfrak{g})$ is a quantization of $\mathcal{S}(\mathfrak{g})$.

Alternatively, let $\hbar$ be a formal parameter. Put $\mathfrak{g}_\hbar = \mathfrak{g} \otimes_k k\,[[\hbar]]$; we rescale the commutator in $\mathfrak{g}_\hbar$ by putting $[X, Y]_\hbar = \hbar\,[X, Y]$. Let $\mathcal{U}\,(\mathfrak{g}_\hbar)$ be the universal enveloping algebra of $\mathfrak{g}_\hbar$, and $\hbar\,\mathcal{U}\,(\mathfrak{g}_\hbar)$ be its maximal ideal consisting of formal series in $\hbar$ with zero constant term; the quotient algebra $\mathcal{U}\,(\mathfrak{g}_\hbar)\,/\hbar\,\mathcal{U}\,(\mathfrak{g}_\hbar)$ is canonically isomorphic to $\mathcal{S}\,(\mathfrak{g})$. Let $p : \mathcal{U}\,(\mathfrak{g}_\hbar) \to \mathcal{S}\,(\mathfrak{g})$ be the canonical projection; then $\{p\,(x)\,, p\,(y)\} = p\left(\hbar^{-1}\,(xy - yx)\right)$ for any $x, y \in \mathcal{U}\,(\mathfrak{g}_\hbar)$. (In the sequel we prefer to set $\hbar = 1$ and so the first definition will be more convenient.)

Let $r \in \operatorname{End}\mathfrak{g}$ be a classical r-matrix. As we have seen, the phase spaces of Lax systems associated with $(\mathfrak{g}, r)$ are coadjoint orbits of $\mathfrak{g}_r$; the Hamiltonians are obtained from the Casimir elements $H \in \mathcal{S}\,(\mathfrak{g})^\mathfrak{g}$ via the isomorphism $\mathcal{S}\,(\mathfrak{g}_r) \to \mathcal{S}\,(\mathfrak{g})$ described in Theorem 1. The quantum counterpart of $\mathcal{S}\,(\mathfrak{g})^\mathfrak{g}$ is the center $\mathcal{Z} \subset \mathcal{U}\,(\mathfrak{g})$ of the universal enveloping algebra. Moreover, under favorable conditions there is a nice correspondence between coadjoint orbits of a Lie algebra and unitary representations of the corresponding Lie group. Thus we may establish the following heuristic vocabulary which describes the correspondence between classical and quantum systems:

Linear Classical Case	Linear Quantum case
$\mathcal{S}(\mathfrak{g})$	$\mathcal{U}(\mathfrak{g})$
Classical r-matrices	*Classical r-matrices*
Casimir functions $(\mathcal{S}\,(\mathfrak{g}))^\mathfrak{g} \subset \mathcal{S}\,(\mathfrak{g})$	*Casimir operators* $\mathcal{Z} \subset \mathcal{U}\,(\mathfrak{g})$
$\mathcal{S}(\mathfrak{g}_r)$	$\mathcal{U}(\mathfrak{g}_r)$
Coadjoint orbits in $\mathfrak{g}_r^*$	*Irreducible* $\mathcal{U}\,(\mathfrak{g}_r)$*-modules*

Notice that, in the linear case, classical r-matrices are used in the quantum case as well!

The following theorem is an exact analogue of Theorem 1

Let $(\mathfrak{g}, r)$ be a Lie algebra equipped with a classical r-matrix $r \in \operatorname{End}\mathfrak{g}$ satisfying (2.3). Let $\mathfrak{g}_r$ be the corresponding Lie algebra (with the same underlying linear space). Put $r_\pm = \frac{1}{2}(r \pm id)$; then (2.3) implies that $r_\pm : \mathfrak{g}_r \to \mathfrak{g}$ are Lie algebra homomorphisms. Extend them to homomorphisms $\mathcal{U}(\mathfrak{g}_r) \to \mathcal{U}(\mathfrak{g})$ which we denote by the same letters. These morphisms also agree with the standard Hopf structure on $\mathcal{U}(\mathfrak{g})$, $\mathcal{U}(\mathfrak{g}_r)$. Define the action

$$\mathcal{U}(\mathfrak{g}_r) \otimes \mathcal{U}(\mathfrak{g}) \to \mathcal{U}\,(\mathfrak{g})$$

by setting

$$x \cdot y = \sum r_+(x_i^{(1)})\, y\, r_-(x_i^{(2)})',\, x \in \mathcal{U}(\mathfrak{g}_r), y \in \mathcal{U}(\mathfrak{g}), \qquad (4.1)$$

where $\Delta x = \sum x_i^{(1)} \otimes x_i^{(2)}$ is the coproduct and $a \mapsto a'$ is the antipode map.

Theorem 7. *(i)* $\mathcal{U}(\mathfrak{g})$ *is a free filtered Hopf* $\mathcal{U}(\mathfrak{g}_r)$*-module generated by* $1 \in \mathcal{U}(\mathfrak{g})$. *(ii) Let* $i : \mathcal{U}(\mathfrak{g}) \to \mathcal{U}(\mathfrak{g}_r)$ *be the induced isomorphism of filtered linear spaces;*

its restriction to $\mathcal{Z} \subset \mathcal{U}(\mathfrak{g})$ is an algebra homomorphism; in particular, $i(\mathcal{Z}) \subset \mathcal{U}(\mathfrak{g}_r)$ is commutative.

Let $\mathfrak{u} \subset \mathfrak{g}_r$ be an ideal, $\mathfrak{s} = \mathfrak{g}_r/\mathfrak{u}$ the quotient algebra, and let $p : \mathcal{U}(\mathfrak{g}_r) \to \mathcal{U}(\mathfrak{s})$ be the canonical projection; restricting p to the subalgebra $I_r = i_r(\mathcal{Z})$ we get a commutative subalgebra in $\mathcal{U}(\mathfrak{s})$; we shall say that the corresponding elements of $\mathcal{U}(\mathfrak{s})$ are obtained by *specialization*.

In particular, let $\mathfrak{g} = \mathfrak{g}_+ \dotplus \mathfrak{g}_-$ be a splitting of $\mathfrak{g}$ into a linear sum of two Lie subalgebras; let $P : \mathcal{U}(\mathfrak{g}) \to \mathcal{U}(\mathfrak{g}_-)$ be the projection onto $\mathcal{U}(\mathfrak{g}_-)$ in the decomposition

$$\mathcal{U}(\mathfrak{g}) = \mathcal{U}(\mathfrak{g}_-) \oplus \mathfrak{g}_+\mathcal{U}(\mathfrak{g}).$$

Corollary 3. *The restriction of P to the subalgebra $\mathcal{Z} \subset \mathcal{U}(\mathfrak{g})$ of Casimir elements is an algebra homomorphism.*

4.1.1 Quantum Reduction.
The quantum analogue of the symplectic induction discussed in Sect. 2.1.1 is the ordinary induction. For completeness we recall the standard definitions (cf., for example, [10]). Let $\mathfrak{g}$ be a Lie algebra. Let us denote by $\mathsf{Rep}_\mathfrak{g}$ the category of $\mathcal{U}(\mathfrak{g})$ -modules. Let $\mathfrak{b} \subset \mathfrak{g}$ be a Lie subalgebra. In complete analogy with Sect. 2.2.1 we construct the induction functor $\mathrm{Ind}_\mathfrak{b}^\mathfrak{g} : \mathsf{Rep}_\mathfrak{b} \rightsquigarrow \mathsf{Rep}_\mathfrak{g}$ which associates to each $\mathcal{U}(\mathfrak{b})$-module a $\mathcal{U}(\mathfrak{g})$-module. Namely, we put $\mathrm{Ind}_\mathfrak{b}^\mathfrak{g}(V) = \mathcal{U}(\mathfrak{g}) \otimes_{\mathcal{U}(\mathfrak{b})} V$, where $\mathcal{U}(\mathfrak{g})$ is regarded as a right $\mathcal{U}(\mathfrak{b})$-module.

[By definition, $\mathcal{U}(\mathfrak{g}) \otimes_{\mathcal{U}(\mathfrak{b})} V$ is the quotient of $\mathcal{U}(\mathfrak{g}) \otimes_\mathbb{C} V$ over the submodule generated by $(u \otimes v) \cdot X$, $u \in \mathcal{U}(\mathfrak{g})$, $v \in V$, $X \in \mathcal{U}(\mathfrak{b})$; the (right) action of $\mathcal{U}(\mathfrak{b})$ on $\mathcal{U}(\mathfrak{g}) \otimes_\mathbb{C} V$ is defined by

$$(u \otimes v) \cdot X = \sum_i uS\left(X^{(i)}\right) \otimes X_{(i)}v,$$

where $\Delta X = \sum_i X^{(i)} \otimes X_{(i)}$ is the coproduct in $\mathcal{U}(\mathfrak{b})$ and S is its antipode. In physical terms passing to the quotient means that we impose 'constraints on the wave functions'; these constraints express the invariance of wave functions with respect to the diagonal action of $\mathcal{U}(\mathfrak{b})$.]

The structure of a $\mathcal{U}(\mathfrak{g})$-module in $\mathrm{Ind}_\mathfrak{b}^\mathfrak{g}(V)$ is induced by the left action of $\mathcal{U}(\mathfrak{g})$ on itself.

Fix a point $F \in \mathfrak{b}^*$ and assume that $\mathfrak{l}_F$ is a Lagrangian subalgebra subordinate to F. In that case F defines a character of $\mathfrak{l}_F$; let V_F be the corresponding 1-dimensional $\mathcal{U}(\mathfrak{l}_F)$-module. Put $V = \mathcal{U}(\mathfrak{b}) \otimes_{\mathcal{U}(\mathfrak{l}_F)} V_F$. Then V is a natural $\mathcal{U}(\mathfrak{b})$-module associated with the coadjoint orbit of F.

Proposition 22. $\mathrm{Ind}_\mathfrak{b}^\mathfrak{g}(V) \simeq \mathrm{Ind}_{\mathfrak{l}_F}^\mathfrak{g}(V_F)$.

Informally, we may say that a $\mathcal{U}(\mathfrak{g})$-module associated with a coadjoint orbit admitting a Lagrangian polarization may be induced from a 1-dimensional module.

Let us now return to the setting of Theorem 7. Let $(\mathfrak{g}, r)$ be a Lie algebra equipped with a classical r-matrix $r \in \mathrm{End}\,\mathfrak{g}$ satisfying (2.3). We regard $\mathcal{U}(\mathfrak{g})$ as a $\mathcal{U}(\mathfrak{g}_r)$-module with respect to the action (4.1). Let V be a $\mathcal{U}(\mathfrak{g}_r)$-module.

Proposition 23. $V \simeq \mathcal{U}(\mathfrak{g}) \otimes_{\mathcal{U}(\mathfrak{g}_r)} V$ *as a linear space.*

Since half of $\mathcal{U}(\mathfrak{g}_r)$ is acting on $\mathcal{U}(\mathfrak{g})$ on the left and the other half on the right, the structure of a $\mathcal{U}(\mathfrak{g})$-module in $\mathcal{U}(\mathfrak{g}) \otimes_{\mathcal{U}(\mathfrak{g}_r)} V$ is destroyed. However, the structure of $\mathcal{Z}$-module in $\mathcal{U}(\mathfrak{g})$ survives tensoring with V over $\mathcal{U}(\mathfrak{g}_r)$.

Proposition 24. *For any* $\zeta \in \mathcal{Z}$, $u \in \mathcal{U}(\mathfrak{g})$, $v \in V$, $\zeta u \otimes v = u \otimes i_r(\zeta) v$.

This gives a new proof of Theorem 7.

Fix $F \in \mathfrak{g}_r^*$ and let $\mathfrak{l}_F \subset \mathfrak{g}_r$ be a Lagrangian subalgebra subordinate to F. Let V_F be the corresponding 1-dimensional $\mathcal{U}(\mathfrak{l}_F)$ -module. Then $\mathcal{U}(\mathfrak{g}) \otimes_{\mathcal{U}(\mathfrak{l}_F)} V_F$ may be identified with V. Let us assume now that $\mathfrak{g} = \mathfrak{g}_+ \dotplus \mathfrak{g}_-$ and the r-matrix is given by (2.5); in that case $\mathcal{U}(\mathfrak{g}_r) \simeq \mathcal{U}(\mathfrak{g}_+) \otimes \mathcal{U}(\mathfrak{g}_-)$. Let W be a $\mathcal{U}(\mathfrak{g}_-)$-module, $\mathbf{W} = \mathcal{U}(\mathfrak{g}) \otimes_{\mathcal{U}(\mathfrak{g}_-)} W$; we regard $\mathbf{W}$ as a left $\mathcal{U}(\mathfrak{g})$-module. There is a canonical embedding $W \hookrightarrow \mathbf{W} : w \mapsto 1 \otimes w$. Let $\mathbf{W}^*$ be the dual module regarded as a right $\mathcal{U}(\mathfrak{g})$-module, and let $W_0^* \subset \mathbf{W}$ be the subspace of $\mathcal{U}(\mathfrak{g}_+)$-invariants.

Lemma 5. W_0^* *is isomorphic to the dual of* W.

This is an easy corollary of the decomposition

$$\mathcal{U}(\mathfrak{g}) \simeq \mathcal{U}(\mathfrak{g}_+) \otimes \mathcal{U}(\mathfrak{g}_-) \simeq \mathfrak{g}_+ \mathcal{U}(\mathfrak{g}) \oplus \mathcal{U}(\mathfrak{g}_-) .$$

Fix a basis $\{e_i\}$ in W, and let $\{e^i\}$ be the dual basis in W_0^*. Let $\Omega = \sum e^i \otimes e_i \in W_0^* \otimes W$ be the canonical element; it defines a natural mapping

$$\mathbf{W} \to W : \varphi \mapsto \langle \varphi \rangle_\Omega = \sum_i \langle e^i, \varphi \rangle e_i.$$

Proposition 25. *For any* $\varsigma \in \mathcal{Z}$, $\varphi \in \mathbf{W}$, $\langle \varsigma \varphi \rangle_\Omega = i_\varsigma \langle \varphi \rangle_\Omega$.

Informally, we may say that quantum Hamiltonians acting in W are *radial parts* of the Casimir operators. (The reasons for this terminology will be obvious from our next example.)

4.1.2 Toda Lattice.

Our first example is the generalized open Toda lattice. We retain the notation of Sect. 2.2.1. Let again $\mathfrak{g}$ be a real split semisimple Lie algebra, $\mathfrak{g} = \mathfrak{k} \dotplus \mathfrak{a} \dotplus \mathfrak{n}$ its Iwasawa decomposition, $\mathfrak{b} = \mathfrak{a} \dotplus \mathfrak{n}$ the Borel subalgebra. Let $\partial_p \subset \mathfrak{b}$ $(p \geq 0)$ be the generalized diagonals defined in (2.12). The subalgebras $\mathfrak{b}^{(p)} = \oplus_{r \geq p} \partial_r$ define a decreasing ad $\mathfrak{b}$-invariant filtration in $\mathfrak{b}$. Put $\mathfrak{p}_k = \oplus_{0 \leq p \leq k} s(\partial_p)$, where $s : \mathfrak{g} \to \mathfrak{p} : X \mapsto \frac{1}{2}(id - \sigma) X$ is the projection onto the subspace of anti-invariants of the Cartan involution σ. Put $\mathfrak{s} = b/\mathfrak{b}^{(2)}$. Let

$L \in \mathfrak{p}_1 \otimes \mathfrak{s} \subset \mathfrak{p}_1 \otimes \mathcal{U}(\mathfrak{s})$ be the canonical element induced by the natural pairing $\mathfrak{p}_1 \times \mathfrak{b}/\mathfrak{b}^{(2)} \to \mathbb{R}$. Let $P : \mathcal{U}(\mathfrak{g}) \to \mathcal{U}(\mathfrak{b})$ be the projection map onto $\mathcal{U}(\mathfrak{b})$ in the decomposition $\mathcal{U}(\mathfrak{g}) = \mathcal{U}(\mathfrak{b}) \oplus \mathcal{U}(\mathfrak{g})\mathfrak{k}$, $p : \mathcal{U}(\mathfrak{b}) \to U(\mathfrak{s})$ the specialization map. Let us set $I_{\mathfrak{s}} = p \circ P(\mathcal{Z})$, where as usual $\mathcal{Z}$ is the center of $\mathcal{U}(\mathfrak{g})$. Fix a faithful representation (ρ, V) of $\mathfrak{g}$ and put $L_V = (\rho \otimes id) L$.

Proposition 26. *(i) $I_{\mathfrak{s}}$ is a commutative subalgebra in $\mathcal{U}(\mathfrak{s})$; moreover, $\mathcal{Z} \to I_{\mathfrak{s}} : z \longmapsto H_z$ is an algebra isomorphism. (ii) $I_{\mathfrak{s}}$ is generated by $H_k = \mathrm{tr}\,_V L_V^k \in \mathcal{U}(\mathfrak{s})$, $k = 1, 2, \dots$*

Recall that $\mathfrak{s} = \mathfrak{a} \ltimes \mathfrak{u}$, where $\mathfrak{u} = \mathfrak{n}/[\mathfrak{n}, \mathfrak{n}]$, and the corresponding Lie group is $S = A \ltimes U$, $U = N/N'$. Let $\mathcal{O}_T$ be the coadjoint orbit of S described in (2.14), $f = \sum_{\alpha \in P}(e_\alpha + e_{-\alpha}) \in \mathcal{O}_T$ the marked point, $L_f = U$ the corresponding Lagrangian subgroup, $V_f = \mathbb{C}$ the 1-dimensional U-module which corresponds to f; we may also regard it as a $\mathcal{U}(\mathfrak{u})$-module.

Proposition 27. *(i) The irreducible $\mathcal{U}(\mathfrak{s})$-module associated with $\mathcal{O}_T$ is the induced module $\mathcal{U}(\mathfrak{s}) \otimes_{\mathcal{U}(\mathfrak{u})} V_f$. (ii) The corresponding unitary representation space $\mathfrak{H}$ may be identified with $L_2(\mathfrak{a})$. (iii) Let $\Delta \in \mathcal{Z}$ be the quadratic Casimir operator which corresponds to the Killing form. Then H_Δ is precisely the Toda Hamiltonian H_T acting in $\mathfrak{H}$; it is given by*

$$H_T = -(\partial, \partial) + \sum_{\alpha \in P} \exp 2\alpha(q), \partial = \left(\frac{\partial}{\partial q_1}, \dots, \frac{\partial}{\partial q_n}\right).$$

In other words, geometric quantization of the Toda Hamiltonian agrees with its 'naive' Schrödinger quantization. What is nontrivial, of course, is the consistent definition of the quantum integrals of motion which correspond to higher Casimirs.

Let us now turn to the description of the Toda lattice based on the reduction procedure. In Sect. 4.1.1 we discussed the reduction using the language of $\mathcal{U}(\mathfrak{g})$-modules; in the present setting we may assume that all these modules are actually integrable, i. e., come from representations of the corresponding Lie groups. In particular, the action of $\mathcal{U}(\mathfrak{g})$ on itself by left (right) multiplications corresponds to the regular representation of G; we may regard $\mathfrak{H} = L_2(G)$ as a result of quantization of T^*G. We are led to the following construction. Let χ_f be the character of N defined by

$$\chi_f(\exp X) = \exp if(X), X \in \mathfrak{n}.$$

Let $\mathfrak{H}_f$ be the space of smooth functions on G satisfying the functional equation

$$\psi(kxn) = \chi_f(n)\psi(x), k \in K, n \in N. \tag{4.2}$$

By Iwasawa decomposition, any such function is uniquely determined by its restriction to $A \subset G$. Thus there is an isomorphism $i : C^\infty(\mathfrak{a}) \to \mathfrak{H}_f : i(\psi)(kan) = \chi_f(n)\psi(\log a)$.

Lemma 6. $\mathfrak{H}_f$ *is invariant with respect to the action of* $\mathcal{Z} \subset \mathcal{U}(\mathfrak{g})$.

For $\zeta \in \mathcal{Z}$ let $\delta(\zeta) \in \operatorname{End} C^\infty(\mathfrak{a})$ be its radial part defined by

$$\zeta i(\psi) = i(\delta(\zeta)\psi), \psi \in C^\infty(\mathfrak{a}).$$

Proposition 28. *We have* $\delta(\zeta) = H_\zeta$, *i. e., the radial parts of the Casimir operators coincide with the quantum integrals of the Toda lattice.*

We are interested in eigenfunctions of the Toda Hamiltonian and higher Toda integrals. By proposition 28 this is equivalent to description of the eigenfunctions of Casimir operators on G which satisfy the functional equation (4.2). At the formal level this problem may be solved as follows. Let $(\pi, \mathfrak{V})$ be an infinitesimally irreducible representation of G, $\mathfrak{V}^*$, the dual representation. Assume that $w \in \mathfrak{V}$ satisfies

$$\pi(n)w = \chi_f(n)w, \text{ for } n \in N. \tag{4.3}$$

(In that case w is called a *Whittaker vector*.) Assume, moreover, that in the dual space $\mathfrak{V}^*$ there is a K-invariant vector $v \in \mathfrak{V}^*$. Put

$$\psi(x) = \langle v, \pi(x)w \rangle. \tag{4.4}$$

Proposition 29. ψ *is an eigenfunction of* $\mathcal{Z}$ *and satisfies the functional equation* (4.2).

Thus ψ is essentially a Toda lattice eigenfunction.

This formal argument can be made rigorous. Since the Toda Hamiltonian describes scattering in a repulsive potential, its spectrum is continuous; so there are no chances that ψ (which is called a *Whittaker function*) is in L_2. The problem is to find representations of G such that ψ has the usual properties of a continuous spectrum eigenfunction (i. e., the wave packets smoothed down with appropriate amplitudes are in L_2). As it appears, the correct class is that of the *spherical principal series representations*.

Definition 4. Let $B = MAN$ be the Borel subgroup in G, $\lambda \in \mathfrak{a}^*$; let $\chi_\lambda : B \to \mathbb{C}$ be the 1-dimensional representation defined by

$$\chi_\lambda(man) = \exp\langle \lambda - \rho, \log a \rangle.$$

The spherical principal series π_λ is the induced representation, $\pi_\lambda = \operatorname{Ind}_B^G \chi_\lambda$.

Principal series representations are infinitesimally irreducible and hence define a homomorphism $\pi : \mathcal{Z} \to P(\mathfrak{a}^*)$ into the algebra of polynomials on $\mathfrak{a}^*$

$$\pi : z \longmapsto \pi_z, \text{ where } \pi_\lambda(z) := \pi_z(\lambda) \cdot id.$$

By a classical Harish Chandra theorem [10], π is actually an isomorphism onto the subalgebra $P(\mathfrak{a}^*)^W \subset P(\mathfrak{a}^*)$ of the Weyl group invariants. Hence the set of principal series representations is sufficiently ample to separate points of the algebraic spectrum of $\mathcal{Z}$. The algebraic theory of Whittaker vectors is exposed in [35], the analytic theory leading to the Plancherel theorem for the Toda lattice is outlined in [47]. A remarkable point is that the eigenfunctions of the quantum system are expressed as matrix coefficients of appropriate representations of the 'hidden symmetry group'. This is a special case of a very general situation. Below we shall discuss another example of this kind, for which the hidden symmetry algebra is infinite-dimensional.

4.1.3 Gaudin Model.

The treatment of the Gaudin model is considerably more difficult, since in this case the hidden symmetry group is infinite-dimensional. Remarkably, the general pattern described in section 4.1.1 is fully preserved in this case as well.

The algebra of observables for the Gaudin model is simply $\mathcal{U}(\mathfrak{g}^N)$. Let V_λ be a finite-dimensional highest weight representation of $\mathfrak{g}$ with dominant integral highest weight λ. Let $\lambda = (\lambda_1, ... \lambda_N)$ be a set of such weights; set $\mathbf{V}_\lambda = \otimes_i V_{\lambda_i}$ (In other words, we associate a 'spin λ_i particle' with the point z_i). The space $\mathbf{V}_\lambda$ is a natural Hilbert space associated with the Gaudin model; so the 'kinematical' part of the quantization problem in this case is fairly simple. Let us fix also an auxiliary representation (ρ, V). By analogy with the classical case, we may introduce the *quantum Lax operator*,

$$L_V(z) \in \operatorname{End} V(z) \otimes \mathfrak{g}^N \subset \operatorname{End} V(z) \otimes \mathcal{U}(\mathfrak{g}^N); \qquad (4.5)$$

the definition remains exactly the same, but now we embed $\mathfrak{g}^N$ into the universal enveloping algebra $\mathcal{U}(\mathfrak{g}^N)$ instead of the symmetric algebra $\mathcal{S}(\mathfrak{g}^N)$. The commutation relations for $L_V(z)$ essentially reproduce the Poisson bracket relations (2.23), but this time the left-hand side is a matrix of true commutators:

$$\left[L_V(u) \overset{\otimes}{,} L_V(v) \right] = \left[r_V(u,v), L_V(u) \otimes 1 + 1 \otimes L_V(v) \right]. \qquad (4.6)$$

The key point in (4.6) is the interplay of the commutation relations in the quantum algebra $\mathcal{U}(\mathfrak{g}^N)$ and the auxiliary matrix algebra $\operatorname{End} V(z)$. Formula (4.6) was the starting point of QISM (as applied to models with *linear* commutation relations). Put

$$S(u) = \frac{1}{2} \operatorname{tr}_V \left(L_V(u) \right)^2; \qquad (4.7)$$

using (4.6), it is easy to check that $S(u)$ form a commutative family of Hamiltonians (called *Gaudin Hamiltonians*) in $\mathcal{U}(\mathfrak{g}^N)$ (see, e.g., [31]). An important property of this commuting family is that it possesses at least one 'obvious' eigenvector $|0\rangle \in \mathbf{V}_\lambda$, the tensor product of highest weight vectors in V_{λ_i}; it is usually called the *vacuum vector*.

One of the key ideas of QISM is to construct other eigenvectors by applying to the vacuum creation operators which are themselves rational functions of z.

This construction is called the *algebraic Bethe Ansatz*. Assume that $\mathfrak{g} = \mathfrak{sl}_2$ and let $\{E, F, H\}$ be its standard basis. Put

$$F(z) = \sum_{z_i \in D} \frac{F^{(i)}}{z - z_i}, \qquad (4.8)$$

where $F^{(i)}$ acts as F in the i-th copy of $\mathfrak{sl}_2$ and as id in the other places. The Bethe vector is, by definition,

$$\mid w_1, w_2, ...w_m \rangle = F(w_1)F(w_2)...F(w_m) \mid 0 \rangle. \qquad (4.9)$$

The Lax matrix (4.5) applied to $\mid w_1, w_2, ...w_m \rangle$ becomes triangular, i. e.,

$$L(u) \mid w_1, w_2, ...w_m \rangle = \begin{pmatrix} a(u, w_1, w_2, ...w_m) & * \\ 0 & d(u, w_1, w_2, ...w_m) \end{pmatrix};$$

after a short computation this yields

$$S(u) \mid w_1, w_2, ...w_m \rangle =$$

$$s_m(u) \mid w_1, w_2, ...w_m \rangle + \sum_{j=1}^{N} \frac{f_j}{u - w_j} \mid w_1, w_2, ..., w_{j-1}, u, w_{j+1}, ..., w_m \rangle,$$

where

$$f_j = \sum_{i=1}^{N} \frac{\lambda_i}{w_j - z_i} - \sum_{s \neq j} \frac{2}{w_j - w_s}$$

and $s_m(u)$ is a rational function,

$$s_m(u) = \frac{c_V}{2} \chi_m(u)^2 - c_V \partial_u \chi_m(u),$$

$$\chi_m(u) = \sum_{i=1}^{N} \frac{\lambda_i}{u - z_i} - \sum_{i=1}^{m} \frac{2}{u - w_j}. \qquad (4.10)$$

(The constant c_V depends on the choice of V.) If all f_j vanish, $\mid w_1, w_2, ...w_m \rangle$ is an eigenvector of $S(u)$ with the eigenvalue $s_m(u)$; equations

$$\sum_{i=1}^{N} \frac{\lambda_i}{w_j - z_i} - \sum_{s \neq j} \frac{2}{w_j - w_s} = 0 \qquad (4.11)$$

are called the *Bethe Ansatz equations*. (Notice that (4.11) is precisely the condition that $s_m(u)$ be nonsingular at $u = w_i$.)

For general simple Lie algebras the study of the spectra of the Gaudin Hamiltonians becomes rather complicated; one way to solve this problem is to treat it inductively by choosing in $\mathfrak{g}$ a sequence of embedded Lie subalgebras of lower rank and applying the algebraic Bethe Ansatz to these subalgebras. An alternative idea (which is completely parallel to the treatment of the Toda lattice in Sect. 4.1) is to interpret the Hamiltonians as radial parts of (infinite-dimensional) Casimir operator of the 'global' Lie algebra $\mathfrak{g}_D$. It is impossible to apply Theorem 7 immediately in the affine case, since the center of $\mathcal{U}(\mathfrak{g}_D)$ is trivial. The point is that the invariants in the (suitably completed) symmetric algebra $\mathcal{S}(\mathfrak{g}_D)$ are infinite series; an attempt to quantize these expressions leads to divergent expressions, unless some kind of ordering prescription is introduced. The commutation relations for the normally ordered operators are already non-trivial and they do not lie in the center of $\mathcal{U}(\mathfrak{g}_D)$. However, the situation can be amended by first passing to the central extension of $\mathcal{U}(\mathfrak{g}_D)$ and then considering the quotient algebra $\mathcal{U}_k(\mathfrak{g}_D) = U(\hat{\mathfrak{g}}_D)/(c - k)$. It is known that for the *critical value* of the central charge $k = -h^\vee$ (here $h^\vee$ is the *dual Coxeter number* of $\mathfrak{g}$) the (appropriately completed) algebra $\mathcal{U}_{-h^\vee}(\mathfrak{g}_D)$ possesses an ample center.

Let us recall first of all the construction of *Sugawara operators*. Let $L\mathfrak{g} = \mathfrak{g}((z))$ be the 'local' algebra of formal Laurent series with coefficients in $\mathfrak{g}$. Let ω be the 2-cocycle on $L\mathfrak{g}$ defined by

$$\omega(X, Y) = \mathrm{Res}_{z=0} \langle X, \partial_z Y \rangle \, dz.$$

It is well known that highest weight representations of $L\mathfrak{g}$ are actually projective and correspond to the central extension $\widehat{L\mathfrak{g}} \simeq L\mathfrak{g} \oplus \mathbb{C} \, c$ of $L\mathfrak{g}$ defined by this cocycle[4]. Let us formally consider the canonical element $J \in L\mathfrak{g} \otimes L\mathfrak{g}$ which corresponds to the inner product

$$(X, Y) = \mathrm{Res}_{z=0} \langle X, Y \rangle \, dz$$

in $L\mathfrak{g}$. Fix a finite-dimensional representation (ρ, V) of $\mathfrak{g}$; it extends to the 'evaluation representation' of $L\mathfrak{g}$ in $V \otimes_\mathbb{C} \mathbb{C}((z))$. Let $(\pi_k, \mathfrak{V})$ be any level k highest weight representation of $\widehat{L\mathfrak{g}}$. (This means that the central element $c \in \widehat{L\mathfrak{g}}$ acts by $\pi_k(c) = k \cdot id$.) Put $J(z) = (\pi_k \otimes \rho) J$. Since J is not a proper element in $L\mathfrak{g} \otimes L\mathfrak{g}$, $J(z)$ is a formal series infinite in both directions; however, its coefficients are well defined. Namely, let $\{e_a\}$ be a basis in $\mathfrak{g}$; the algebra $L\mathfrak{g}$ is spanned by the vectors $e_a(n) = e_a \otimes z^n$, $n \in \mathbb{Z}$. It is easy to see that the coefficients of $J(z)$ are *finite* linear combinations of the vectors $\pi_k(e_a(n))$.

Put

$$T(u) = \frac{1}{2} : \mathrm{tr}\, J(u)^2 : \quad , \quad T(u) = \sum_{n=-\infty}^{\infty} T_n u^n, \tag{4.12}$$

---

[4] The value of the central charge associated with a given projective representation of $L\mathfrak{g}$ depends on the normalization of this cocycle, or, equivalently, of the inner product in the Lie algebra $\mathfrak{g}$. The standard choice (which leads to the value $k = -h^\vee$ for the critical level) is fixed by the condition that the square length of the long roots of $\mathfrak{g}$ is equal to 2.

where the normal ordering $: :$ is defined in the following way:

$$: e_a(n) e_b(m) := \begin{cases} e_b(m) e_a(n), & \text{if } n < 0, m \geq 0, \\ e_a(n) e_b(m), & \text{otherwise.} \end{cases}$$

Due to the normal ordering the coefficients of $T(u)$ are well defined operators in End $\mathfrak{V}$. It is well known that

$$\begin{aligned} [T_m, T_n] &= (k + h^\vee) \left[(m - n) T_{m+n} + \frac{k \dim \mathfrak{g}}{12} \left(n^3 - n \right) \delta_{n,-m} \right], \\ [T_m, \pi_k(e_a(n))] &= -(k + h^\vee) \pi_k(e_a(n+m)). \end{aligned} \tag{4.13}$$

Therefore, if $k \neq -h^\vee$, T_n generate the Virasoro algebra; for $k = -h^\vee$ the elements T_n, $n \in \mathbb{Z}$, are central.

One can show that for $\mathfrak{g} = \mathfrak{sl}_2$ the center of $\mathcal{U}_{-h^\vee}(\widehat{L\mathfrak{g}})$ is generated by T_n, $n \in \mathbb{Z}$; for higher rank algebras there are other Casimirs. An attempt to construct these higher Casimir elements by considering 'higher Sugawara currents' $: \operatorname{tr} J(u)^n :$ for arbitrary $n \geq 2$ runs into trouble. However, by applying a different technique [21], it was proved that for the critical central charge the algebra of Casimirs is very ample: essentially, there exists a Casimir element with prescribed symbol and hence there is an isomorphism $\mathcal{S}(L\mathfrak{g})^{L\mathfrak{g}} \longrightarrow \mathcal{Z}\left(\mathcal{U}_{-h^\vee}\left(\widehat{L\mathfrak{g}}\right)\right)$. The situation with the Lie algebra $\mathfrak{g}_D$ is basically the same as described above. This allows to realize the Gaudin Hamiltonians as radial parts of appropriate Casimir operators.

Let us describe this construction (due to [22]) more precisely. Let the Lie algebras $\mathfrak{g}_D$, $\mathfrak{g}_D^+$, $\mathfrak{g}_D^{++}$, $\mathfrak{g}(D)$ be as above. It will be convenient to add one more point $\{u\}$ to the divisor D and to attach to it the trivial representation V_0 of $\mathfrak{g}$. (This will not affect the Hilbert space of our model, since $\otimes_{z_i \in D} V_{\lambda_i} \otimes V_0 = (\otimes_{z_i \in D} V_{\lambda_i}) \otimes \mathbb{C} \simeq \otimes_{z_i \in D} V_{\lambda_i}.$) Thus we write $D' = D \cup \{u\}$, etc. Let ω be the 2-cocycle on $\mathfrak{g}_{D'}$ defined by

$$\omega(X, Y) = \sum_{z_i \in D'} \operatorname{Res}(X_i, dY_i). \tag{4.14}$$

Let $\hat{\mathfrak{g}}_{D'} = \mathfrak{g}_{D'} \oplus \mathbb{C}c$ be the central extension of $\mathfrak{g}_{D'}$ defined by this cocycle. Note that since the restriction of ω to the subalgebra $\mathfrak{g}(D') \subset \mathfrak{g}_{D'}$ is zero, the algebra $\mathfrak{g}(D')$ is canonically embedded into $\hat{\mathfrak{g}}_{D'}$. Put $\hat{\mathfrak{g}}_{D'}^+ = \mathfrak{g}_{D'}^+ \oplus \mathbb{C}c$. Fix a highest weight representation $V_{(\lambda,0)} = \otimes_{z_i \in D} V_{\lambda_i} \otimes V_0$ of the Lie algebra $\mathfrak{g}^{N+1} = \mathfrak{g}_{D'}^+ / \mathfrak{g}_{D'}^{++}$ as above and let $V_{(\lambda,0)}^k$ be the associated representation of $\hat{\mathfrak{g}}_{D'}^+$ on which the center $\mathbb{C}c$ acts by multiplication by $k \in \mathbb{Z}$.. Let $\mathbf{V}_{(\lambda,0)}^k$ be the induced representation of $\hat{\mathfrak{g}}_{D'}$,

$$\mathbf{V}_{(\lambda,0)}^k = \mathcal{U}(\hat{\mathfrak{g}}_D) \otimes_{\mathcal{U}(\hat{\mathfrak{g}}_{D'}^+)} V_{(\lambda,0)}^k. \tag{4.15}$$

There is a canonical embedding

$$V_{(\lambda,0)}^k \hookrightarrow \mathbf{V}_{(\lambda,0)}^k : v \longmapsto 1 \otimes v.$$

Let $\left(\mathbf{V}^k_{(\lambda,0)}\right)^*$ be the dual of $\mathbf{V}_{(\lambda,0)}$, $\mathbf{H}^k_{(\lambda,0)} \subset \left(\mathbf{V}^k_{(\lambda,0)}\right)^*$ the subspace of $\mathfrak{g}(D')$-invariants. Decomposition (2.19) together with the obvious isomorphism $V_{(\lambda,0)} \simeq V_\lambda$ immediately implies that $\mathbf{H}_{(\lambda,0)}$ is canonically isomorphic to V_λ^*. Let $\Omega \in \mathbf{H}_\lambda \otimes V_\lambda$ be the canonical element; it defines a natural mapping

$$\mathbf{V}^{k,}_{(\lambda,0)} \longrightarrow V_\lambda : \varphi \longmapsto \langle\varphi\rangle_\Omega .$$

(In Conformal Field Theory $\langle\varphi\rangle_\Omega$ are usually called *correlation functions*.) For any $x \in U(\hat{\mathfrak{g}}_D)$, let $\gamma_\lambda(x) \in \operatorname{End} V_\lambda$ be the linear operator defined by the composition mapping $v \mapsto \langle x(1 \otimes v)\rangle_\Omega$, $v \in V_\lambda$.

Now suppose that $k = -h^\vee$; let again J be the canonical element which corresponds to the inner product (2.20) in $\mathfrak{g}((z))$, $J(z) = (id \otimes \rho)J$, $T(z) = {:}\operatorname{tr} J(z)^2{:}$. Let us embed $\mathcal{U}_{-h^\vee}(\mathfrak{g}((z)))$ into $\mathcal{U}_{-h^\vee}(\mathfrak{g}_D)$ sending it to the extra place $\{u\} \subset D'$ to which we attached the trivial representation of $\mathfrak{g}$.

Proposition 30. $S(u) = \gamma_\lambda(J_V(-2))$ *coincides with the Gaudin Hamiltonian.*

Remark 13. Besides quadratic Hamiltonians associated with the Sugawara current there are also higher commuting Hamiltonians which may be obtained using the methods of [21].

In the calculation above we started with a generalized Verma module $\mathbf{V}^k_{(\lambda,0)}$; however, any representation of the critical level will do. The problem is to find a sufficiently ample class of critical level representations which will account for the spectrum of the Gaudin model. Note that this construction is exactly similar to the description of the Toda eigenfunctions in the finite-dimensional case. In the Toda case the correct class consisted of the spherical principal series representations; an important point is that principal series representations separate points of the algebraic spectrum of $\mathcal{Z}$. Now, at the critical level the algebra of the Casimir operators is extraordinarily rich; hence the generalized Verma modules which are parametrized by the dual of the extended Cartan subalgebra $\hat{\mathfrak{a}} = \mathfrak{a} \oplus \mathbb{C}$ (i. e. depend on a finite number of parameters) cannot be used. Remarkably, there is another class of representations of $\mathfrak{g}_D$, the *Wakimoto modules*, which play the role of the principal series representations. Roughly, the idea is to use the loop algebra $L\mathfrak{a} = \mathfrak{a} \otimes \mathbb{C}((z))$ as the substitute for the Cartan subalgebra, to extend its characters trivially to $L\mathfrak{n}$ and to take the induced $L\mathfrak{g}$ -module. Due to the normal ordering (which is necessary, since after central extension $L\mathfrak{a}$ becomes a Heisenberg algebra), this construction should be modified (among other things, this leads to a shift of the level, i. e., of the value of the central charge, which becomes $-h^\vee$).

For $\mathfrak{g} = \mathfrak{sl}_2$ the Wakimoto module of the critical level has the following explicit realization [23]. Let $\mathbf{H}$ be the Heisenberg algebra with generators a_n, a_n^*, $n \in \mathbb{Z}$, and relations

$$[a_n, a_m^*] = \delta_{n,-m}. \tag{4.16}$$

Let $\mathfrak{F}$ be the Fock representation of H with vacuum vector v satisfying $a_n v = 0$, $n \geq 0$, $a_n^* v = 0$, $n > 0$. Put

$$a\left(u\right) = \sum_{n \in \mathbb{Z}} a_n u^{-n-1}, a^*\left(u\right) = \sum_{n \in \mathbb{Z}} a_n^* u^{-n}. \tag{4.17}$$

Let $\{E, F, H\}$ be the standard basis of $\mathfrak{sl}_2$. Put $E\left(n\right) = E \otimes z^n$, $F\left(n\right) = F \otimes z^n$, $H\left(n\right) = H \otimes z^n$; let

$$E\left[u\right] = \sum_{n \in \mathbb{Z}} E\left(n\right) u^n, F\left[u\right] = \sum_{n \in \mathbb{Z}} F\left(n\right) u^n, H\left[u\right] = \sum_{n \in \mathbb{Z}} F\left(n\right) u^n$$

be the corresponding 'generating functions'. For any formal power series $\chi\left(u\right) = \sum_{n \in \mathbb{Z}} \chi_n u^{-n-1}$ define the (projective) action of $L\mathfrak{g}$ on F by

$$E\left[u\right] = a\left(u\right), H\left[u\right] = -2 : a\left(u\right) a^*\left(u\right) : + \chi\left(u\right), \tag{4.18}$$
$$F\left[u\right] = - : a\left(u\right) a^*\left(u\right) a^*\left(u\right) : -2\partial_u a^*\left(u\right) + \chi\left(u\right) a^*\left(u\right).$$

Thus we get an $L\mathfrak{g}$-module structure in $\mathfrak{F}$ which depends on $\chi\left(u\right)$; this is the Wakimoto module of the critical level $k = -2$ (usually denoted by $W_{\chi(u)}$).

The Virasoro generators $T_n, n \in \mathbb{Z}$, are scalar in $W_{\chi(u)}, T_n = q_n \cdot id$. Put $q\left(u\right) = \sum_{n \in \mathbb{Z}} q_n u^n$. One can show that

$$q\left(u\right) = \frac{1}{4} \chi\left(u\right)^2 - \frac{1}{2} \partial_u \chi\left(u\right), \tag{4.19}$$

i. e., q is the Miura transform of χ. This means of course that

$$\partial_u^2 - q\left(u\right) = \left(\partial_u - \frac{1}{2} \partial_u \chi\right)\left(\partial_u + \frac{1}{2} \partial_u \chi\right);$$

as a matter of fact, due to the normal ordering, if we make a change of coordinates on the (formal) disk, q transforms as a projective connection. Now a comparison with formula (4.10) suggests that the Bethe Ansatz is related to a Wakimoto module with an appropriate choice of χ. Namely, we consider the following global rational function

$$\lambda\left(u\right) = \sum_{i=1}^{N} \frac{\lambda_i}{u - z_i} - \sum_{j=1}^{m} \frac{2}{u - w_j}.$$

Denote by $\lambda_i\left(u - z_i\right)$ its expansion at the points z_i, $i = 1, ..., N$, and by $\mu_j\left(u - w_j\right)$ its expansion at the points w_j, $j = 1, ..., m$. We have

$$\lambda_i\left(t\right) = \frac{\lambda_i}{t} + ..., \mu_j\left(t\right) = -\frac{2}{t} + \mu_j\left(0\right) + ...,$$

where

$$\mu_j\left(0\right) = \sum_{i=1}^{N} \frac{\lambda_i}{w_j - z_i} - \sum_{s \neq j} \frac{2}{w_j - w_s}.$$

We want to attach the Wakimoto modules to the points on the Riemann sphere. To this end let us observe that creation and annihilation operators (4.17) define a projective representation of the 'local' algebra $\Gamma = \mathbb{C}\left(\left(u\right)\right) \oplus \mathbb{C}\left(\left(u\right)\right) du$ consisting of formal series and formal differentials with the cocycle

$$\Omega_0\left(\left(\varphi_1, \alpha_1\right), \left(\varphi_2, \alpha_2\right)\right) = \mathrm{Res}_{u=0}\left(\varphi_1 \alpha_2 - \varphi_2 \alpha_1\right).$$

The 'global' algebra is the direct sum of the local algebras with the cocycle

$$\Omega\left(\left(\varphi_1, \alpha_1\right), \left(\varphi_2, \alpha_2\right)\right) = \sum_i \mathrm{Res}_{z_i}\left(\varphi_1 \alpha_2 - \varphi_2 \alpha_1\right).$$

Now let us consider the tensor product of the Wakimoto modules attached to the points z_i, w_j,

$$\mathbb{W} = \bigotimes_{i=1}^N W_{\lambda_i(t)} \bigotimes_{j=1}^m W_{\mu_j(t)}.$$

Let $\mathbb{H}$ be the corresponding 'big' Heisenberg algebra. The eigenfunctions of the Gaudin Hamiltonians may be constructed as the appropriate correlation functions associated with $\mathbb{W}$. To formulate the exact statement we need one more definition. Let V be a $\mathfrak{g}\left(\left(t\right)\right)$-module. For $X \in \mathfrak{g}, n \in \mathbb{Z}$ we put $X\left(n\right) = X \otimes t^n \in \mathfrak{g}\left(\left(t\right)\right)$. A vector $v \in V$ is called a *singular vector of imaginary weight* if $X\left(n\right)v = 0$ for all $X \in \mathfrak{g}, n \in \mathbb{Z}, n \geq 0$. Let $v_j \in W_{\mu_j(t)}$ be the vacuum vector; put $w_j = a_{-1}v_j$, where a_{-1} is the creation operator introduced in (4.16).

Lemma 7. *The vector w_j is a singular vector of imaginary weight if and only if $\mu_j\left(0\right) = 0$.*

Note that this condition coincides with (4.11). Let $\tilde{M}_{\lambda_i}$ be the subspace of $W_{\lambda_i(t)} \simeq \mathcal{F}$ generated from the vacuum vector by the creation operators a_n^*.

Lemma 8. *$\tilde{M}_{\lambda_i}$ is stable with respect to the subalgebra $\mathfrak{g} \subset L\mathfrak{g}$ of constant loops and is isomorphic to the dual of the Verma module M_{λ_i} over $\mathfrak{g} = \mathfrak{sl}_2$ with the highest weight $\lambda_i = \mathrm{Res}_{t=0}\lambda_i\left(t\right)$.*

This assertion is immediate from the definition (4.18) of the $\widehat{\mathfrak{sl}}_2$-action on $\mathcal{F}$. Let $\mathbb{W}^*$ be the dual of $\mathbb{W}$. Let $\mathfrak{h}_{z,w}$ be the algebra of rational functions with values in the Cartan subalgebra $\mathfrak{h} \subset \mathfrak{sl}_2$ which are regular outside $\{z_1, ..., z_N, w_1, ..., w_m\}$. There is a natural embedding of $\mathfrak{h}_{z,w}$ into the 'big' Heisenberg algebra $\mathbb{H}$ defined via expansion at each point of the divisor $\{z_1, ..., z_N, w_1, ..., w_m\}$. Let $\mathcal{H}_{z,w}$ be the maximal abelian subalgebra of $\mathbb{H}$ which contains $\mathfrak{h}_{z,w}$.

Lemma 9. *The space of $\mathcal{H}_{z,w}$-invariants in $\mathbb{W}^*$ is 1-dimensional; it is generated by a functional τ whose value on the tensor product of the vacuum vectors of $W_{\lambda_i(t)}, W_{\mu_j(t)}$ is equal to 1; the restriction of τ to $\otimes_{i=1}^N \tilde{M}_{\lambda_i} \otimes w_1 \otimes ... \otimes w_m$ is nontrivial.*

Let $\psi \in \otimes_{i=1}^{N} \tilde{M}_{\lambda_i}$. We may write $\tau(\psi) = \langle \psi, \phi \rangle$ where $\phi \in \otimes_{i=1}^{N} M_{\lambda_i}$ is a vector in the tensor product of Verma modules over $\mathfrak{sl}_2$.

Theorem 8. *[22] If the Bethe equations (4.11) are satisfied, $\phi \in \otimes_{i=1}^{N} M_{\lambda_i}$ is an eigenvector of the Gaudin Hamiltonians in $\otimes_{i=1}^{N} M_{\lambda_i}$ with the eigenvalue (4.10).*

Remark 14. If the weights λ_i are dominant integral, there is a natural projection

$$\pi : \bigotimes_{i=1}^{N} M_{\lambda_i} \longrightarrow \bigotimes_{i=1}^{N} V_{\lambda_i}.$$

It is easy to see that π maps the eigenvectors of the Gaudin Hamiltonians in $\otimes_{i=1}^{N} M_{\lambda_i}$ onto those in $\otimes_{i=1}^{N} V_{\lambda_i}$.

The construction described above admits a generalization to arbitrary semi-simple Lie algebras. We shall only write down the generalized Bethe equations. To parametrize a Bethe vector first choose a set $\{w_1, ..., w_m\}$, $w_j \in \mathbf{C}$, and assign to each w_j a set of simple roots $\{\alpha_{i_j}\}_{i=1}^{N}$, one for each copy of $\mathfrak{g}$ in $\mathfrak{g}^N$. Let $F_{i_j}^i$ be the corresponding Chevalley generator of $\mathfrak{g}$ acting nontrivially in the i-th copy of $\mathfrak{g}$. The straightforward generalization of the Bethe creation operator (4.8) to the higher rank case is

$$F(w_j) = \sum_{i=1}^{N} \frac{F_{i_j}^i}{w_j - z_i} \tag{4.20}$$

[4]. The problem with (4.20) is that these operators no longer commute with each other. Hence one cannot use a string of creation operators to produce an eigenvector as is done in (4.9). The correct way to decouple them is to use the Wakimoto modules $W_{\mu_j(t)}$ which correspond to the poles of the creation operator; as above, the eigenvectors are generated by the singular vectors of imaginary weight which exist if and only if the constant term in the expansion of $\mu_j(t)$ satisfies certain orthogonality conditions. This leads to the *generalized Bethe equations*:

$$\sum_{i=1}^{N} \frac{(\lambda_i, \alpha_{i_j})}{w_j - z_i} - \sum_{s \neq j} \frac{(\alpha_{i_s}, \alpha_{i_j})}{w_j - w_s} = 0, \ j = 1, \dots, m. \tag{4.21}$$

Bethe vectors are again computed as correlation functions.

Let us end this section with a brief remark on the Knizhnik-Zamolodchikov equations. Recall that this is a system of equations satisfied by the correlation functions for an *arbitrary* value of the central charge; the critical value $c = -h^\vee$ corresponds to the semiclassical limit for the KZ system (small parameter before the derivatives). The outcome of this is two-fold: first, the Bethe vectors for the Gaudin model appear naturally in the semiclassical asymptotics of the solutions of the KZ system [43]. Moreover, the exact integral representation of the solutions (for any value of the central charge) also involves the Bethe vectors [22].

4.2   Quadratic Case. Quasitriangular Hopf Algebras

For an expert in Quantum Integrability, the Gaudin model is certainly a sort of limiting special case. The real thing starts with the quantization of *quadratic* Poisson bracket relations (3.7). This is a much more complicated problem which eventually requires the whole machinery of Quantum Groups (and has led to their discovery). The substitute for the Poisson bracket relations (3.30) is the famous relation

$$R(u\ v^{-1})L^1(u)L^2(v)R(u\ v^{-1})^{-1} = L^2(v)L^1(u), \qquad (4.22)$$

where $R(u)$ is the quantum R-matrix satisfying the quantum Yang-Baxter identity

$$R_{12}(u)R_{13}(u\ v)R_{23}(v) = R_{23}(v)R_{13}(u\ v)R_{12}(u). \qquad (4.23)$$

To bring a quantum mechanical system into Lax form one has to arrange quantum observables into a Lax matrix $L(u)$ (which is a rational function of u) and to find an appropriate R-matrix satisfying (4.22), (4.23). The first examples of quantum Lax operators were constructed by trial and error method; in combination with the Bethe Ansatz technique this has led to the explicit solution of important problems ([18,53], Faddeev [14,15]).

The algebraic concept which brings order to the subject is that of *quasitriangular Hopf algebra* [12]. The main examples of quasitriangular Hopf algebras arise as q-deformations of universal enveloping algebras associated with Manin triples. Remarkably, the general pattern represented by Theorems 1, 7 survives q-deformation. The standard way to describe quantum deformations of simple finite-dimensional or affine Lie algebras is by means of generators and relations generalizing the classical Chevalley–Serre relations [12, 30]. We shall recall this definition below in Sect. 4.2.2 in the finite-dimensional case, and in Sect. 4.2.3 for the quantized universal enveloping algebra of the loop algebra $L(\mathfrak{sl}_2)$. A dual approach, due to [17], is to construct quantum universal enveloping algebras as deformations of coordinate rings on Lie groups (regarded as linear algebraic groups). Of course, the construction of a quantum deformation of the Poisson algebra $\mathcal{F}(G)$ was one of the first results of the quantum group theory and is, in fact, a direct generalization of the Baxter commutation relations $RT^1T^2 = T^2T^1R$. A nontrivial fact, first observed by Faddeev, Reshetikhin and Takhtajan, is that the dual Hopf algebra (usually described as a q-deformation of the universal enveloping algebra) may also be regarded as a deformation of a Poisson algebra of functions on the dual group. More generally, the FRT construction is related to the *quantum duality principle*, which we shall now briefly discuss.

Let $(\mathfrak{g}, \mathfrak{g}^*)$ be a factorizable Lie bialgebra, G, G^* the corresponding dual Poisson groups, $\mathcal{F}(G), \mathcal{F}(G^*)$ the associated Poisson-Hopf algebras of functions on G, G^*, and $\mathcal{F}_q(G), \mathcal{F}_q(G^*)$ their quantum deformations. (For simplicity we choose the deformation parameter q to be the same for both algebras.) The quantum duality principle asserts that these algebras are dual to each other as Hopf algebras. More precisely, there exists a nondegenerate bilinear pairing

$$\mathcal{F}_q(G) \otimes \mathcal{F}_q(G^*) \to \mathbb{C}[[q]]$$

which sets the algebras $\mathcal{F}_q(G), \mathcal{F}_q(G^*)$ into duality as Hopf algebras. Hence, in particular, we have, up to an appropriate completion,

$$\mathcal{F}_q(G^*) \simeq \mathcal{U}_q(\mathfrak{g}).$$

In the dual way, we have also

$$\mathcal{F}_q(G) \simeq \mathcal{U}_q(\mathfrak{g}^*).$$

For factorizable Lie bialgebras the quantum deformations $\mathcal{F}_q(G), \mathcal{F}_q(G^*)$ may easily be constructed once we know the corresponding quantum R-matrices. An equivalence of this formulation to the definition of Drinfeld and Jimbo is not immediate and requires the full theory of universal R-matrices. Namely, starting with Drinfeld's definition of a quasitriangular Hopf algebra we may construct 'quantum Lax operators' whose matrix coefficients generate the quantized algebras of functions $\mathcal{F}_q(G), \mathcal{F}_q(G^*)$. Using explicit formulæ for universal R-matrices one can, in principle, express these generators in terms of the Drinfeld-Jimbo generators. In the context of integrable models, the FRT formulation has several important advantages: it allows stating the quantum counterpart of the main commutativity theorem as well as a transparent correspondence between classical and quantum integrable systems. The analogue of the FRT realization in the affine case is nontrivial, the key point being the correct treatment of the central element which corresponds to the central extension; it was described by [41].

4.2.1 Factorizable Hopf Algebras and the Faddeev-Reshetikhin-Takhtajan Realization of Quantized Universal Enveloping Algebras

Definition 5. Let $\mathcal{A}$ be a Hopf algebra with coproduct Δ and antipode S; let Δ' be the opposite coproduct in $\mathcal{A}$; $\mathcal{A}$ is called *quasitriangular* if

$$\Delta'(x) = \mathcal{R}\Delta(x)\mathcal{R}^{-1} \tag{4.24}$$

for all $x \in \mathcal{A}$ and for some distinguished invertible element $\mathcal{R} \in \mathcal{A} \otimes \mathcal{A}$ (*the universal R-matrix*) and, moreover,

$$(\Delta \otimes id)\,\mathcal{R} = \mathcal{R}_{13}\mathcal{R}_{23}, \quad (id \otimes \Delta)\,\mathcal{R} = \mathcal{R}_{13}\mathcal{R}_{12}. \tag{4.25}$$

Identities (4.25) imply that $\mathcal{R}$ satisfies the Yang-Baxter identity

$$\mathcal{R}_{12}\mathcal{R}_{13}\mathcal{R}_{23} = \mathcal{R}_{23}\mathcal{R}_{13}\mathcal{R}_{12}. \tag{4.26}$$

(We use the standard tensor notation to denote different copies of the spaces concerned.)

Let $\mathcal{A}^0$ be the dual Hopf algebra equipped with the opposite coproduct (in other words, its coproduct is dual to the opposite product in $\mathcal{A}$). Put $\mathcal{R}^+ = \mathcal{R}, \mathcal{R}^- = \sigma(\mathcal{R}^{-1})$ (here σ is the permutation map in $\mathcal{A} \otimes \mathcal{A}$, $\sigma(x \otimes y) = y \otimes x$) and define the mappings

$$R_{\pm} : \mathcal{A}^0 \longrightarrow \mathcal{A} : f \longmapsto \langle f \otimes id, \mathcal{R}^{\pm} \rangle;$$

axioms (4.25) imply that $R_{\pm}$ are Hopf algebra homomorphisms. Define the action $\mathcal{A}^0 \otimes \mathcal{A} \to \mathcal{A}$ by

$$f \cdot x = \sum_i R^+(f_i^{(1)}) \, x \, S(R^-(f_i^{(2)})), \quad \text{where} \quad \Delta^0 f = \sum_i f_i^{(1)} \otimes f_i^{(2)}. \qquad (4.27)$$

Definition 6. $\mathcal{A}$ is called *factorizable* if $\mathcal{A}$ is a free $\mathcal{A}^0$-module generated by $1 \in \mathcal{A}$.

(Let us denote the corresponding linear isomorphism $\mathcal{A}^0 \to \mathcal{A}$ by F for future reference.)

There are several important examples of factorizable Hopf algebras:

- Let $\mathcal{A}$ be an arbitrary Hopf algebra; then its Drinfeld double $D(\mathcal{A})$ is factorizable [42].
- Let $\mathfrak{g}$ be a finite-dimensional semisimple Lie algebra; then $\mathcal{A} = \mathcal{U}_q(\mathfrak{g})$ is factorizable.
- Let $\hat{\mathfrak{g}}$ be an affine Lie algebra; then $\mathcal{A} = \mathcal{U}_q(\hat{\mathfrak{g}})$ is factorizable.

(Observe that the last two cases are actually special cases of the first one; up to now, the double remains the principal (if not the only) source of factorizable Hopf algebras.)

Let $\mathfrak{g}$ be a simple Lie algebra. Let $\mathcal{A} = \mathcal{U}_q(\mathfrak{g})$ be the corresponding quantized enveloping algebra, and let $\mathcal{A}^0$ be the dual of $\mathcal{A}$ with the opposite coproduct. Let $\mathcal{R}$ be the universal R-matrix of $\mathcal{A}$. Let V, W be finite-dimensional irreducible representations of $U_q(\mathfrak{g})$. Let $\mathcal{L} \in \mathcal{A} \otimes \mathcal{A}^0$ be the canonical element. Set

$$L_V = (\rho_V \otimes id)\mathcal{L}, \quad R_{\pm}^{VW} = (\rho_V \otimes \rho_W)\mathcal{R}_{\pm}. \qquad (4.28)$$

We may call $L_V \in \operatorname{End} V \otimes \mathcal{A}^0$ the *universal quantum Lax operator* (with auxiliary space V). Property (4.24) immediately implies that

$$L_W^2 \, L_V^1 = R_+^{VW} \, L_V^1 \, L_W^2 \, \left(R_+^{VW}\right)^{-1}. \qquad (4.29)$$

Proposition 31. *The associative algebra $\mathcal{F}_q(G)$ generated by the matrix coefficients of L_V satisfying the commutation relations (4.29) and with the matrix coproduct*

$$\Delta L_V = L_V \dot{\otimes} L_V$$

is a quantization of the Poisson-Hopf algebra $\mathcal{F}(G)$ with the Poisson bracket (3.6), (3.8).[5]

The dual algebra $\mathcal{F}_q(G^*)$ is described in the following way. Let again $\mathcal{L} \in \mathcal{A} \otimes \mathcal{A}^0$ be the canonical element. Put

$$T_V^\pm = (\rho_V \otimes R^\pm) \mathcal{L} \in \text{End}\, V \otimes \mathcal{A}, \tag{4.30}$$

$$T_V = T_V^+ \, (id \otimes S)\, T_V^-. \tag{4.31}$$

Proposition 32. *(i) The matrix coefficients of $T_V^\pm$ satisfy the following commutation relations:*

$$T_W^{\pm 2} T_V^{\pm 1} = R^{VW}\, T_V^{\pm 1}\, T_W^{\pm 2}\, \left(R^{VW}\right)^{-1},$$

$$T_W^{-2} T_V^{+1} = R^{VW} T_V^{+1} T_W^{-2} \left(R^{VW}\right)^{-1}. \tag{4.32}$$

(ii) The associative algebra $\mathcal{F}_q(G^*)$ generated by the matrix coefficients of $T^{\pm V}$ satisfying the commutation relations (4.32) and with the matrix coproduct

$$\Delta T^{\pm V} = T^{\pm V} \dot{\otimes} T^{\pm V}$$

is a quantization of the Poisson-Hopf algebra $\mathcal{F}(G^*)$ with Poisson bracket (3.9), (3.13).

(iii) $T_V = (id \otimes F)L_V$. The matrix coefficients of T_V satisfy the following commutation relations:

$$\left(R_+^{VW}\right)^{-1} T_W^2 R_+^{VW} T_V^1 = T_V^1 \left(R_-^{VW}\right)^{-1} T_W^2 R_-^{VW}. \tag{4.33}$$

(iv) The associative algebra generated by the matrix coefficients of T_V satisfying the commutation relations (4.32) is a quantization of the Poisson-Hopf algebra $\mathcal{F}(G^*)$ with Poisson bracket (3.13).

(v) The pairing

$$\langle T_V^{\pm 1}, L_W^2 \rangle = R_+^{VW}$$

sets the algebras $\mathcal{F}_q(G)$, $\mathcal{F}_q(G^*)$ into duality as Hopf algebras.

Formulae (4.29, 4.32, 4.33) are the exact quantum analogues of the Poisson bracket relations (3.8, 3.9, 3.13), respectively.

4.2.2 Quantum Commutativity Theorem and Quantum Casimirs. In complete analogy with the linear case the appropriate algebra of observables which is associated with quantum Lax equations is not the quasitriangular Hopf algebra $\mathcal{A}$ but rather its dual $\mathcal{A}^0$; the Hamiltonians arise from the Casimir

---

[5] As usual, $\Delta L = L \dot{\otimes} L$ is the condensed notation for $\Delta L_{ij} = \sum_k L_{ik} \otimes L_{kj}$.

elements of $\mathcal{A}$. We may summarize this picture in the following heuristic correspondence principle.

Quadratic Classical Case	Quadratic Quantum case
$\mathcal{F}(G)$	$\mathcal{A}^0 = \mathcal{U}_q\left(\mathfrak{g}^*\right) \simeq \mathcal{F}_q\left(G\right)$
$\mathcal{F}\left(G^*\right)$	$\mathcal{A} = \mathcal{U}_q\left(\mathfrak{g}\right) \simeq \mathcal{F}_q\left(G^*\right)$
Classical r-matrices	*Quantum R-matrices*
Casimir functions in $\mathcal{F}\left(G^*\right)$	*Casimir operators* $\mathcal{Z} \subset \mathcal{U}_q\left(\mathfrak{g}\right)$
Symplectic leaves in G	*Irreducible representations of* $\mathcal{A}^0$
Symplectic leaves in G^*	*Irreducible representations of* $\mathcal{A}$

According to this correspondence principle, in order to get a 'quantum Lax system' associated with a given factorizable Hopf algebra A we may proceed as follows:

- *Choose a representation* $(\pi, \mathfrak{V})$ *of* $\mathcal{A}^0$.
- *Choose a Casimir element* $\zeta \in \mathcal{Z}\left(\mathcal{A}\right)$ *and compute its inverse image* $F^{-1}\left(\zeta\right)$ $\in \mathcal{A}^0$ *with respect to the factorization map* $F : \mathcal{A}^0 \to \mathcal{A}$.
- *Put* $\mathcal{H}_\zeta = \pi\left(F^{-1}\left(\zeta\right)\right)$.

Moreover, we expect that the eigenvectors of the quantum Hamiltonian $\mathcal{H}_\zeta$ can be expressed as matrix coefficients of appropriate representations of $\mathcal{A}$. In order to describe quantum Casimirs explicitly we have to take into account the effect of twisting. We shall start with their description in the finite-dimensional case. The results stated below are the exact quantum counterparts of those described in Sect. 3.2.2.

Let $\mathfrak{g}$ be a finite-dimensional simple Lie algebra, $\mathcal{A} = \mathcal{U}_q(\mathfrak{g})$ the quantized universal enveloping algebra of $\mathfrak{g}$. We shall recall its standard definition in terms of the Chevalley generators [12, 30]. Let P be the set of simple roots of $\mathfrak{g}$. For $\alpha_i \in P$ we set $q_i = q^{\langle \alpha_i, \alpha_i \rangle}$. We denote the Cartan matrix of $\mathfrak{g}$ by A_{ij}.

Definition 7. $\mathcal{U}_q(\mathfrak{g})$ is a free associative algebra with generators k_i, k_i^{-1}, e_i, f_i, $i \in P$, and relations

$$k_i \cdot k_i^{-1} = k_i^{-1} k_i = 1, \quad [k_i, k_j] = 0,$$

$$[a_i, e_j] = q_i^{A_{ij}} e_j, \qquad [a_i, f_j] = q_i^{-A_{ij}} e_j,$$

$$[e_i, f_j] = \delta_{ij} \frac{\left(k_i^2 - k_i^{-2}\right)}{q_i^2 - q_i^{-2}};$$

moreover, we assume the following relations *(q-deformed Serre relations)*:

$$\sum_{n=0}^{1-A_{ij}} (-1)^n \begin{bmatrix} 1 - A_{ij} \\ n \end{bmatrix}_{q_i^2} e_i^{1-A_{ij}-n} e_j e_i^n = 0,$$

$$\sum_{n=0}^{1-A_{ij}} (-1)^n \begin{bmatrix} 1 - A_{ij} \\ n \end{bmatrix}_{q_i^2} f_i^{1-A_{ij}-n} f_j f_i^n = 0.$$

(Here $\begin{bmatrix} m \\ n \end{bmatrix}_q$ are the q-binomial coefficients.)

Theorem 9. *[12] $\mathcal{U}_q(\mathfrak{g})$ is a quasitriangular Hopf algebra.*

As usual, we denote by $\mathcal{R} \in \mathcal{U}_q(\mathfrak{g}) \otimes \mathcal{U}_q(\mathfrak{g})$ its universal R-matrix and put $R_{VW} = (\rho_V \otimes \rho_W)\,\mathcal{R}$.

We recall the following well known corollary of definition 7. Let A be the 'Cartan subgroup' generated by the elements $k_i, i \in P$, $\mathbb{C}[A]$ its group algebra. We denote by $\mathcal{U}_q(\mathfrak{n}_\pm)$, $\mathcal{U}_q(\mathfrak{b}_\pm) \subset \mathcal{U}_q(\mathfrak{g})$ the subalgebras generated by e_i, f_i (respectively, by k_i, e_i and by k_i, f_i).

Proposition 33. *(i) $\mathbb{C}[A]$, $\mathcal{U}_q(\mathfrak{n}_\pm), \mathcal{U}_q(\mathfrak{b}_\pm)$ are Hopf subalgebras in $\mathcal{U}_q(\mathfrak{g})$. (ii) $\mathcal{U}_q(\mathfrak{n}_\pm)$ is a two-sided Hopf ideal in $\mathcal{U}_q(\mathfrak{b}_\pm)$ and $\mathcal{U}_q(\mathfrak{b}_\pm)/\mathcal{U}_q(\mathfrak{n}_\pm) \simeq \mathbb{C}[A]$.*

Let $\pi_\pm : \mathcal{U}_q(\mathfrak{b}_\pm) \to \mathbb{C}[A]$ be the canonical projection. Consider the embedding

$$i : \mathbb{C}[A] \to \mathbb{C}[A] \otimes \mathbb{C}[A] : h \longmapsto (id \otimes S)\,\Delta h.$$

(Here Δ is the coproduct in the commutative and cocommutative Hopf algebra $\mathbb{C}[A]$ and S is its antipode.)

Proposition 34. *The dual of $\mathcal{U}_q(\mathfrak{g})$ may be identified with the subalgebra*

$$\mathcal{A}^0 = \{x \in \mathcal{U}_q(\mathfrak{b}_+) \otimes \mathcal{U}_q(\mathfrak{b}_-); \ \pi_+ \otimes \pi_-(x) \in i\,(\mathbb{C}[A])\}\,.$$

Lemma 10. *For any $h \in A$, $h \otimes h\,\mathcal{R} = \mathcal{R}\,h \otimes h$.*

The algebra $\mathcal{A}$ admits a (trivial) family of deformations $\mathcal{A}_h = \mathcal{U}_q(\mathfrak{g})_h, h \in A$, defined by the following prescription:

$$e_i \longmapsto h e_i, f_i \longmapsto h f_i, k_i \longmapsto h k_i. \tag{4.34}$$

It is instructive to look at the commutation relations of $\mathcal{A}_h$ in the FRT realization. Let $V, W \in \mathsf{Rep}\,\mathcal{A}$; we define the 'generating matrices' $T_V^\pm \in \mathcal{A} \otimes \mathrm{End}\,V, T_W^\pm \in \mathcal{A} \otimes \mathrm{End}\,W$, as in (4.31).

Proposition 35. *(i) The deformation (4.34) maps $T_V^\pm$ into $\rho_V(h^{\mp 1})T_V^\pm$; the commutation relations in the deformed algebra $\mathcal{A}_h$ amount to*

$$\begin{aligned}
T_W^{\pm 2}\,T_V^{\pm 1} &= R^{VW}\,T_V^{\pm 1}\,T_W^{\pm 2}\,\left(R^{VW}\right)^{-1}, \\
T_W^{-2}T_V^{+1} &= R_h^{VW}T_V^{+1}T_W^{-2}\left(R^{VW}\right)^{-1},
\end{aligned} \tag{4.35}$$

where $R_h^{VW} = \rho_V \otimes \rho_W\,(\mathcal{R}_h)$ and $\mathcal{R}_h = h \otimes h^{-1}\mathcal{R}h^{-1} \otimes h$.

(ii) Put $T_h = \rho_V\,(h^2)\,T_V^+\,\left(T_V^-\right)^{-1}$; the 'operator-valued matrix' T_h satisfies the following commutation relations

$$\left(R_+^{VW}\right)^{-1} T_W^2 R_+^{VW} T_V^1 = T_V^1\,\left(R_-^{VW}\right)^{-1} T_W^2 R_-^{VW}. \tag{4.36}$$

Let $L_V \in \mathcal{A}^* \otimes \mathrm{End}\,V$ be the 'universal Lax operator' introduced in (4.28). Fix $h \in A$ and put

$$l_V^h = \mathrm{tr}\,_V \rho_V(h^2)L^V\,. \tag{4.37}$$

Elements $l_V^h \in \mathcal{A}^0$ are usually called (twisted) *transfer matrices*. The following theorem is one of the key results of QISM (as applied to the finite-dimensional case).

Theorem 10. *For any representations V, W and for any $h \in A$,*

$$l_V^h \, l_W^h = l_W^h \, l_V^h \, . \tag{4.38}$$

The pairwise commutativity of transfer matrices is a direct corollary of the commutation relations (4.29); twisting by elements $h \in A$ is compatible with these relations, since

$$(h \otimes h) \, \mathcal{R} = \mathcal{R} \, (h \otimes h) \, .$$

We want to establish a relation between the transfer matrices and the Casimirs of $\mathcal{U}_q(\mathfrak{g})_{h'}$ (where $h' \in A$ may be different from h). To account for twisting we must slightly modify the factorization map.

Let $A \times A \to A$ be the natural action of the 'Cartan subgroup' A on $\mathcal{A}$ by right translations, and let $A \times \mathcal{A}^0 \to \mathcal{A}^0$ be the contragredient action. The algebra $\mathcal{A} = \mathcal{U}_q(\mathfrak{g})$ is factorizable; let $F : \mathcal{A}^0 \to \mathcal{A}$ be the isomorphism induced by the action (4.27); for $h \in A$ put $F^h = F \circ h$. Let us compute the image of the universal Lax operator $L_V \in \operatorname{End} V \otimes \mathcal{A}^0$ under the 'factorization mapping' $(id \otimes F^h)$. It is easy to see that

$$(id \otimes F^h)(L_V) := T_V^{h\pm} = (id \otimes h^{\pm 1}) T_V^{1\pm}$$

Put

$$t_V^h = tr_V T_V^{h+} \left(T_V^{h-}\right)^{-1} = tr_V \rho_V(h)^2 T_V^1; \tag{4.39}$$

clearly, we have $t_V^h = F(l_V^h)$.

Theorem 11. *(i) [17] Suppose that $h = q^{-\rho}$, where $2\rho \in \mathfrak{h}$ is the sum of the positive roots of $\mathfrak{g}$. Then all coefficients of t_V^h are central in $\mathcal{A} = \mathcal{U}_q(\mathfrak{g})$. (ii) The center of $\mathcal{A}$ is generated by $\{t_V^h\}$ (with V ranging over all irreducible finite-dimensional representations of $\mathcal{U}_q(\mathfrak{g})$). (iii) For any $s \in A$ we have $l_V^s = F^{sh^{-1}}(t_V^h)$.*

Theorem 11 is an exact analogue of Theorem 7 for finite-dimensional factorizable quantum groups. Its use is twofold: first, it provides us with commuting quantum Hamiltonians; second, the eigenfunctions of these Hamiltonians may be constructed as appropriate matrix coefficients ('correlation functions') of irreducible representations of $\mathcal{U}_q(\mathfrak{g})$. Again, to apply the theorem to more interesting and realistic examples we must generalize it to the affine case.

4.2.3 Quantized Affine Lie Algebras.

Two important classes of infinite-dimensional quasitriangular Hopf algebras are quantum affine Lie algebras and the full Yangians (that is, the doubles of the Yangians defined in [12]). For concreteness I shall consider only the first class. To avoid technical difficulties we shall consider the simplest nontrivial case, that of the affine $\mathfrak{sl}_2$. The standard definition of $\mathcal{U}_q(\widehat{\mathfrak{sl}_2})$ is by means of generators and relations:

Definition 8. $\mathcal{U}_q(\widehat{sl_2})$ is a free associative algebra over $\mathbb{C}\left[q, q^{-1}\right]$ with generators $e_i, f_i, K_i, K_i^{-1}, i = 0, 1$, which satisfy the following relations:

$$
\begin{aligned}
K_i K_j &= K_j K_i, \\
K_i e_j &= q^{A_{ij}} e_j K_i, \quad K_i f_j = q^{-A_{ij}} f_j K_i, \\
[e_i, f_j] &= \delta_{ij} \left(q - q^{-1}\right) \left(K_i - K_i^{-1}\right),
\end{aligned}
\tag{4.40}
$$

where A_{ij} is the Cartan matrix of the affine sl_2, $A_{00} = A_{11} = -A_{01} = -A_{10} = 2$. In addition to (4.40) the following q-*Serre* relations are imposed:

$$
\begin{aligned}
e_i^3 e_j - \left(1 + q + q^{-1}\right)\left(e_i^2 e_j e_i - e_i e_j e_i^2\right) - e_j e_i^3 &= 0, \\
f_i^3 f_j - \left(1 + q + q^{-1}\right)\left(f_i^2 f_j f_i - f_i f_j f_i^2\right) - f_j f_i^3 &= 0.
\end{aligned}
\tag{4.41}
$$

The Hopf structure in $\mathcal{U}_q(\widehat{sl_2})$ is defined by

$$
\begin{aligned}
\Delta K_i &= K_i \otimes K_i, \\
\Delta e_i &= e_i \otimes 1 + K_i \otimes e_i, \\
\Delta f_i &= f_i \otimes 1 + K_i \otimes e_i.
\end{aligned}
\tag{4.42}
$$

The element $K_0 K_1$ is central in $\mathcal{U}_q(\widehat{sl_2})$; the quotient of $\mathcal{U}_q(\widehat{sl_2})$ over the relation

$$
K_0 K_1 = q^k, k \in \mathbb{Z},
$$

is called the level k quantized universal enveloping algebra $\mathcal{U}_q(\widehat{sl_2})_k$.

An alternative realization of $\mathcal{U}_q(\widehat{sl_2})_k$ is based on the explicit use of the quantum R-matrix; in agreement with the quantum duality principle, this realization may be regarded as an explicit quantization of the Poisson algebra of functions $\mathcal{F}_q(G^*)$, where G^* is the *Poisson dual* of the loop group $L(SL_2)$. The relevant Poisson structure on G^* has to be *twisted* so as to take into account the central charge k; moreover, in complete analogy with the non-deformed case, the center of $\mathcal{U}_q(\widehat{sl_2})_k$ is nontrivial only for the critical value of the central charge $k_{crit} = -2$. (The critical value $k = -2$ is a specialization, for $\mathfrak{g} = sl_2$, of the general formula $k = -h^\vee$; thus the value of k_{crit} is not affected by q-deformation.)

Introduce the R-matrix

$$
R(z) = \begin{pmatrix} 1 & 0 & 0 & 0 \\ 0 & \frac{1-z}{q-zq^{-1}} & \frac{z\left(q-q^{-1}\right)}{q-zq^{-1}} & 0 \\ 0 & \frac{q-q^{-1}}{q-zq^{-1}} & \frac{1-z}{q-zq^{-1}} & 0 \\ 0 & 0 & 0 & 1 \end{pmatrix}.
\tag{4.43}
$$

Let $\mathcal{U}_q'(\widehat{gl_2})_k$ be an associative algebra with generators $l_{ij}^{\pm}[n], i, j = 1, 2, n \in \pm\mathbb{Z}_+$. In order to describe the commutation relations in $\mathcal{U}_q'(\widehat{gl_2})_k$ we introduce the generating series $L^{\pm}(z) = \left\| l_{ij}^{\pm}(z) \right\|$,

$$l_{ij}^{\pm}(z) = \sum_{n=0}^{\infty} l_{ij}^{\pm}[\pm n] z^{\pm n}; \tag{4.44}$$

we assume, moreover, that $l_{ij}^{\pm}[0]$ are upper (lower) triangular: $l_{ij}^{+}[0] = l_{ji}^{-}[0] = 0$ for $i < j$. The defining relations in $\mathcal{U}_q'(\widehat{\mathfrak{gl}}_2)_k$ are

$$l_{ii}^{+}[0]\, l_{ii}^{-}[0] = l_{ii}^{-}[0]\, l_{ii}^{+}[0] = 1, \; i = 1, 2,$$
$$R\left(\frac{z}{w}\right) L_1^{\pm}(z) L_2^{\pm}(w) = L_2^{\pm}(w) L_1^{\pm}(z) R\left(\frac{z}{w}\right), \tag{4.45}$$
$$R\left(\frac{z}{w}q^{-k}\right) L_1^{+}(z) L_2^{-}(w) = L_2^{-}(w) L_1^{+}(z) R\left(\frac{z}{w}q^{k}\right)$$

(we again use the standard tensor notation). Relations (4.45) are understood as relations between formal power series in $\frac{z}{w}$. Observe the obvious parallels between (4.45) and (4.35): the shift in the argument of R reflects the effects of the central extension; in the finite-dimensional case this extension is of course trivial and the shift may be eliminated by a change of variables. The commutation relations (4.45) are the exact quantum analogue of the Poisson bracket relations (3.26) with $p = q^k$.

Lemma 11. *The coefficients of the formal power series*

$$q\det(L^{\pm}(z)) := l_{11}^{\pm}(zq^2)\left(l_{22}^{\pm}(z) - l_{21}^{\pm}(z) l_{11}^{\pm}(z)^{-1} l_{12}^{\pm}(z)\right)$$

are central in $\mathcal{U}_q'(\widehat{\mathfrak{gl}}_2)_k$.

Theorem 12. *The quotient of* $\mathcal{U}_q'(\widehat{\mathfrak{gl}}_2)_k$ *over the relations* $q\det(L^{\pm}(z)) = 1$ *is isomorphic to* $\mathcal{U}_q(\widehat{\mathfrak{sl}}_2)_k$.

The accurate proof of this theorem (and of its generalization to arbitrary quantized affine Lie algebras) is far from trivial (see [9]); it relies on still another realization of quantized affine Lie algebras, the so-called *Drinfeld's new realization* [13]; the important point for us is that the realization described in Theorem 12 is adapted to the explicit description of the Casimir elements.

Let us set

$$L(z) = L^{+}\left(q^{-k/2}z\right) L^{-}\left(zq^{k/2}\right)^{-1} \tag{4.46}$$

(this is the so-called *full quantum current*)

Theorem 13. *[41] (i) The quantum current (4.46) satisfies the commutation relations*

$$R\left(\frac{z}{w}\right) L_1(z) R^{-1}(\frac{q^{-k/2}z}{w})\; L_2(w) = \tag{4.47}$$
$$L_2(w) R(\frac{q^{k/2}z}{w}) L_1(z) R\left(\frac{z}{w}\right)^{-1}.$$

(ii) Suppose that $k = -2$; then the coefficients of the formal series

$$t(z) = \text{tr } q^{2\rho} L(z) \tag{4.48}$$

are central elements in the algebra $\mathcal{U}_q(\widehat{sl_2})_k$.

As usual, ρ stands for half the sum of the positive roots. In the present setting we have simply $q^{2\rho} = \begin{pmatrix} q & 0 \\ 0 & q^{-1} \end{pmatrix}$. The critical value $k = -2$ is a special case of the general formula $k = -h^\vee$ (the dual Coxeter number); thus the critical value remains the same as in the non-deformed case.

The definition of the quantum current and formulæ (4.47,4.48) are in fact quite general and apply to arbitrary quantized loop algebras. There are many reasons to expect that the center of the quotient algebra $A_{-h^\vee} = A/(c + h^\vee)$ for the critical value of the central charge is generated by (4.48) (with V ranging over all irreducible finite-dimensional representations of $\mathcal{U}_q(\hat{\mathfrak{g}})$) (cf. [8,23]). There are two main subtle points:

1. Our construction involved the 2-dimensional 'evaluation representation' $\mathcal{U}_q(sl_2) \to \text{End } \mathbb{C}((z))$. In the general case we must describe finite-dimensional representations $V(z)$ of $\mathcal{U}_q(\mathfrak{g})$ with spectral parameter; this requires additional technical efforts.
2. In the sl_2 case it was possible to use the generating series (4.44) as an exact alternative to the Drinfeld-Jimbo definition; in the general case the analogues of relations (4.45) are still valid but there may be others; anyway, the generating series $L^\pm_{V(z)}$ may be defined as the image of the canonical element:

$$L^\pm_{V(z)} = \left(id \otimes \rho_{V(z)}\right) \circ (R_\pm \otimes id)L;$$

the definition of the quantum current and Theorem 13 are still valid.

We may now use Theorem 13 to relate the elements of the center to quantum Hamiltonians. Our main pattern remains the same. For concreteness let us again assume that $A = \mathcal{U}_q(\widehat{sl_2})$. Let us take its dual A^0 as the algebra of observables. Let $\mathcal{L} \in A \otimes A^0$ be the canonical element. For any finite-dimensional sl_2-module V there is a natural representation of A in $V((z))$ (evaluation representation) Let W be another finite-dimensional sl_2-module. Set $L^V(z) = (\rho_{V(z)} \otimes id)\mathcal{L}$, $L^W(z) = (\rho_{W(z)} \otimes id)\mathcal{L}$, $R^{VW}(z) = (\rho_{V(z)} \otimes \rho_{W(z)})\mathcal{R}$. We may call $L^V(z) \in \text{End } V((z))$ $\otimes A^0$ the *universal quantum Lax operator* (with auxiliary space V). Fix a finite-dimensional representation π of A^0; one can show that $(id \otimes \pi) L_V(z)$ is *rational* in z. (Moreover, it was shown in [55] how to use this dependence on z to classify finite-dimensional representations of A^0.) Property (4.24) immediately implies that

$$L^W_2(w)L^V_1(v) = R^{VW}(vw^{-1})L^V_1(v)L^W_2(w)\left(R^{VW}(vw^{-1})\right)^{-1}. \tag{4.49}$$

Moreover, $R^{VW}(z)$ satisfies the Yang-Baxter identity (4.23).

Let A be a Cartan subgroup of G. Fix $h \in A$ and put

$$l_V^h(v) = \mathrm{tr}\,_V \rho_V(h^2) L^V(v),$$

then $l_V^h(v) l_W^h(w) = l_W^h(w) l_V^h(v)$. Elements $l_V^h(v)$ are called (twisted) *transfer matrices*.

It is convenient to extend $\mathcal{A}$ by adjoining to it group-like elements which correspond to the extended Cartan subalgebra $\hat{\mathfrak{a}} = \mathfrak{a} \oplus \mathbb{C}d$ in $\hat{\mathfrak{g}}$. Let $\hat{A} = A \times \mathbb{C}^*$ be the corresponding 'extended Cartan subgroup' in $\mathcal{A}$ generated by elements $h = q^\lambda, \lambda \in \mathfrak{h}, t = q^{kd}$. Let $\hat{A} \times \mathcal{A} \to \mathcal{A}$ be its natural action on $\mathcal{A}$ by right translations, $\hat{A} \times \mathcal{A}^0 \to \mathcal{A}^0$ the contragredient action. The algebra $\mathcal{A} = \mathcal{U}_q(\hat{\mathfrak{g}})$ is factorizable; let $F : \mathcal{A}^0 \to \mathcal{A}$ be the isomorphism induced by the action (4.27); for $h \in \hat{H}$ put $F^h = F \circ h$,

$$L_V^h(z)^\pm = \left(\,id \otimes R^\pm \circ h\right) L_V(z) \in \mathrm{End}\,V\left[\left[z^{\pm 1}\right]\right] \otimes \mathcal{A},$$
$$L_V^h(z) = L_V^h(z)^+ \,(id \otimes S)\,L_V^h(z)^-, \quad h \in \hat{H}. \tag{4.50}$$

It is easy to see that

$$L_V^h(z)^\pm = \left(id \otimes h^{\pm 1}\right) L_V^1(z)^\pm$$

and hence

$$T_V^h(z) = (id \otimes h)\,T_V^1(z)\,(id \otimes h)\,;$$

moreover, if $h = s^{-1}t$, where $s \in H$ and $t = q^{kd}$, we have also

$$T_V^h(z) = (\rho_V(s) \otimes id)\,T_V^t(z)\,(\rho_V(s) \otimes id)\,.$$

Put $t_V^h = \mathrm{tr}\,_V T_V^h(z) = \mathrm{tr}\,_V \rho_V(s)^2 T_V^t(z)$; clearly, we have $t_V^h(z) = F^t(l_V^s(z))$. Let 2ρ be the sum of the positive roots of $\mathfrak{g} = \mathfrak{sl}_2$, $h^\vee = 2$ its dual Coxeter number (our notation again hints at the general case).

Theorem 14. *(i) [41] Suppose that $h = q^{-\rho}q^{-h^\vee d}$. Then all coefficients of $t_V^h(z)$ are central in U. (ii) For any $s \in H$ we have $l_V^s(z) = F^{sh^{-1}}(t_V^h(z))$.*

Thus the duality between Hamiltonians and Casimir operators holds for quantum affine algebras as well. This allows us to anticipate connections between the generalized Bethe Ansatz, the representation theory of quantum affine algebras at the critical level and the q-KZ equation [24,54]. The results bearing on these connections are already abundant [8,23,56], although they are still not in their final form.

As we have already observed in the classical context, the Hopf structure on $\mathcal{A}^0$ is perfectly suited to the study of lattice systems. Let

$$\Delta^{(N)} : \mathcal{A}^0 \longrightarrow \overset{N}{\bigotimes} \mathcal{A}^0$$

be the iterated coproduct map; algebra $\otimes^N \mathcal{A}^0$ may be interpreted as the algebra of observables associated with a 'multiparticle system'. Set

$$\hat{t}_V^h(z) = \Delta^{(N)} t_V^h(z);$$

the Laurent coefficients of $\hat{t}_V^h(z)$ provide a commuting family of Hamiltonians in $\otimes^N \mathcal{A}^0$. Let $i_n : \mathcal{A}^0 \to \otimes^N \mathcal{A}^0$ be the natural embedding, $i_n : x \mapsto 1 \otimes ... \otimes x \otimes ... \otimes 1$; set $L_V^n = (id \otimes i_n) L_V$. Then

$$\hat{t}_V^h(z) = \text{tr}_V \left(\rho_V(h) \prod_n \overset{\frown}{L_V^n} \right). \tag{4.51}$$

Formula (4.51) has a natural interpretation in terms of lattice systems: L_V^n may be regarded as 'local' Lax operators attached to the points of a periodic lattice $\Gamma = \mathbb{Z}/N\mathbb{Z}$; commuting Hamiltonians for the big system arise from the monodromy matrix $M_V = \overset{\frown}{\prod} L_V^n$ associated with the lattice. Finally, the twist $h \in H$ defines a quasiperiodic boundary condition on the lattice. The study of the lattice system again breaks into two parts: (a) *Find the joint spectrum of $\hat{t}_V(z)$.* (b) *Reconstruct the Heisenberg operators corresponding to 'local' observables and compute their correlation functions.* This is the *Quantum Inverse Problem* (profound results on it are due to [54]).

Acknowledgements

The present lectures were prepared for the CIMPA Winter School on Nonlinear Systems which was held in Pondicherry, India, in January 1996. I am deeply grateful to the Organizing Committee of the School and to the staff of the Pondicherry University for their kind hospitality. I would also like to thank Prof. Y. Kosmann-Schwarzbach for her remarks on the draft text of these lectures.

References

1. Adler M. (1979). On a trace functional for formal pseudodifferential operators and the symplectic structure for the KdV type equations, *Invent. Math.* **50**, 219–248
2. Baxter R.J. (1982). Exactly Solved Models in Statistical Mechanics. Academic Press, London
3. Belavin A.A. and Drinfeld V.G. (1983). Triangle equations and simple Lie algebras, *Sov. Sci. Rev. C* **4**, 93–165
4. Babujian H.M. and Flume R. (1994). Off-shell Bethe ansatz equations for Gaudin magnets and solutions of KZ equations, *Mod. Physics Lett. A* **9**, 2029–2039
5. Bogolyubov N.M., Izergin A.G., and Korepin V.E. (1994). Correlation Functions of Integrable Systems and Quantum Inverse Scattering Method. Cambridge University Press, Cambridge
6. Bayen F., Flato M., Fronsdal C., Lichnerowicz A., and Sternheimer D. (1977). Deformation theory and quantization, I, II. *Ann. Phys.* **111**, 61–151

7. Cartier P. (1994). Some fundamental techniques in the theory of integrable systems, in: *Lectures on integrable systems*, O. Babelon, P. Cartier, and Y. Kosmann-Schwarzbach, eds. World Scientific

8. Ding J. and Etingof P. (1994). Center of a quantum affine algebra at the critical level. *Math. Res. Lett.* **1**, 469–480

9. Ding J. and Frenkel I. (1993). *Commun. Math. Phys.* **156**, 277–300

10. Dixmier J. (1974). Algèbres enveloppantes. *Cahiers Scientifiques, fasc. XXXVII.* Gauthier-Villars, Paris

11. Drinfeld V.G. (1983). Hamiltonian structures on Lie groups, Lie bialgebras and the geometric meaning of the classsical Yang-Baxter equation. *Sov. Math. Dokl.* **27**, 68–71

12. Drinfeld V.G. (1987). Quantum Groups. *Proc. ICM-86 (Berkeley)*, vol. 1, p. 798. AMS

13. Drinfeld V.G. (1987). *Sov. Math. Dokl.* **36**, 212–216

14. Faddeev L.D. (1980). Quantum Completely Integrable Systems in Quantum Field Theory. *Sov. Sci. Rev. C* **1**, 107–155

15. Faddeev L.D. (1984). Integrable Models in (1+1)-dimensional Quantum Field Theory. *Les Houches* 1982. Elsevier Science Publ.

16. Faddeev L.D. (1995). Algebraic aspects of Bethe ansatz, Int. J. Mod. Phys. A**10**, 1845–1878, *e-print archive: hep-th*/9404013

17. Faddeev L.D., Reshetikhin N.Y., and Takhtajan L.A. (1989). Quantization of Lie groups and Lie algebras. *Leningrad Math. J.* **1**, 178–207

18. Faddeev L.D. and Takhtajan L.A. (1979). Quantum Inverse Scattering Method. *Russian Math. Surveys* **34**:5, 11–68

19. Faddeev L.D. and Takhtajan L.A. (1987). Hamiltonian Methods in the Theory of Solitons. Springer-Verlag

20. Faddeev L.D. and Volkov A.Y. (1993). Abelian current algebra and the Virasoro algebra on the lattice. *Phys. Lett. B* **315**, 311–318

21. Feigin B. and Frenkel E. (1992). Affine Kac-Moody algebras at the critical level and Gelfand-Dikij algebras, *Int. J. Mod. Phys. A* **7**, Suppl. 1A, 197–215

22. Feigin B., Frenkel E., and Reshetikhin N. (1994). Gaudin model, Bethe ansatz and correlation functions at the critical level, *Commun. Math. Phys.* **166**, 27–62

23. Frenkel E. (1995). Affine algebras, Langlands duality and Bethe ansatz, *Proc. International Congr. of Math. Physics,* Paris 1994. International Press, pp. 606–642

24. Frenkel I.B. and Reshetikhin N.Y. (1992). Quantum affine algebras and holonomic difference equations, *Commun. Math. Phys.* **146**, 1–60

25. Gardner C.S., Greene J.M., Kruskal M.D., and Miura R.M. (1967). Method for solving the KdV equation, *Phys. Rev. Lett.* **19**, 1095–1097

26. Gaudin M. (1983). La Fonction d'Onde de Bethe. Masson, Paris

27. Goodman R. and Wallach N.R. (1982). Classical and quantum-mechanical systems of Toda lattice type. I, *Commun. Math. Phys.* **83**, 355–386

28. Goodman R. and Wallach N.R. (1984). Classical and quantum-mechanical systems of Toda lattice type. II, *Commun. Math. Phys.* **94**, 177–217

29. Harnad J. and Winternitz P. (1995). Classical and quantum integrable systems in $\tilde{\mathfrak{gl}}(2)^{+*}$ and separation of variables, *Commun. Math. Phys.* **172**, 263–285

30. Jimbo M. (1986). A q-difference analogue of $U(\mathfrak{g})$ and the Yang-Baxter equation. *Lett. Math. Phys.* **10**, 62–69

31. Jurčo B. (1989). Classical Yang-Baxter equation and quantum integrable systems, *J. Math. Phys.* **30**, 1289–1293

32. Kazhdan D., Kostant B., and Sternberg S. (1978). Hamiltonian group actions and dynamical systems of Calogero type. *Commun. Pure Appl. Math.* **31**, 481–508
33. Kirillov A.A. (1976). Elements of representation theory. Springer Verlag, Berlin a.o.
34. Kirillov A.N. and Reshetikhin N.Y. (1990). q-Weyl group and a multiplicative formula for the universal R-matrix. *Commun. Math. Phys.* **134**, 421–431
35. Kostant B. (1978). On Whittaker vectors and representation theory. *Inventiones Math.* **48**, 101–184
36. Kostant B. (1979). Quantization and Representation Theory. In:*Representation Theory of Lie Groups, Proc. SRC/LMS Res. Symp., Oxford 1977. London Math. Soc. Lecture Notes Series* **34**, 287–316
37. Kulish P.P. and Sklyanin E.K. (1981). Quantum Spectral Transform Method. In: Integrable Quantum Field Theories. *Lecture Notes in Physics* **151**, 61–119, Springer.
38. Kuznetsov V. (1994). Equivalence of two graphical calculi. *J. Phys. A* **25**, 6005–6026
39. Lax P.D. (1968). Integrals of nonlinear equations and solitary waves. *Commun. Pure Appl. Math.* **21**, 467–490
40. Olshanetsky M.A. and Perelomov A.M. (1994). Integrable systems and finite-dimensional Lie algebras. In: *Encyclopaedia of Math. Sciences*, vol. 16. Springer-Verlag, Berlin a.o.
41. Reshetikhin N.Y. and Semenov-Tian-Shansky M.A. (1990). Central extensions of quantum current groups. *Lett. Math. Phys.* **19**, 133–142
42. Reshetikhin N.Y. and Semenov-Tian-Shansky M.A. (1991). Quantum R-matrices and factorization problems. In: *Geometry and Physics, essays in honour of I.M. Gelfand*, S. Gindikin and I.M. Singer, eds. North Holland, Amsterdam – London – New York, pp. 533–550
43. Reshetikhin N. and Varchenko A. (1993). Quasiclassical asymptotics of the KZ equations. Preprint
44. Reyman A.G. (1996). Poisson structures related to quantum groups. In: *International School of Physics "E. Fermi", Course 127, Varenna, 1994. Quantum Groups and their Applications in Physics*, L. Castellani and J. Wess, eds. Amsterdam
45. Reyman A.G. and Semenov-Tian-Shansky M.A. (1994). Group-theoretical methods in the theory of finite-dimensional integrable systems. In: *Encyclopaedia of Math. Sciences*, vol. 16. Springer-Verlag, Berlin a.o.
46. Semenov-Tian-Shansky M.A. (1983). What is the classical r-matrix? Funct. Anal. Appl. **17**, 259–272
47. Semenov-Tian-Shansky M.A. (1994a) Quantization of open Toda lattices. In: *Encyclopaedia of Math. Sciences*, vol. 16. Springer-Verlag, Berlin a.o.
48. Semenov-Tian-Shansky M.A. (1994b). Lectures on r-matrices, Poisson Lie groups and integrable systems. In: *Lectures on integrable systems*, O. Babelon, P. Cartier, and Y. Kosmann-Schwarzbach, eds. World Scientific
49. Semenov-Tian-Shansky M.A. (1994c). Poisson Lie groups, quantum duality principle and the quantum double. *Contemporary Math.* **175**, 219–248
50. Sklyanin E.K. (1979) On complete integrability of the Landau-Lifshitz equation. *Preprint LOMI E-3-79*, Leningrad
51. Sklyanin E.K. (1992). Quantum Inverse Scattering Method. Selected Topics. In: *Quantum Groups and Quantum Integrable Systems*, Ge M.L., ed. World Scientific, Singapore

52. Sklyanin E.K. (1995). Separation of variables. New trends. In: *QFT, Integrable Models and beyond*, T. Inami and R. Sasaki, eds. Progr. Theor. Phys. Suppl. **118**, 35–60

53. Sklyanin E.K., Takhtajan L.A., Faddeev L.D. (1979). The quantum inverse scattering method. I. *Theor. Math. Phys.* **101**, 194–220

54. Smirnov F.A. (1992). Form Factors in Completely Integrable Models of Quantum Field Theory. *Adv. Series in Math. Phys.* **14**. World Scientific, Singapore

55. Tarasov V.O. (1985). Irreducible monodromy matrices for the R-matrix of the XXZ-model and lattice local quantum Hamiltonians. *Theor. Math. Phys.* **63**, 440–454

56. Tarasov V.O. and Varchenko A. (1995). Jackson integral representations for solutions of the quantized KZ equation. *St. Petersburg Math. J.* **6**, no. 2, 275–313

57. Weinstein A. (1994). Deformation quantization. *Séminaire Bourbaki*, exposé n° 789

Lecture Notes in Physics

For information about Vols. 1–591
please contact your bookseller or Springer-Verlag
LNP Online archive: springerlink.com